HANDBOOK
OF
ANATOMY AND PHYSIOLOGY
FOR STUDENTS
OF
MEDICAL RADIATION TECHNOLOGY

—

COMPILED BY
M. MALLETT, M.D.
Radiologist, retired, Edmonton, Canada

—

ILLUSTRATED BY
M. J. BABIUK, M.D.
Director of Radiological Services
General Hospital, Edmonton

This Handbook is dedicated to students of Radiological Technology everywhere. It is hoped that some students may acquire an acceptable knowledge of Anatomy and Physiology from a study of the material included in this Text, and that they will develop an interest in these two most fascinating subjects.

THE BURNELL COMPANY / PUBLISHERS, INC.
821 NORTH SECOND STREET • P. O. BOX 304
MANKATO, MINNESOTA 56002

ACKNOWLEDGEMENTS

Dr. M. J. Babiuk, F.R.C.P.(Canada), a graduate in medicine of the University of Alberta, and at the time of this edition Director of Radiological Services at the Edmonton General Hospital spent many hours in drawing the illustrations used in the Handbook. The excellence of his work is displayed in the diagrams. Dr. Babiuk refused to accept any recompense for his efforts. Thank you Dr. Babiuk.

Miss Shelia M. Robinson, A.C.P., B.Ed., a teacher of English and Art with the Edmonton Public School Board, at Grandview Heights Junior High School, Edmonton, very thoroughly reviewed the manuscript and made many invaluable suggestions.

Mrs. Naomi Hite, R.T.R., contributed in the correction of errors in the text.

Miss Joan M. Graham, R.N., R.T.R., assisted in many ways in finalizing the publication.

As in the earlier Editions of the Handbook the compiler, the illustrator and all others connected with this publication have no financial interest in it and will not receive any royalty therefrom.

PREFACE

The first Edition of the Handbook (1952) was compiled to supply students of radiological technolgy with sufficient information in anatomy and physiology to perform x-ray examinations. Very elementary information in these subjects **was** added to supply a basic knowledge of general anatomy and physiology. The **text**books for student nurses and other paramedical groups did not satisfactorily supply the information that our students should know. For instance, the information on bones and joints was limited, while that in some other areas was superfluous. Our text was therefore timely.

There have been so many recent advances in the application of radiology to medicine that the technologist must now have a more complete knowledge of both anatomy and physiology in order to perform the required examinations. This is particularly true of the co-called special procedures.

In addition, as each revision of the Syllabus of training is published more and more detailed knowledge of these two sciences is demanded. Much of this material has no direct application to radiology.

In anatomy it is simply a matter of including details that were previously required knowledge only for medical students. In gross anatomy there have not been many changes over the years, other than changes in the terminology, the exclusion of proper names for anatomical structures, etc.

Our knowledge of microscopic anatomy has advanced a great deal since the application of the electron microscope has made visible structures that were not previously visible or which were visible without sufficient precision to determine details. Concepts that were accepted when the magnification by light microscopes was used have been discarded or drastically changed, and this trend will continue as research progresses.

Physiology, unlike gross anatomy, has shown marked advances in our knowledge, and many functions have been proven or at least suggested. Present concepts may be proven incorrect tomorrow or may be radically revised. Possibly physiology should have a separate textbook even for the paramedical group in order to allow for frequent revision.

The plan of this new Edition follows that adopted in earlier ones. The subjects have been covered in Sections, a section in many instances being limited to a single topic. The definition method has again been used, and the structure to be described is printed in bold type, at the beginning of the definition. Many chapters are introduced by a list of the structures to be studied. This is followed by a list of the important parts of each structure, for reference in review. A detailed study of each part is then undertaken. Finally anatomical terms used in the chapter are listed. Instructors who disagree with this method of presentation should disregard what they consider superfluous.

THE STUDENT'S USE OF THE HANDBOOK

The arrangement of the material contained in this Handbook has been planned with a definite purpose in mind. The objective has been to make it as easy as possible for the student to find the information, and to acquire the knowledge required by a study of the subject matter. Explanations and drawings have been made simple but inclusive.

The text has been divided into twenty-four chapters, most of which include the study of a single system. The skeletal system however has been subdivided into six chapters, each dealing with some part of this system.

Each chapter has been further divided into Sections. Each Section deals with a single topic where possible. It should be possible, by designating the section, to specify what material is to be studied or reviewed.

The drawings have been labelled as F. 1-1, F. 1-2 etc. The first figure denotes the Chapter "1", the second figure indicates the number of the drawing in that first Chapter "1", "2", etc.

Where possible the definition method of presentation has been employed, and the structure defined has been printed in bold type and conspicuously in the first part of the definition.

Anatomy is largely a memory course. Great numbers of terms must be sorted out and remembered. The terms are unfamiliar, and sometimes difficult to pronounce. The student must correlate the explanation of each term with a drawing wherever possible. The new terms should be pronounced over and over and written down.

Many of the drawings have been simplified purposely. Each drawing should be studied so that a visual impression is acquired.

In any study there may be three approaches: the student may read a text, may listen to a lecture on the topic, or may study drawings. When listening to a lecture a student may jot down the essential and important facts and may disregard the padding. In any event, whether it is in reading, listening, or in studying drawings, the student should learn to select the essential facts and disregard the nonessentials. For example the definition of a prefix should contain three facts — (1) it consists of one or more syllables, (2) it is placed before a word, (3) it modifies or alters the meaning of the term. Most definitions or brief descriptions may be treated similarly. Remember — what we hear we forget, what we see we remember, and what we do we understand.

In defining any anatomical structure the information listed should include:
—what it is
—where it is (the location)
—what it does (its functions)
  e.g. the pancreatic duct:
—is a small passage
—extends through the length of the pancreas and joins the common bile duct
—conveys pancreatic juice into the duodenum to mix with food to help digest the food

Students must acquire some method of study, and must eliminate bad habits such as studying in a room where other people are talking, where a radio is blasting away, or where a television is in operation.

SOME SUGGESTIONS FOR INSTRUCTORS

As each part of the body is studied specimens, models, charts, pictures and drawings should be utilized in the preliminary lecture or in review. In the study of bones an attempt should be made to provide each student with a specimen of each bone to be studied. Remember, a visual image is much more valuable than any written or spoken description.

Since anatomical terms are like a foreign language each term must be spoken clearly, repeated often, and written out. Students should be encouraged to use the terms. Correct pronunciation may be acquired if the terms are used frequently by the instructors. Following an adequate study of prefixes and suffixes students should be taught to analyse each medical term where possible.

Radiographs should be employed to illustrate anatomical structures when it is possible, and no study should be considered complete until radiographs have been demonstrated, e.g. bones and joints.

In using the Handbook some routine method should be employed. Each system of the body might be introduced by naming the parts, as in the list introducing the chapter. A general statement of the functions of the system might follow in order that the student may understand the purpose for which the system was designed. Following this introduction a detailed study of each part of the system should be made, with the normal physiology of each part. In reviewing each structure the list of the important parts could be utilized or the student's attention might be directed to it.

Students seem to have a tendency to ignore the drawings, and to memorize the detailed description. This is unfortunate if it is true that what we see we remember while what we hear or read, we forget. Including some diagrams on the examination papers may induce students to look at and know illustrations.

Because students seem to have been brought up on a diet of tests it may be necessary to give frequent tests in order indirectly to get students to study the material included. While critics may condemn this method as "teaching for exams" it may prove to be the whip that works.

While some instructors will have a tendency to include more details than the curriculum requires, and each instructor will have a tendency to favor certain parts of the course, unnecessary details must be omitted. With no provision for dissection of the body, or opportunity to view the dissected body, students have a real problem in attempting to memorize all of the structures listed in the course.

REFERENCE BOOKS

Morris' Human Anatomy, McGraw Hill Book Co., Toronto
Gray's Anatomy, American Edition, Lea & Febiger
Gray's Anatomy, English Edition, Longmans Canada, Don Mills
Anatomy, Meschan, Saunders Ltd., Toronto
Ciba Collection Medical Illustrations, Netter
Basic Physiology & Anatomy, Taylor, Macmillan Co., Canada.

— CONTENTS —

CONTENTS OF HANDBOOK — LISTED
BY SECTIONS

LIST OF ILLUSTRATIONS

F.1-1, F.1-2, etc. F = figure, 1 = chapter 1, 1 = number of diagram in chapter

1. ANATOMICAL TERMS

S. 1   BASIC DEFINITIONS

ANATOMY is the study of the structure and form of an organism; human anatomy deals with the structure and form of the various parts of the human body (anatomy - ana + tome = to cut apart).

The term structure, in the sense in which it is used here, would include a study of such things as:
— the composition or materials composing a part;
— the manner in which these are put together;
— whether the unit is hollow with coverings, or is solid;
— the parts or units making up the whole;
— the number of such units;
— the size, shape, and position in the body.

The term "structure" is sometimes used with reference to any unit or part of the body not generally considered to be an organ, such as a cell, gland, hair, nerve, part of an organ, etc., hence the heading "Organs and other structures of ————".

GROSS OR MACROSCOPIC ANATOMY is the study of any part of the body as seen with the naked eye. This study is frequently undertaken in an anatomy laboratory upon a corpse. (Makros = large + skopeo = to see).

HISTOLOGY OR MICROSCOPIC ANATOMY is the study of the structure of minute parts such as cells or tissues that are not visible as separate entities to the naked eye. A lens, a light microscope, or an electron microscope is therefore used to magnify them. (Mikros = small + skopeo = to see).

EMBRYOLOGY OR DEVELOPMENTAL ANATOMY is the study of the structural changes that take place in the fertilized ovum up to the time of birth.

SURFACE OR REGIONAL OR TOPOGRAPHICAL ANATOMY (for radiographers) would be a study of what part of the surface of the body corresponds to the various organs and structures within the body. It would also include a study of what structures are responsible for the prominences and depressions visible or palpable (able to be felt) through the skin.

COMPARATIVE ANATOMY is the study of the similarities and differences in structure of the various parts of the body, among different species of living things. Often man is compared to some species of animal.

PHYSIOLOGY is the study of the function or behavior of the organs or other structures of the body. In the Handbook it will be limited to the normal functions of the normal structures.

PATHOLOGY is the study of disease, its causes and its effects upon the body.
In this presentation of anatomy emphasis has been placed upon the gross and surface anatomy, with a brief introduction to embryology, and of microscopic anatomy only when considered essential.

S. 2   THE NOMINA ANATOMICA

The names of anatomical structures described in the Handbook are based upon those listed in the NOMINA ANATOMICA. This is a list of the names of the parts of the body compiled by the International Anatomical Nomenclature Committee, and published in booklet form. The object was to produce a list of anatomical names that would be acceptable internationally to replace the various names employed. The fourth edition of the booklet was published in 1977, and includes sections on Histology and also on Embryology. In this Handbook of Anatomy Nomina Anatomica is listed as (NA). In the booklet all names are in the Latin form, in order to make them universally acceptable.

Eponyms, proper names used as names for anatomical structures, have been replaced by descriptive names, e.g., Parotid duct replaces Stensen's duct, glomerular capsule replaces Bowman's capsule.
Diphthongs, two vowels occurring together and pronounced as a single sound such as ae, oe, etc., have been eliminated. Cecum replaces caecum, fetus replaces foetus, hemorrhage replaces haemorrhage, etc.
Hyphens between vowels coming together in the middle of words, such as a prefix and its stem have been dropped, e.g. infra-orbital becomes infraorbital, intra-uterine becomes intrauterine.
Hyphens in compound words have also been discontinued, e.g. postero-anterior becomes posteroanterior.
Alternate names have been retained in some instances, e.g. calcaneus and os calcis; visceral pleura and pulmonary pleura.
In some instances where two or more names were employed one of them has been selected, e.g. the scaphoid bone of the wrist (not navicular).
In other instances the meaning of a term has been restricted, e.g. extremity now refers to the end of some long structure, not to the limb.
Undoubtedly older medical personnel will resist these changes, so the original terms will continue to be used by some people.

S. 3   THE ORGIN OF ANATOMICAL NAMES

The names of anatomical structures have been derived from Greek (G), Latin (L), AngloSaxon (AS) and from some other languages:

From Greek	-kephale, cephale = the head; kondylos, condyle = knuckle; karpos, carpus = wrist
From Latin	-scapula = shoulder blade femur = thigh bone clavicula, clavicle = collar bone
AngloSaxon	-sceaft = shaft heorte = heart hype = hip lungen = lung

Sometimes derivatives from both Greek and Latin have been used in reference to a single structure, hence renal from ren (L), and nephritic from nephros (G), both refer to the kidney; cor (L) and cardiac from Kardia (G) the heart.
Ancient anatomists frequently recognized some similarity between an anatomical structure and some common object. They named the structure after the thing it resembled, e.g. clavicula = clavicle, the collar bone, from its resemblance to an old fashioned key; pelvis from its likeness to a basin; pisiform - shaped like a pea; cuneiform - wedge shaped; arachnoid - spiderlike; deltoid - like the Greek letter delta - triangular.

S. 4   ANATOMICAL TERMINOLOGY IN THE HANDBOOK

Older medical personnel doctors, technicians, etc., have learned and used older names. It has been necessary therefore to list these along with the names recognized in

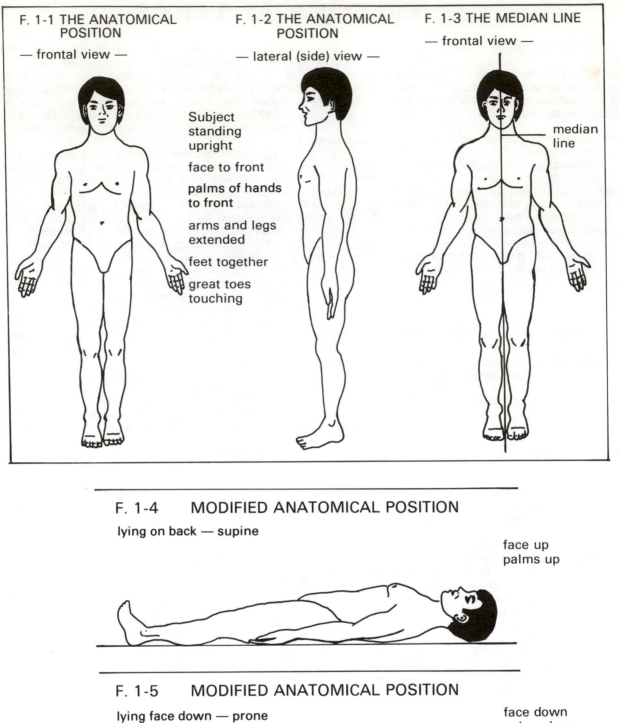

F. 1-1 THE ANATOMICAL POSITION

— frontal view —

F. 1-2 THE ANATOMICAL POSITION

— lateral (side) view —

Subject standing upright

face to front

palms of hands to front

arms and legs extended

feet together

great toes touching

F. 1-3 THE MEDIAN LINE

— frontal view —

median line

F. 1-4   MODIFIED ANATOMICAL POSITION

lying on back — supine

face up
palms up

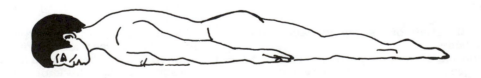

F. 1-5   MODIFIED ANATOMICAL POSITION

lying face down — prone

face down
palms down

the (NA). Students must become familiar with each name listed since any of them may be used on requisitions to the department.

Whenever two or more names are commonly used for any structure the (NA) and older names are listed. If an eponym is frequently used it has been listed along with the descriptive anatomical name (NA). Very occasionally some name other than the (NA) term has been listed, e.g. proximal and distal tibiofibular joints.

While Latin has been used in the Nomina Anatomica for all terms modern texts in anatomy have not adhered to this principle. Because modern students do not have a knowledge of Latin or Greek, there is an increasing tendency to use English terms based upon AngloSaxon names or upon interpretations of the Latin or Greek. The result is that in most anatomical texts there is a conglomeration of terms.

Many Latin names for structures have been retained unchanged in modern tests, e.g. humerus, ulna, radius, femur, tibia, fibula, etc.

In other instances the Latin has been translated into English, e.g. clavicle from clavicula; notch from incisura; head from caput, etc. In some cases the AngloSaxon name has been utilized, e.g. elbow, not cubitus; kidney, not ren; liver, not hepar; lung, not pulmo; heart, not cor or cardia.

Frequently, while the Latin name itself has not been used, an Anglicized derivative (especially an adjective) has been accepted; hence hepatic from hepar (liver); gastric from gaster (stomach); enteric from enteron (intestine); pulmonary from pulmo (lung); and cephalic for head.

S. 5 PLURAL FORMS AND ADJECTIVES

In English most plurals are formed by adding "s" or "es" to the singular form. There are also some irregular forms such as ox - oxen; knife - knives.

In the Latin names that have been retained, the plural form varies with the ending of the singular form:-
1. Words ending in "a" change the "a" to "ae", e.g. ulna = ulnae; fibula = fibulae; tibia = tibiae.
2. Some names ending in "us" change the "us" to "i", e.g. radius = radii; humerus = humeri; nucleus = nuclei. Others have the same plural form as the singular e.g. manus = manus; fetus = fetus, etc.
3. Terms ending in "um" change the "um" to "a", e.g. ilium = ilia; ischium = ischia; ovum = ova.
4. Some terms appear to be different; phalanx = phalanges; meninx = meninges; spermatozoon = spermatozoa.

There is a modern tendency to discard the Latin plural form and substitute an English ending; ulna = ulnas; tibia = tibias; fibula = fibulas. Some names, such as humerus, radius, etc., appear awkward if "s" or "es" is added.

Adjectives are frequently Anglicized forms of the Latin names, e.g. ulna = ulnar; radius = radial; axilla = axillary, etc.

Other adjectives, as stated above, are derived from rarely used Latin or Greek nouns, such as pulmonary from pulmo; pneumonic from pneumon; renal from ren; cardiac from kardia, etc.

S. 6 SPELLING

The student should refer to S. 2 under Nomina Anatomica for recently adopted spellings for diphthongs such as fetus, hemorrhage, and anemia.

Hyphens as noted have been eliminated, e.g. infraorbital, intrauterine, and in compound terms, e.g. superoinferior, occipitomental, etc.

Calix has been suggested instead of calyx, and anulus instead of annulus.

Possibly some of these changes may not prove to be acceptable, and the older ones may be retained. Students must therefore learn both.

S. 7 (1) GENERAL ANATOMICAL TERMS F. 1-1,2,4,5.

1. **THE ANATOMICAL POSITION;** the subject stands upright, with the limbs extended, face to the front, palms of the hands facing the front, and the feet together. In descriptions of the position of a part of the body it is assumed that the body is in this position, an awkward one to maintain. In the **supine position,** the subject lies on the back, the face and palms are directed upwards: the other conditions would be as described above. In the **prone position,** or face down position, the face and palms would be directed downwards. Often radiography is done in the supine or prone positions and the conditions listed must be met whenever practical.

2. **THE MEDIAN LINE OF THE BODY;** F.1-3, a line drawn vertically through the center of the forehead, nose, chin, neck, chest, abdomen, and between the legs. The body would be divided into right and left halves. See also the median plane below. Sometimes an anterior and posterior median line are described.

S. 7 (2) PARTS OF ORGANS & OTHER STRUCTURES

Head; caput (L), the expanded end of a part of the body, e.g. head of pancreas, head of radius, etc.
Little Head; capitulum (L) a small head (from caput = head + ulum = little) e.g. capitulum of humerus, since it already has one head.

Neck; cervix or collum (L), the constricted part of a structure adjacent to the head of it, e.g. neck of radius; neck (cervix) of the uterus.

Body; corpus (L), the principal part of a structure or the shaft of a long bone; body of pancreas; body of a vertebra; radius, etc; corpse = body.

Shaft; sceaft (AS), the principal part or body of a long bone, a rodlike part, e.g. shaft of radius.

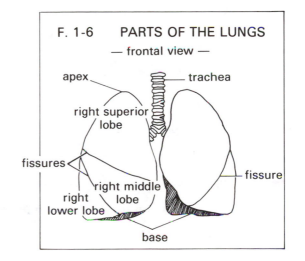

F. 1-6   PARTS OF THE LUNGS
— frontal view —

apex — trachea — right superior lobe — fissures — right middle lobe — right lower lobe — fissure — base

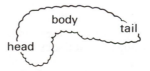

F. 1-7   PARTS OF THE PANCREAS

body — tail — head

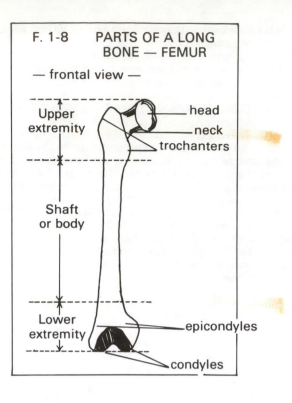

F. 1-8 PARTS OF A LONG BONE — FEMUR

— frontal view —

- Upper extremity
- head
- neck
- trochanters
- Shaft or body
- Lower extremity
- epicondyles
- condyles

Tail; cauda(L), the tapered end of a structure, the tail of the pancreas, etc.

Lobe; lobus(L), a subdivision of an organ or structure; e.g. a lobe of the lung, liver, etc.

Lobule; lobulus(L), a subdivision of a lobe (lobe + ulus = little), e.g. lobule of lung, liver, etc.

Segment; segmentum(L), a unit of structure, separated from other segments, with its own blood supply etc. e.g. segment of a lobe of a lung, with its own bronchus, a branch of a lobar bronchus, and its own blood vessels.

Extremity; extremitas(L), the end of a long structure, e.g. extremity of a kidney, of a bone, etc. In (NA) not used referring to a limb of the body.

Apex; the pointed end of a structure e.g. of lung.

Base; the broad flattened end of a structure, often its lower end, e.g. base of the lung.

S. 7 (3) OPENINGS INTO OR WITHIN ORGANS

Aperture	—a hole or opening.
Foramen	—a hole or opening, pl. = foramina.
Hiatus	—an opening or gap; esophageal hiatus.
Orifice	—an opening; mitral orifice of heart.
Os	—an opening or mouth; oral hygiene, pl, ossa.
Ostium	—an opening; cardiac ostium of stomach, pl. ostia.
Lumen	—a window or opening; usually refers to the cavity of a hollow organ, stomach, artery, etc.
Porus	—an opening, pore, foramen, e.g. external acoustic (auditory) meatus.

S. 7 (4) DEPRESSIONS OF ANATOMICAL STRUCTURES

Fissure; fissura, a narrow slit, cleft or groove; e.g. fissures of the lung, liver, brain.

Fossa; (L), a depression or hollow below the normal surface of a structure, a trench, ditch, e.g. putuitary fossa, cranial fossa. (pl. fossae)

Hilus; hilum; an indentation or depression on the surface of an organ, where vessels enter or leave the organ e.g., hilum of lung, kidney, lymph node (pl. hila).

Sulcus; (L) a groove or furrow, broader than a fissure sometimes has a tendon or artery, etc., running along it e.g. intertubercular (bicipital) groove of humerus, central sulcus of cerebrum, pl. sulci.

Sinus; three meanings:-
a **cavity** within an organ, bone, etc., the paranasal sinuses, sinus of kidney (cavity within it)
a **channel** for blood, etc; e.g. venous sinuses within the cranium.
a **canal** or passage from one organ to another, or to the outside by which pus etc., escapes.

S. 7 (5) ADDITIONAL DESCRIPTIVE TERMS

Incisura; (L) a notch or cleft, e.g. incisura angularis or angular notch of stomach, pl. incisurae.

Meatus; (L) a canal or tubelike passage, e.g. the external acoustic (auditory) meatus of the ear.

Ramus; (L) a branch or a division of a vessel, nerve or bone; e.g. ramus of the mandible, anterior ramus of a spinal nerve, etc. pl. rami.

Septum; (L) a partition separating two cavities, e.g. nasal septum, interatrial septum of heart.

Note: students must realize that the terms defined in the preceding sections are used in connection with many organs in addition to bones. For this reason they have been transferred to this general section. See additional terms, S. 47, bones.

S. 8 TERMS RELATING TO POSITION AND LOCATION F. 1-9.10.11

Medial or mesial; that part of any structure or organ that lies nearest to the median line of the body; medial malleolus of the tibia.
Note: middle has a different meaning, and refers to a structure lying between two other structures, such as the middle lobe of the right lung.

Lateral; that part of an organ or structure that lies farthest away from the median line of the body e.g. lateral malleolus of fibula.

Anterior; towards the front or in the front part of the body, an organ or other structure, e.g., the lower anterior teeth.

Posterior; towards the back, or in the back part of the body or a part of it, e.g. lower posterior teeth.

Ventral; in human anatomy refers to the front or anterior part - the same as anterior above.

Dorsal; in human anatomy refers to the back or the posterior part, e.g. dorsal surface of the forearm.

Palmar; refers to the front or palm of the hand.

Plantar; refers to the sole of the foot.

Volar; either the palm of the hand, or sole of foot.

Dorsal; REFERRING TO HAND = the back of the hand.
Dorsal; REFERRING TO FOOT = its upper surface.

Superior; refers to the upper part, or that part towards the head end, assuming the body to be in the anatomical position, e.g. superior lobe of a lung.
Inferior; refers to the lower part, or that part away from the head end, e.g. inferior lobe of a lung.

Cephalic; an adjective from Kephale (G) head; refers to the head or head end of the body or an organ.
Caudal; an adjective from cauda = tail; refers to the tail or tail end of the body, or of an organ;
Cranial; frequently used instead of cephalic.
Cephalad; towards the head; ad = towards; Caudad - towards the tail end.

Proximal; that part of a structure closest to its source or origin, its attached end e.g. proximal end of humerus.
Distal; terminal; that part farthest away from the source or origin or point of attachment, e.g. the distal end of the humerus.

Supine; the position assumed when lying upon the back with the face up, or if referring to the hands, with the palms up. Note - remember S & S = lying upon the sacrum, i.e. supine.
Prone; lying face down, or if referring to the hands, with the palms directed downwards. Note: P & P = lying upon the pubis - prone.

Longitudinal; lengthwise, i.e. along the long axis or length of a structure.
Transverse; crossways, or at right angles to the long axis or length of a structure, e.g. a cross section.

Vertical; perpendicular or at right angles to the horizon, assuming that the subject is standing up.
Horizontal; parallel to the horizon, subject upright.

Central; the inner part, farthest from the surface.
Peripheral; on or near the surface, outer part.

Superficial; on or near the surface, e.g. superficial veins of the forearm.
Deep; (L profundus) far from the surface, e.g. deep veins of the forearm.

Major; the larger or greater of two.
Minor; the smaller or lesser of two.

Internal; on the inside of the body, or a part of.
External; on the outside, the outer, e.g. external acoustic meatus of the ear.

Intrinsic; part of an organ itself.
Extrinsic; originating outside an organ.

Visceral; refers to some organ; from viscus = an organ; visceral pleura, visceral pericardium.
Parietal; refers to a wall, relating to the wall of a structure; from paries (G) = wall, e.g. parietal pleura, parietal pericardium.

Note: **lateral** is frequently used as referring to the side of the body or a part of it. Latus (L) side; **bilateral** refers to both sides, e.g. the bones of the upper limbs are bilaterally similar.

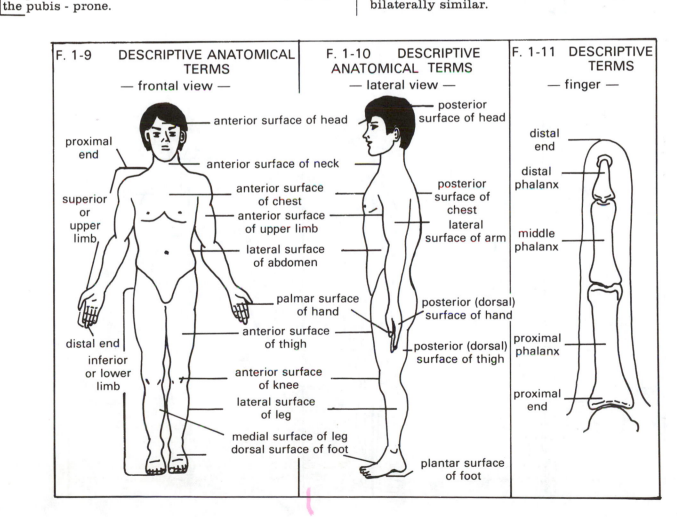

F. 1-9 DESCRIPTIVE ANATOMICAL TERMS
— frontal view —

proximal end
superior or upper limb
distal end
inferior or lower limb
anterior surface of head
anterior surface of neck
anterior surface of chest
anterior surface of upper limb
lateral surface of abdomen
palmar surface of hand
anterior surface of thigh
anterior surface of knee
lateral surface of leg
medial surface of leg
dorsal surface of foot

F. 1-10 DESCRIPTIVE ANATOMICAL TERMS
— lateral view —

posterior surface of head
posterior surface of chest
lateral surface of arm
posterior (dorsal) surface of hand
posterior (dorsal) surface of thigh
plantar surface of foot

F. 1-11 DESCRIPTIVE TERMS
— finger —

distal end
distal phalanx
middle phalanx
proximal phalanx
proximal end

S. 9 ANATOMICAL PLANES AND SECTIONS
F.1-12,13.

A PLANE: from planus (L) = flat; a real or imaginary flat surface. This might be made by taking a saw, cutting through the body or a part of it, then turning the part so as to view its cut flat surface. The cut is usually lengthwise or transverse. (A plane is a carpenter's tool used to make wood flat and smooth).

1. **A longitudinal plane** is a flat surface made by cutting lengthwise or along the long axis of the body or a part of it, then turning this so that the cut surface is visible. The cut could be made from side to side, or from front to back, and with the subject standing upright or lying down, or in any other position.

2. **A transverse plane** is a flat surface made by cutting through the body or part of it crossways, or at right angles to the long axis, then turning this so that the flat surface is visible. In this case the cut from front to back, or from side to side would have the same result. The subject might be standing or lying down.

3. **A vertical plane** is a **longitudinal** plane made with the subject upright, and the cut made perpendicular to the horizon, either from front to back or from side to side.

4. **A horizontal plane** is a **transverse** plane made with the subject upright, and the cut made parallel to the horizon.

Planes made in the longitudinal or transverse directions at specific locations have acquired specific names denoting their position in the body. Most planes are made in the longitudinal or transverse directions. The longitudinal planes might be made by cutting from front to back or from side to side. Some of these planes are defined below:

(1) **A midsagittal or median plane** is a longitudinal plane made by cutting from front to back along the median line of the body, and along the sagittal suture of the skull. This suture is a joint that passes from front to back on the top (vertex) of the skull along the median line hence sagittal.

(2) **A sagittal plane** is a longitudinal plane made by cutting from front to back to one or the other side of the sagittal suture, and parallel to the midsagittal plane (defined above).

(3) **A coronal plane**, or frontal plane, is a longitudinal plane made by cutting lengthwise and from side to side through the head and body or part of it along the coronal suture, or parallel to it. The coronal suture is a joint that extends across the skull from side to side behind the forehead.

(4) **A subcostal plane** is a transverse plane made by cutting across the upper abdomen at right angles to the long axis of the body, and opposite the 10th costal cartilages, the lowest limit of the thoracic cage.

(5) **A transpyloric plane** is a transverse plane made by cutting across the body from one side to the other at the level of the 9th costal cartilages or half way between the upper end of the sternum and the symphysis (the joint between the anterior ends of the two pubic bones). This should cut across the pyloric part of the stomach, hence the name.

A SECTION is a slice made by making two parallel cuts close to a plane, frequently one of those defined above, and similar to a slice of bread or beefsteak. The section is named according to the plane through which it passes - longitudinal, or transverse:

longitudinal section
transverse section

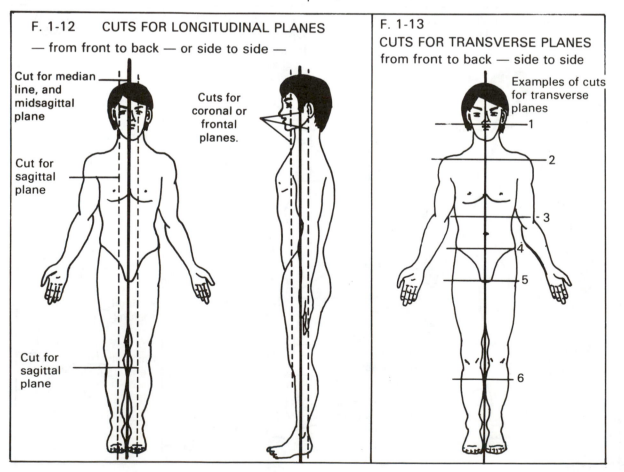

F. 1-12   CUTS FOR LONGITUDINAL PLANES
— from front to back — or side to side —

Cut for median line, and midsagittal plane

Cuts for coronal or frontal planes.

Cut for sagittal plane

Cut for sagittal plane

F. 1-13
CUTS FOR TRANSVERSE PLANES
from front to back — side to side

Examples of cuts for transverse planes
1
2
3
4
5
6

cross section
vertical section
horizontal section
median section or

midsagittal section
sagittal section
frontal section or
coronal section
subcostal section
transpyloric section

Many textbooks have illustrations of sections from various parts of the body.

PREFIXES AND SUFFIXES

Many anatomical and other medical terms have been built up from a root or stem and one or more syllables placed before or after the root or stem. These syllables are called prefixes (S. 10) or suffixes (S. 12). A knowledge of these will be of considerable value in determining the meaning of any term, e.g. duco means to lead, or bring, and from it many words are formed; abduct, adduct, induct, reduce, conduct, seduce, ductile, ductless, ductitis, etc.

S. 10   PREFIXES - AN ALPHABETICAL LIST

A prefix is one or more syllables placed before a name to modify, alter, or limit its meaning. The adjective clavicular is formed from clavicula, (collar bone), and from it are formed supraclavicular, above the clavicle; infraclavicular, below the clavicle, etc. From sternal, the adjective of sternum (breast bone) are formed retrosternal, behind the sternum; substernal, below the sternum, etc. Many anatomical terms contain prefixes and a large number of medical terms also have prefixes.

In the list catalogued below the number in the brackets refers to the number in Section 11 in which the prefixes are grouped and defined with examples.

a, ab	= away from (1)
a, ad	= to, towards (5)
a, an	= without (2)
ambi	= both sides, or around (6)
amphi	= both sides, or around (7)
ana	= up, towards, apart (8)
ante	= before, in front of (9)
anti	= against, opposed to (12)
auto	= self (16)
bi, bis	= two, twice, double (14)
circum	= around (17)
contra	= against, opposed to (13)
de -	= away, take away (19)
di, dis	= two, twice, double (15)
dia -	= through, across (20)
dys -	= difficult, bad (23)
ecto	= outer, outside (24)
en -	= in, inside, within (27)
endo	= in, inside, within (28)
ento	= in, inside, within (29)
epi -	= upon, on, over (32)
e, ex	= out of, outside (25)
extra	= outside, beyond (26)
hemi	= half (33)
homo	= the same (15)
hydro	= water (36)
hyper	= above, over (37)
hypo	= below, under (40)
in	= not (3)
in, intra	= in, inside, within (30 & 31)
infra	= below, under (41)

inter	= between or among (43)
leuko, leuco	= white (44)
macro	= large (45)
mal -	= bad, faulty, poor (47)
meta	= beyond, after (48)
micro	= small (46)
ortho	= straight (50)
pan -	= all (51)
para	= beside (52)
per -	= through, across (21)
peri	= around (21)
poly	= much, many (53)
Post	= after (54)
pre -	= before, in front of (10)
pro -	= before, in front of (11)
pseudo	= false (55)
re -	= again, back (56)
retro	= behind, backwards (57)
semi	= half (34)
sub -	= below, under (42)
super	= above, over (38)
supra	= above, over (39)
sym -	=together, with (60)
syn -	=together, with (61)
telo	= end, fulfilment (58)
trans	= across, through (22)
ultra	= beyond, after (49)
un -	= not (4) See also in
uni -	= one, single, one form (59)

S. 11 PREFIXES GROUPED DEFINED, ILLUSTRATED

Prefixes with similar meanings are grouped together, and nonmedical examples are given as the students may not have acquired a medical vocabulary.

1. a, ab = away from;
abnormal, away from, i.e., not normal
abduct, away from the midline + duco

2. a, an = not, or without;
anonymous, nameless, without a name;
atheist, without a God, (Theos = God),
anhydrous, without water
anemia, without (poor) blood (haema)
anuria, without urine, (uria = urine)

3. in = not, or without;
insoluble, not soluble, indestructible, cannot be destroyed
inefficient, not fully capable
inactive, not active
inoperable, cannot be operated upon

4. un =not;
unbalanced, not balanced
unbroken, not broken
unfair, not fair
unsanitary, not sanitary
unorthodox, not according to the accepted standards
unhealthy, not healthy.

5. ad = to, towards;
adduct, to bring towards the midline

6. ambi = both sides, around
ambidexterous, using both hands

7. amphi = both sides, around
amphibian, on land and water
amphitheater, seats surround stage
amphiarthrosis, a type of joint

8. ana = up, towards, apart;
anaphase, a stage in cell division

9. ante = before, in front of;
anteroom, a room in front of another
antedated, to date before present date
antebrachium, forearm (brachium = arm)

10. pre	= before, in front of;	

10. pre = before, in front of;
preamble, introduction to a document
precede, to go before
precocious, maturing early
premolar, in front of molar teeth

11. pro = before;
prologue, introduction to a play
prophase, first stage cell division

12. anti = against, opposed to;
anticommunist, opposed to communism
antisocial, against society
antiseptic, against, counteracting dirt

13. contra = against, opposed to;
contraception, to prevent pregnancy
contraband, prohibited import
contralateral, opposite side of

14. bi, bis = two, twice, double;
bilingual, two languages (lingua, tongue)
biplane, aeroplane with two wings
bipolar, two poles or ends
bifocal, two focuses, (near & far)
bipartite, two parts, bipartite patella

15. di, dis = two, twice, double;
dioxide, two atoms of oxygen
dicotyledonous, two cotyledons
dicephalic, two heads, two headed monster

16. auto = self;
automobile, self moving vehicle
autonomic, self regulating
automatic, self running

17. circum = around;
circumnavigate, sail around (world)
circumflex, to bend around

18. peri = around;
perimeter, the distance around
periscope, to see around
pericardium, covering around heart

19. de = take away, delouse, take away (get rid of) lice
defrost, take away (remove) frost
decapitate, take away the head, behead

20. dia = across, through;
diameter, distance through, (Meters)
diarrhoea, to flow through, rhoia (to flow)
diarthrosis, movable joint, with cavity

21. per = through, across;
perennial, through the years
percutaneous, through the skin

22. trans = across, through;
trans-Canada, across Canada
transcontinental, across a continent
transparent, to let light through
transposition, changing places, e.g. organ to opposite side

23. dys = difficult, bad, painful;
dysfunction, bad, impaired function
dysphagia, difficult swallowing
dysmenorrhoea, painful monthly periods

24. ecto = outer, outside;
ectoderm, outer layer primitive embryo;

25. e, ex = outer, outside;
eject, to force out, expel
exhale, to breathe out
export, to send out of the country
evert, to turn out

26. extra = outside, beyond
extramural, outside the walls
extracurricular, outside curriculum
extracellular, outside of a cell

27. en = in, inside, within;
encircle, to place within
engulf, to take inside, to swallow up
encephalic, brain (inside of the head)

28. endo = in, inside, within;
endoscope, instrument to look inside
endotheliium, lining membrane (inside);
endosteum, lining of cavity in a long bone

29. ento = in, inside, within;
entoderm, inner layer primitive embryo

30. in = in, inside, within;
inhale, to breathe in
inhabit, to live within

31. intra = in, inside, within;
intramural, within walls (of a school)
intravenous, inside a vein

32. epi = upon;
epigraph, inscription on a building
epitaph, inscription on a tombstone
epicondyle, upon a condyle (knuckle)

33. hemi = half;
hemisphere, half a sphere
hemithorax, half the thorax

34. semi = half;
semicivilized, half civilized
semicircle, half a circle
semilunar, half moon shaped
semiannual, half yearly, twice a year

35. homo = the same;
homogeneous, the same throughout
homosexual, the same sex

36. hydro = water;
hydroelectric, electricity by water-power
hydroplane, a plane able to land on water
hydrocephalus, water in the head
(actually in the ventricles of the brain)

37. hyper = above, over;
hypersensitive, unusually sensitive
hypercritical, critical above normal
hypertension, above normal blood pressure

38. super = above, over;
superman, an above normal man
supervoltage, above usual voltage
superficial, above, close to skin

39. supra = above, over;
suprarenal, above the kidney
supraorbital, above the orbit

40. hypo = below, under;
hypochondriac, below normal health
hypodermic, under the skin (dermis)

41. infra = below, under;
infrared, below the red rays (spectrum)
infraorbital, below the orbit

42. sub = below, under;
sub-basement, basement below basement
substandard, below normal standard
subway, a way (passage below a street)
submental, under the chin (mentum)

43. inter = between two, among if more than two;
interprovincial, between provinces
interstate, between states
intermarriage, between relatives
intercostal, between ribs (costae)

44. leuko leuco = white;
leukocyte, white blood cell
leukoplacia, a white plac (patch)

45. macro = large;
macroscopic, large enough to be seen with the naked eye
macrocephalic, a large head

46. micro = small;
microorganisms, very small living bodies
microphone, instrument for intensifying small sounds

47. mal = bad, poor, faulty;
malnutrition, poor nutrition
maladjusted, poorly adjusted

48. meta = beyond, after;
metacarpal, beyond the carpals
metaphase, a stage in cell division

49. ultra = beyond, after;
ultramodern, excessively modern
ultraviolet, beyond the violet rays

50. ortho = straight;
orthodox, correct (accepted) doctrine
orthography, correct writing
orthopedics, correction of deformities
of children (pais = child)

51. pan = all;
panAmerican, entirely American
panchromatic, sensitive to all colors
pansinusitis, inflammation all sinuses

52. para = beside;
paramour, illicit partner of married man
or woman
parathyroid, beside the thyroid
paratyphoid, in addition to typhoid
parasympathetic, beside the sympathetic

53. poly = much or many;
polygamy, having more than one spouse
polygon, a many sided figure
polyuria, passing much urine

54. post = after;
postgraduate, after graduation
postdated, dated after present date
postmortem, after death

55. pseudo = false;
pseudonym, false name
pseudoAmerican, not actually American
pseudoarthrosis, false joint

56. re - = again, back;
recall, to call or bring back
rebuild, to build over again
recapture, to capture again
reduce, to put back into normal position

57. retro = behind, backwards;
retroactive, effective before present
retrospect, looking backwards
retrograde, directed backwards
retrosternal, behind the sternum

58. telo = end, fulfilment;
telophase, final stage of cell division

59. uni = one, single, one form;
unilateral, confined to one side
uniform, similar garments
unicellular, one cell only

60. sym = together, with;
symposium, a set of writings
symphysis, a joint (growing together)

61. syn = together, with;
syndicate, group of persons
synthetic, simple substances put together
to form complex unit

PREFIXES WITH OPPOSITE MEANINGS PAIRED

ante(before)	- post(after)
endo(inside)	- ecto(outside)
hypo(below)	- hyper(above)
in(inside)	- ex(outside of)
intra(within)	- extra(outside)
macro(large)	- micro(small)
super(above)	- sub(under)
supra(above)	- infra(below)

PREFIXES WITH OPPOSITE MEANINGS GROUPED

ante, pre, pro	- post
e, ex, extra, ecto	- en, endo, ento, in, intra
hypo, infra, sub	- hyper, supra, super

S. 12 SUFFIXES GROUPED, DEFINED, & ILLUSTRATED

A suffix is one of more syllables added to the end of a word to modify, alter, or limit its meaning. Some anatomical terms have suffixes.

1. Algia = pain;
cephalalgia, headache
neuralgia, nerve pain

2. cele = swelling, hernia, tumor;
meningocele, a hernia of membranes
of the brain or spinal cord

3. centesis = a tapping or a puncture;
thoracocentesis, tapping thoracic
cavity to obtain fluid

4. ectomy = to cut out, to excise, remove;
appendectomy, removal of appendix
gastrectomy, removal of stomach
cholecystectomy, removal of gallbladder

5. otomy = an incision, a cut into;
cholecystotomy, cut into gallbladder
gastrotomy, cut into the stomach

6. ostomy = to make an opening or mouth;
gastrostomy, to make artificial opening in
stomach

7. graph = a writing, tracing, drawing, picture;
gram bronchograph, bronchogram

8. graphy = whole procedure of making tracing, e.g.
bronchography, all procedures

9. iasis = a condition, state, presence of;
cholelithiasis, presence of gallstones

10. osis = a condition, state, presence of;
diverticulosis, having diverticula

11. less = without;
hairless, without hair
spotless, without a spot, clean
ductless, without a duct

12. itis = an inflammation;
appendicitis, tonsillitis, bronchitis

13. oma = a tumor;
lipoma, a tumor of fatty tissue
osteoma, a tumor of bone

14. lith = a stone;
phlebolith, stone in wall of a vein
cholelithiasis, stones in gall bladder

15. pathy = disease;
adenopathy, disease of lymph nodes

16. ptosis = a falling down;
nephroptosis, a low kidney

17. oid or = like or resembling;
oides coronoid, resembling a crow's beak
condyloid, like a condyle

18. form = shape, shaped like;
pisiform, pea shaped
cuneiform, wedge shaped

19. logia = the science or study of;
logy zoology, the study of animals
biology, the study of living things

20. ulus = small;
lobulus, lobule, a small lobe

21. olus = small;
malleolus, a little hammer

22. culus = small;
corpusculum, a small corpus, body

23. uria = referring to the urine;
(G. ouron = urine)
dysuria, difficult urination

S. 13 ANATOMICAL TERMS FROM CHAPTER 1

anatomy	macroscopic anatomy
structure	histology or
gross anatomy or	microscopic anatomy

9

embryology or developmental anatomy
surface, (regional or topographic) anatomy
comparative anatomy
physiology
pathology
anatomical position
median line, plane
Nomina Anatomica (NA)
head or caput
little head, capitulum
neck, cervex, collum
body or corpus
shaft
tail or cauda
lobe or lobus
lobule or lobulus
segment or segmentum
extremity or extremitas
apex, apices
base
See under bones, prominences, etc.
aperture
foramen or foramina, pl

hiatus
orifice or orificium (L)
os, adj. oral
ossa, pl.
ostium
ostia, pl.
lumen
porus

fissure
fissura (L)
fossa
fossae, pl.
hilum or hilus
hila, pl.
sulcus
sulci, pl.
sinus
notch or incisura (L)
canal or meatus
branch or ramus
septum or partition

Terms describing location, etc.

medial, mesial
lateral
anterior
posterior
ventral
dorsal
palmar
plantar
volar
dorsal - hand
dorsal - foot
superior
inferior

cephalic
cranial
caudal
cephalad
caudad
proximal
distal
supine
prone
longitudinal
transverse
vertical
horizontal

central
peripheral
superficial
deep(profundus)
major, greater
minor, lesser
internal
external
intrinsic
extrinsic
visceral
parietal
bilateral

Anatomical planes and section;

vertical plane (section)
horizontal plane (section)
longitudinal plane (section)
transverse plane (section)
median plane (section)
midsagittal plane (section)
sagittal plane (section)
coronal plane (section)
frontal plane (section)
subcostal plane (section)
transpyloric plane (section)

Prefixes

a, ab	de	hydro	micro	restro
a, ad	di, dis	hyper	ortho	semi
a, an	dia	hypo	pan	sub
ambi	dys	in	para	super
amphi	ecto	in(or)	per	supra
ana	endo	intra	peri	sym
ante	ento	infra	poly	syn
anti	epi	inter	post	telo
auto	e, ex	leuko	pre	trans
bi, bis	extra	macro	pro	ultra
circum	hemi	mal	pseudo	un
contra	homo	meta	re	uni

Suffixes

algia	ostomy	osis	pathy	logia
cele	graph	less	ptosis	ulus
centesis	gram	itis	oid	olus
ectomy	graphy	oma	oides	culus
otomy	iasis	lith	form	uria

Examples of anatomical terms with prefixes;

abduct	exocrine	paratyphoid
abductor	hemisphere	percutaneous
adduct	hemithorax	pericardium
adductor	hypogastrium	peridental
adrenal	infraorbital	periosteum
amphiarthrosis	inhale	premolar
anaphase	interatrial	prepatellar
autonomic	intercarpal	prophase
bicuspid	intercondylar	retrograde
circumduct	intercostal	semilunar
diaphysis	interphalangeal	subarachnoid
ectoderm	interspinous	subclavian
endocardium	intertarsal	subcostal
endocrine	intertrochanteric	subcutaneous
endometrium	intertubercular	sublingual
endosteum	interventricular	submandibular
endothelium	intervertebral	subphrenic
entoderm	intracranial	superficial
epicardium	intravenous	supraorbital
epicondyle	mesoderm	suprarenal
epidermis	mesothelium	suprascapular
epididymis	metacarpal	supraspinous
epigastrium	metaphase	symphysis
epiglottis	metaphysis	synarthrosis
epiphysis	metatarsal	telophase
epithelium	paranasal	transpyloric
exhale	parasympathetic	

NOTES

— ORGANIZATION OF BODY STRUCTURES —

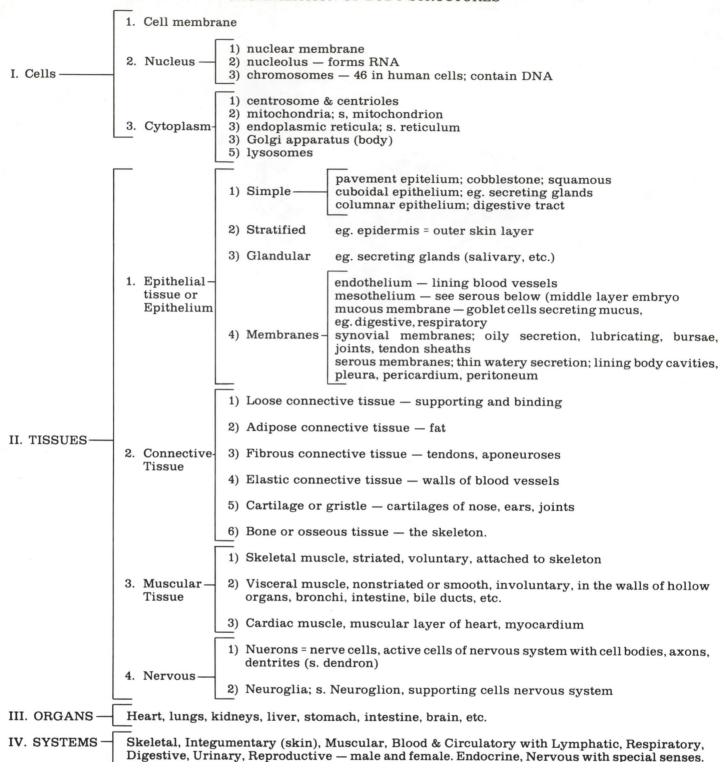

I. Cells

1. Cell membrane

2. Nucleus
   1) nuclear membrane
   2) nucleolus — forms RNA
   3) chromosomes — 46 in human cells; contain DNA

3. Cytoplasm
   1) centrosome & centrioles
   2) mitochondria; s, mitochondrion
   3) endoplasmic reticula; s. reticulum
   3) Golgi apparatus (body)
   5) lysosomes

II. TISSUES

1. Epithelial tissue or Epithelium
   1) Simple
      - pavement epitelium; cobblestone; squamous
      - cuboidal epithelium; eg. secreting glands
      - columnar epithelium; digestive tract
   2) Stratified — eg. epidermis = outer skin layer
   3) Glandular — eg. secreting glands (salivary, etc.)
   4) Membranes
      - endothelium — lining blood vessels
      - mesothelium — see serous below (middle layer embryo
      - mucous membrane — goblet cells secreting mucus, eg. digestive, respiratory
      - synovial membranes; oily secretion, lubricating, bursae, joints, tendon sheaths
      - serous membranes; thin watery secretion; lining body cavities, pleura, pericardium, peritoneum

2. Connective Tissue
   1) Loose connective tissue — supporting and binding
   2) Adipose connective tissue — fat
   3) Fibrous connective tissue — tendons, aponeuroses
   4) Elastic connective tissue — walls of blood vessels
   5) Cartilage or gristle — cartilages of nose, ears, joints
   6) Bone or osseous tissue — the skeleton.

3. Muscular Tissue
   1) Skeletal muscle, striated, voluntary, attached to skeleton
   2) Visceral muscle, nonstriated or smooth, involuntary, in the walls of hollow organs, bronchi, intestine, bile ducts, etc.
   3) Cardiac muscle, muscular layer of heart, myocardium

4. Nervous
   1) Nuerons = nerve cells, active cells of nervous system with cell bodies, axons, dentrites (s. dendron)
   2) Neuroglia; s. Neuroglion, supporting cells nervous system

III. ORGANS — Heart, lungs, kidneys, liver, stomach, intestine, brain, etc.

IV. SYSTEMS — Skeletal, Integumentary (skin), Muscular, Blood & Circulatory with Lymphatic, Respiratory, Digestive, Urinary, Reproductive — male and female. Endocrine, Nervous with special senses.

2. UNITS OF BODY STRUCTURE

S. 14 DIVISIONS OF THE BODY F. 2-1

The divisions of the body are;

1. The head
2. The neck
3. The trunk

   (1) Thorax or chest
   (2) Abdomen
   (3) Pelvis

4. The limbs or members (L. membrana):

   (1) Superior or upper limb or member,
   (2) Inferior or lower limb or member.

S. 15 THE BODY CAVITIES F. 2-2,3

The main body cavities are developed as the fundamental ventral (anterior), and dorsal (posterior) cavities.

The ventral cavity is formed as the celom in front of the vertebral column. It is divided into the thoracic, abdominal, and pelvic cavities.

(1) **The thoracic cavity** occupies the upper part of the trunk and is enclosed by the chest wall. Between it and the abdomen is the double dome-shaped muscular partition, the diaphragm. This muscle is attached at its base to the inner surface of the chest wall. The thoracic cavity is further divided by the vertical partition, the mediastinum, into the right and left pleural cavities.

(2) **The abdominal cavity** extends from the inferior surface of the diaphragm to the pelvic bones.

(3) **The pelvic cavity**, lying within the bony pelvis, is actually continuous with the abdominal cavity with no separating partition between the two.

Its upper limit is marked by a ridge of bone that passes around the inner surface of the pelvic bones. Its lower boundary is the floor of the pelvis, formed by muscles and ligaments. With no partition present to separate the abdomen and pelvis, part of an organ may lie in the abdomen and the remainder in the pelvis. Further, the position of an organ may vary, for instance in the upright and supine positions.

The term abdominopelvic cavity, since it suggests a single cavity, might be a more suitable term for the abdominal and pelvic cavities.

The false pelvis, an (NA) term is misleading. It is that part of the abdominal cavity that lies above the pelvic brim, with the upper parts of the pelvic bones behind it, and no bones in front. It does not lie within the pelvis. (See the Section describing the pelvic bones).

The dorsal cavity lies within the skull and the vertebral column. It forms two cavities:

(1) **The cranial cavity** lies within the cranium, and contains the brain.

(2) **The spinal canal** extends lengthwise within the spinal column and contains the spinal cord.

S. 16 SOME REGIONAL SURFACE AREAS DEFINED

Many areas of the body have been given names, a few of which have been defined below.

1. **The axilla** is the armpit, the space between the medial surface of the upper arm and the adjacent lateral chest wall.

2. **The groin** is the area of the oblique crease on the front of the body where the lower limb joins the trunk in front

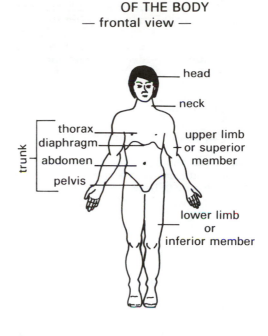

F. 2-1   DIVISIONS (PARTS) OF THE BODY
— frontal view —

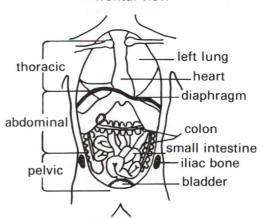

F. 2-2   THE VENTRAL BODY CAVITIES
— frontal view —

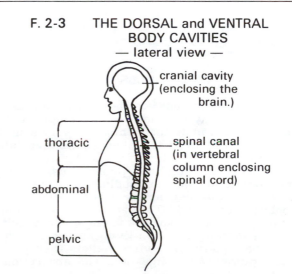

F. 2-3   THE DORSAL and VENTRAL BODY CAVITIES
— lateral view —

of the hip. This crease marks the location of the inguinal ligament, a ligament extending from the anterior superior spine of the ilium to the pubis.

3. The loin is the lateral side of the abdomen between the lowest rib and the upper margin of the ilium.

4. The lumbar region is one of the nine regions into which the abdomen is divided. It sometimes refers to the posterior abdominal wall adjacent to the lumbar vertebrae.

5. The buttock is the prominent area of the rump on either side lying lateral to and behind the hip. It is formed by the gluteal muscles, hence the name gluteal region.

6. The perineum is the space between the upper medial surfaces of the thighs, and extending from the anus behind to the pubic arch in front. It becomes visible when the thighs are separated. It contains the anus, the vaginal opening (in the female), and the opening of the urinary passage, the urethra.

S. 17 THE ORGANIZATION OF BODY STRUCTURES

The many parts of the body are composed of cells, which are organized into tissues. The tissues form organs which in turn form the systems of the body.

1. CELLS: the cell is the unit of structure and of function of the body. Cells are so small that they cannot be seen individually without the aid of a microscope. The entire body is composed of many trillions of cells of varying shapes and sizes. Each cell has its own function such as support, contraction, transportation, conduction of impulses, etc. These functions will be described in subsequent chapters.

2. TISSUES: groups of **similar** cells. As many cells combine to form a tissue it is usually visible to the naked eye. There are four types of tissues: epithelial tissue or epithelium, connective tissue, muscular tissue, and nerve tissue. These will be described in later sections.

3. ORGANS: groups of tissues organized into a unit and concerned with some **specific** function. For example the heart is an organ designed to pump blood to all parts of the body. It is composed of muscle, connective tissue, and epithelium. Other examples include the lungs, kidneys, spleen, liver, etc.

4. SYSTEMS: groups of tissues and organs arranged to perform some function. For example, the cardiovascular system is made up of the heart, arteries, veins and lymphatics. They work together to transport blood to all parts of the body, and to return it to the heart. Other examples include the skeletal system, muscular system, digestive system, etc.

S. 18 COMPOSITION OF CELLS:

Protoplasm is a name used to indicate living matter. Each cell of the body is composed of protoplasm. It is made of compound molecules which are very complex and varied in structure. The molecules are made up of atoms of oxygen, carbon, hydrogen, and nitrogen, with lesser amounts of sulphur, phosphorus, calcium, chlorine, iodine, magnesium, sodium, and potassium. The compound molecules include proteins, nucleic acid, carbohydrates, and lipids. The nature of their structure enables protoplasmic molecules to engage in activities not possible in inorganic (non-living) matter. By these activities they qualify as living matter. They include respiration, circulation, digestion, and absorption of food to produce energy, or to build up new complex compounds. Included also are the secretion of new compounds, excretion of waste, reproduction, and irritability. The latter results in conduction of impulses, and contraction with movement.

Body cells, being composed of protoplasm, breathe, circulate contents, digest and absorb food, break down complex molecules to produce energy, and build up new compounds to replace those used up. They form secretions, excrete waste, conduct impulses and cause movement. The composition and the arrangement of the various compounds vary with the cell function.

DNA or DEOXYRIBOSE NUCLEIC ACID consists of complex molecules located in the chromosomes of the nuclei of body cells. Each contains thousands of compounds with a nitrogen base, a sugar with 5 carbon atoms, deoxyribose, and a phosphate. The nitrogen base, the arrangement and number of the three constituents, and the number of units in the molecule may all vary. These various combinations are responsible for the multiple genes, and the transmission of hereditary traits.

RNA or RIBOSE NUCLEIC ACID is a compound molecule located in the nucleoli of cell nuclei, or in the cytoplasm of cells. It contains nitrogen, ribose, a sugar with 5 carbon atoms and a phosphate. It has ribose rather than the deoxyribose sugar of DNA. The number and arrangement of the units may vary.

S. 19 INSTRUMENTS USED IN MICROSCOPY

Since cells are microscopic they must be studied under a microscope that magnifies very minute structures, rendering them visible as separate entities.

The conventional or "light" microscope utilizes light and may magnify objects up to about 2000 times their actual size.

The electron microscope, utilizing electrons is capable of magnifying objects thousands of times their actual size. Our knowledge of the structure of cells has been revolutionized as a result of electron microscopy, and many previous conceptions of cell structure have had to be drastically revised. Structures that were previously invisible or only barely visible by light microscopy may now be examined and analysed. Undoubtedly our present concepts will have to be still further revised as further studies increase our knowledge. Students should study the diagrams and photos from electron microscopy.

For microscopic study of cell structure very thin slices (sections) of the structure are shaved off. These are dipped in dyes that stain the various parts different colors. The stained sections are mounted on glass slides and examined microscopically.

UNITS OF MICROSCOPIC MEASUREMENTS

Micron = 1/1000 millimeter, (1/25000 of an inch),
Millimicron = 1/1000 micron, (1/25,000,000 inch)
Angstrom unit = 1/10 millimicron, (1/250,000,000 inch)

S. 20 STRUCTURE OF CELLS - MICROSCOPIC ANATOMY
F. 2-5,6.

Cells consist of three parts. A cell membrane, a nucleus, and cytoplasm.

1. THE CELL MEMBRANE, also called the plasma membrane, is a very thin covering that surrounds each cell. It forms a wall to contain the cell contents but is permeable to water and many molecules, allowing them to pass in or out of the cell. The structure of this membrane may be quite complex, and it may contain pores (openings).

2. THE NUCLEUS is a centrally-placed rounded part of the cell that is separated from the remainder of the cell by the **nuclear membrane**. This membrane regulates the passage of materials into and out of the nucleus. The nucleus controls cellular activity and contains chromosomes and a small rounded granule, the nucleolus. (Nucleus, L. = a little nut)

(1) **The chromosomes** are long threadlike filaments contained within the nucleus. Using a light microscope they appear to consist of a single continuous thread or skein. Investigation with an electron microscope, however, shows that there are 46 separate chromosomes in each nucleus of the cells of the human body. When cells are not dividing, stained specimens demonstrate many small colored granules forming part of each chromosome - the chromosome granules. Chromosomes contain proteins and DNA molecules that carry genes, possibly 20,000 per chromosome. In some way these genes transmit parental characteristics in cell division. Chromosomes are important factors in cell division.

(2.) **The nucleolus** is a rounded body within the nucleus. It contains RNA molecules and proteins. Possibly it manufactures RNA that is expelled into the cytoplasm. (Nucleolus, L.= nucleus + olus, hence a little nucleus).

3. THE CYTOPLASM is that part of a cell that lies outside of the nucleus. Under light microscopy many minute granules are observed in the cytoplasm. Electron microscopy has shown these granules to be actually complex units each containing some type of membrane. These units, also named organelles, include :

mitochondria - sausage shaped bodies (s. mitochondrion) the power plants, with enzymes producing chemical reactions that produce energy.

endoplasmic reticula - minute canals (s. reticulum) that build up proteins (synthesize proteins),

Golgi bodies - minute vesicles, i.e. small sacs that concentrate or condense intracellular materials,

lysosomes - minute droplets that digest proteins, etc.

centrosomes - small spherical bodies containing two cylinders which take part in cell division,

intracellular fluid - the fluid within the cytoplasm,

other granules - of proteins, carbohydrates, fats, pigments etc., in the cytoplasm.

THE CELL BODY consists of the nucleus and surrounding cytoplasm.

CELL PROCESSES are projections that extend out from the cell body in some types of cells, such as nerve and bone cells. They may be short, long, threadlike or thick, multiple or single.

SOMATIC CELLS is a term used to include all the cells of the body except the genetic (reproductive) cells.

GENETIC CELLS OR **REPRODUCTIVE CELLS** include those cells that produce an embryo, the ovum, or egg in the female, and the spermatozoon, or sperm, in the male.

S. 21 TISSUES; MICROSCOPIC AND GROSS ANATOMY F. 2-7

As defined previously a tissue is a group of cells that are **similar** in structure and function, and which are organized into a unit.

The cellular structure of tissues must be studied under a microscope, although the tissue mass itself is usually large enough to be examined by the eye without magnification.

THERE ARE FOUR BASIC TISSUES

1. Epithelial tissue or epithelium
2. Connective tissue (see below)**
3. Muscular tissue
4. Nervous tissue - nerve tissue

Each of these four basic types is further subdivided into several varieties that have basic similarities, but also have differences. For example, muscular tissue includes skeletal muscle, visceral muscle, and cardiac muscle.
** Some authorities divide connective tissue into two main groups - connective tissue, and sclerous (skeletal) tissue, hence 5 groups not four.

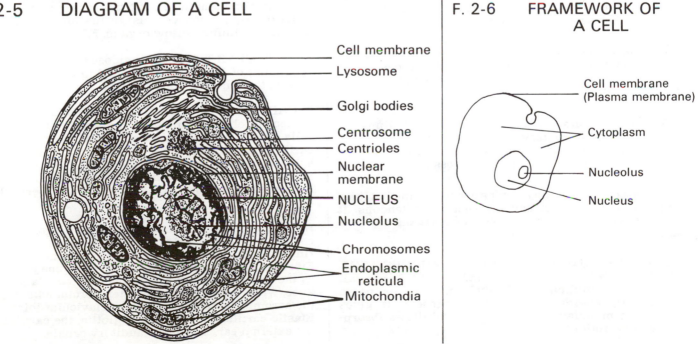

F. 2-5   DIAGRAM OF A CELL

- Cell membrane
- Lysosome
- Golgi bodies
- Centrosome
- Centrioles
- Nuclear membrane
- NUCLEUS
- Nucleolus
- Chromosomes
- Endoplasmic reticula
- Mitochondia

F. 2-6   FRAMEWORK OF A CELL

- Cell membrane (Plasma membrane)
- Cytoplasm
- Nucleolus
- Nucleus

Intercellular substance is the material that lies between adjacent cells in any tissue. It may be minimal - a thin layer of "cement" between cells. It may be abundant when the cells are somewhat separated from each other. It may be fluid or jellylike, fibrous, or even solid as in bone.

S. 22 EPITHELIAL TISSUE - EPITHELIUM F. 2-7

Epithelium: a thin sheet of tissue composed of cells cemented together to form a covering or lining membrane, such as the skin, covering of a lung, or lining of a blood vessel or the intestine, etc.

Pavement epithelium - single layer of flat cobblestone or tilelike cells.

Cuboidal epithelium - a layer of cube-shaped cells cemented together.

Columnar epithelium - a layer of cylindrical cells joined together.

Polyhedral epithelium - many-sided cells are cemented together.

Goblet cells, shaped like wine goblets, are found in columnar epithelium. They secrete mucus, a clear sticky colorless fluid. They are found in mucous membranes of the digestive tract, etc.

Simple epithelium consists of a single layer of cells cemented together to form a continuous sheet. It may be composed of flat, cuboidal, columnar, or polyhedral cells. It forms the linings of blood vessels and the intestine, the coverings of the heart, lungs, etc.

Stratified epithelium consists of several layers of cells cemented together to form a membrane. The mouth, esophagus, vagina, skin, and many other structures have this type of epithelium. (Stratum = a layer.)

In stratified squamous epithelium the outer cells are scalelike, having lost their cellular form.

Endothelium is a type of simple epithelium with a single layer of cells forming the lining of blood vessels. (Endo = inside). It is derived from the mesothelial layer of the embryo.

Mesothelium is another type of simple single-layered epithelium that lines the body cavities, the pleura, pericardium, and peritoneum. The name "meso" denotes its origin from the middle or mesodermal layer of the embryo. (See serous membrane).

Mucous membrane is a type of simple or stratified epithelium that contains goblet cells and secretes mucus. It lines many organs including the respiratory organs the mouth, stomach and intestine.

A serous membrane is a type of simple epithelium that secretes a thin watery colorless fluid. It is the mesothelium of the pleura, pericardium, and the peritoneum. (Serous = whey, a colorless fluid).

A synovial membrane is a type of epithelium that lines joint cavities, bursae, and the sheaths of tendons. It secretes an oily substance that lubricates the adjacent surfaces. (Synovia = joint oil).

ENDOTHELIUM, MESOTHELIUM, MUCOUS MEMBRANE, AND SYNOVIAL MEMBRANES are all varieties of epithelium. Epithelium contains nerve endings (receptors) but no blood vessels. It gives protection, manufactures secretions, and allows absorption and excretion.

S. 23 CONNECTIVE TISSUES

The connective tissues are the supporting and binding tissues of the body. Their functions are to support, connect or bind other important tissues and to fill in spaces within and about organs. Connective tissues have cell bodies that are often well separated by an intercellular substance or matrix that varies with the type of connective tissue. There are several types:-

1. LOOSE (AREOLAR) CONNECTIVE TISSUE consists of a semiliquid or jellylike matrix surrounding well separated cells called fibroblasts (spindle shaped), histoplasts, plasma, and mast cells. The matrix has a network of loosely arranged fibers. These fibers include wavy white bundles of fibers containing collagen that yield gelatin with boiling, and yellow elastic fibers that are stretchable. These fibers are woven into a loosely knit tissue. This kind of tissue is located under the skin, between it and adjacent muscles, leaving the skin freely movable. It also separates muscles, and binds together muscle bundles. It binds together the several tissues in an organ, and surrounds many structures. F. 2-7

2. ADIPOSE (FATTY) TISSUE is modified connective tissue. The nucleus of each cell is pushed to one side by fat that is deposited in the cytoplasm. The fat cell resembles a signet ring. Some fat will be found wherever there is connective tissue. There is a layer of fat under the skin that acts as insulation. Fat forms a layer about some organs such as the kidney. It may form deposits in any organ containing connective tissue. F. 2-7

3. FIBROUS TISSUE, white fibrous tissue is a type of connective tissue in which the matrix consists of bundles of collagen fibers that either lie parallel to each other or form a crisscross of fibers when in sheets. These bundles form strong cords or sheets of tissue. Tendons are composed of the white fibrous bundles with a few cells at the margins of the bundles. Ligaments and various aponeuroses (sheets of tissue) are also fibrous tissue. These cords or sheets are strong and break or tear with difficulty. F. 2-7

4. ELASTIC TISSUE, ELASTIC CONNECTIVE TISSUE, is largely composed of elastic fibers that are contractile and stretchable. There are scattered cell bodies. This type of tissue is found in structures that must expand and contract. For example, elastic tissue is contained in the walls of large blood vessels, in the lungs, and in the membranes lining hollow organs. F.2-7

5. RETICULAR TISSUE is composed of a fine network (a reticulum) of fibers and is located in the lymph nodes, spleen, thymus, and bone marrow.

6. CARTILAGE OR GRISTLE consists of oval shaped nucleated cells surrounded by a matrix that cements the cells into a firm but flexible structure. The cell bodies of cartilage are often paired. In cartilage the intercellular substance or matrix forms the bulk of the tissue. The matrix may be clear (hyaline) or may have white collagen fibers (fibrocartilage) or elastic fibers (elastic cartilage).

Hyaline cartilage forms articular cartilages at joints, costal cartilages, nasal and laryngeal cartilages, and the ringlike cartilages of the trachea and bronchi.
Fibrocartilage forms articular discs at some joints such as the intervertebral discs, semi-lunar cartilages at the knee, wrist, symphysis pubis, temporomandibular, and the acromioclavicular and sternoclavicular joints.
Elastic cartilage forms the epiglottis, the cartilages of the external ear, and of the auditory canals.

F. 2-7 TISSUES

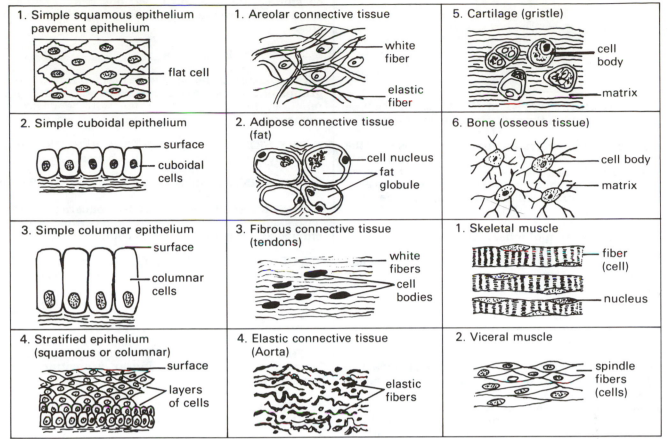

1. Simple squamous epithelium pavement epithelium — flat cell	1. Areolar connective tissue — white fiber, elastic fiber	5. Cartilage (gristle) — cell body, matrix
2. Simple cuboidal epithelium — surface, cuboidal cells	2. Adipose connective tissue (fat) — cell nucleus, fat globule	6. Bone (osseous tissue) — cell body, matrix
3. Simple columnar epithelium — surface, columnar cells	3. Fibrous connective tissue (tendons) — white fibers, cell bodies	1. Skeletal muscle — fiber (cell), nucleus
4. Stratified epithelium (squamous or columnar) — surface, layers of cells	4. Elastic connective tissue (Aorta) — elastic fibers	2. Viceral muscle — spindle fibers (cells)

7. BONE OR OSSEOUS TISSUE is modified connective tissue. It consists of nucleated cells with many processes extending out from the cell bodies like the legs of a spider. The spaces about the processes and between cells are impregnated with calcium phosphate forming a rigid matrix. In compact bone, the bone cells form concentric layers around a central canal - the Haversian canal. This canal contains minute blood vessels and nerves. The microscopic appearance is that of a central canal surrounded by a circular layer of calcium phosphate then a circular layer of bone cells and processes, a further layer of calcium phosphate outside of the cellular layer, another layer of bone cells, etc. All of these successive layers encircle the central canal and form a Haversian system. Bones being rigid, supply a framework upon which the body is built, and in certain locations, such as the skull, they afford protection to many vital structures. F. 2-7

S.24 MUSCULAR TISSUE: MUSCLE TISSUE F.2-7 is composed of elongated, cylindrical or spindle-shaped cells cemented together to form bundles or sheets. Muscle cells are also called **muscle fibers**. Minute fibrils (little fibers) or myofibrils extend from one end of a muscle fiber to the other in the cytoplasm. These fibrils by contraction can shorten the muscle. Bundles of muscle cells are bound together, side by side, and end to end. The bundles of fibers are also bound together by connective tissue. The whole muscle, consisting of many bundles, is often enclosed in a connective tissue sheath or covering. There are three kinds of muscles, skeletal, visceral, and cardiac.

1. SKELETAL MUSCLE: VOLUNTARY OR STRIATED MUSCLE F.2-7 is usually attached to bones across a joint. It has sensory and motor nerves supplying it. It contracts in response to messages transmitted along motor nerves from the brain. It is called "voluntary", because it may be made to contract at will. It has cross markings that are visible microscopically hence the name striated. (Stria = a line, band). Each fiber or cell is cylindrical in shape and may be 3.2 cm or 1.5 inches in length, or much shorter. Each slender cylinder is cemented by its cell membrane to other fibers, side by side and end to end. Each cell or fiber has several nuclei located along the margins - so multinuclear. Contraction of skeletal muscle results in movement.

2. VISCERAL MUSCLE: INVOLUNTARY: NON-STRIATED or smooth muscle is found in the walls of many hollow organs such as the stomach, intestine, gall bladder, blood vessels, etc., hence the name visceral, from viscus = an organ. F. 2-7

Since these organs are controlled by the autonomic nervous system, and not by the cerebrum the muscle is termed involuntary. The muscle fibers (cells) are spindle-shaped with pointed ends. Each has a single nucleus and many myofibrils that run lengthwise; their contraction shortens the fibers. The longer fibers are said to measure about 0.5 mm or 1|50 inch in length. The fibers are joined side to side and end to end to make up sheets of muscle. Fibers do not have cross markings so are labelled nonstriated or smooth muscle.

There are frequently two layers of visceral muscle in the covering of a hollow organ. The inner **circular layer** has its fibers encircling the organ while the outer **longitudinal layer** has its fibers running lengthwise. Contraction of the circular layer will cause a decrease in the size of the cavity. The visceral muscle layers are responsible for peristalsis. Visceral muscle is capable of considerable distension to accommodate the contents of an organ filled with gas, fluid, etc.

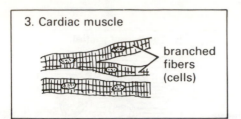

3. Cardiac muscle

branched fibers (cells)

3. **CARDIA MUSCLE** OR **THE MYOCARDIUM** is a type of muscle found only in the wall of the heart. (Cardia, G.= heart). F. 2-7

The fibers are cylindrical similar to those of the skeletal muscle, but branched. Each fiber has a single nucleus, and has cross markings, so is termed striated. The fibers are joined together side to side and end to end. The arrangement of the fibers encircling the heart is quite complex.

Note: the classification of muscular tissue according to location seems to be the most logical one since there is no overlapping — skeletal, visceral, and cardiac. When classified as to striations or as voluntary or involuntary there is definite overlapping of the three groups.

S. 25 **NERVOUS TISSUE**

The cells of the nervous system are of two types nerve cells or neurons, and neuroglia.

1. **NEURONS,** or nerve cells, consist of a cell body with a central nucleus and two sets of processes, an axon, and dendrites.
The axon is a single slender process extending out from the cell body. It may be very short or may be two or three feet in length. It conducts nerve impulses away from the cell body.
Dendrites are several processes that often extend out from the opposite pole of the cell body to the axon. They carry impulses towards the cell body. They are called dendrites because they resemble the branches of a tree. (Dendron = branched tree). s. dendron; pl. dendrites.

Neurons are of two kinds - sensory or motor. Sensory neurons convey sensory impulses from the skin or other structures to the spinal cord and brain. Sensations of touch, pressure, pain, heat and cold, as well as of sight, hearing, smell, and taste, are transmitted.
Receptors are minute structures at the distal ends of dendrites that pick up sensory impulses.
Motor neurons convey impulses from the brain and spinal cord out to muscles or secreting glands thereby initiating contraction or secretion.
Effectors are minute plates or branching fibrils at the distal ends of axons of motor neurons that transmit impulses to muscle fibers or glands.

2. **NEUROGLIA** are the supporting cells of the nervous system and are placed among and around neurons. There are several types that do not concern the technician, except when tumors arise from them. Students should refer to the Nervous System in a later chapter.

S. 26. **ORGANS**

These were defined in S. 17, No. 3 as groups of tissues organized into units concerned with some specific bodily function. Frequently they contain epithelial, connective and muscular tissues and nerve tissue. The heart was listed as an example of a pump. A further excellent example is the lung, with an epithelial covering and lining, visceral muscle, cartilaginous rings in its bronchi, and elastic and other connective tissues. Its blood vessels have assorted tissues in their walls. Since so many tissues are involved, a tumor of the lung could be composed of any one of them, not just the epithelial lining. The human body has many organs and these will be studied along with the various body systems. Examples cited are the lungs, kidneys, spleen, liver, brain, etc.

S. 26 A **SYSTEMS**

The systems of the body were defined in S. 17 No. 4, as groups of tissues and organs arranged to perform some function. In the Handbook ten will be described in sequence.

1. **THE SKELETAL** OR **OSSEOUS SYSTEM,** including adjacent joints as well as all the bones;

2. **THE SKIN, CUTIS,** OR **INTEGUMENTARY SYSTEM;**

3. **THE MUSCULAR SYSTEM,** concerned primarily with skeletal muscles. Visceral and cardiac muscles are studied along with the organs containing them.

4. **THE BLOOD** AND **CIRCULATORY SYSTEM,** including the heart, arteries, veins, lymphatics and the blood forming organs;

5. **THE RESPIRATORY SYSTEM,** including the respiratory passages, lungs, and pleura;

6. **THE DIGESTIVE SYSTEM,** with the mouth, throat, esophagus, stomach, small and large intestine, liver, gall bladder, pancreas, and bile ducts.

7. **THE URINARY SYSTEM,** including the kidneys, the ureters, bladder, and urethra.

8. **THE REPRODUCTIVE SYSTEM**

    (1) **Female:** ovaries, uterine tubes, uterus, vagina, external genitals, breasts.

    (2) **Male:** testes, seminal vesicles, prostate gland, epididymides, deferent ducts, seminal ducts, ejaculatory ducts, and penis.

9. **THE ENDOCRINE GLANDS:** including the ductless glands; - pituitary, pineal, thryroid, parathyroid, pancreas, suprarenal, testes, ovaries, and some hormonal glands in the digestive system.

10. **THE NERVOUS SYSTEM:** including the brain, and spinal cord, cranial and spinal nerves, and the autonomic nervous system.

Caution: frequently the term "system" is used in reference to part of a system, e.g. the arterial system, venous system, central nervous system, the autonomic nervous system, etc. The lymphatics are often classified as a separate system.

S. 27 **PHYSIOCHEMICAL PROCESSES IN CELLS & TISSUES**

Cell membranes and nuclear membranes of body cells have minute pores and are semipermeable. That is to say they will allow water and substances with small molecules dissolved in water to pass in and out through them. They will not permit substances having large molecules to pass through. Thus undigested food proteins, fats and carbohydrates are not passed.
The very thin walls of the blood and lymph capillaries, of the minute tubules of the kidney, of the lining membrane of the lungs and intestine are also semipermeable. These also allow water, small molecules of dissolved substances, and gases in solution, to pass in and out.

There is a continual flow of water and these solutes into cells or out of cells, into intercellular spaces surrounding cells, as well as into and out of capillaries.

This exchange is accomplished by processes of filtration, diffusion and osmosis. These have been defined briefly below.

FILTRATION is a process by which water, with its dissolved smaller molecules, passes through cell or vessel walls in the same concentration as in the original solution. This passage is due to a difference in pressure on the two sides of the membrane. At the arterial end of a capillary the pressure within the vessel is high, forcing the solution out. At the venous end of the capillary the capillary pressure is lower than that of the intercellular spaces forcing the solution into the capillary. The kidneys provide a good example of filtration. The fluid part of the blood, and its smaller dissolved substances, pass from the renal capillaries into the minute tubules of the kidney in the same concentration as was present in blood. The larger protein molecules are not filtered out.

DIFFUSION is the movement of molecules from an area of greater concentration to one of lesser concentration within a cell or through a semipermeable membrane. For instance, oxygen breathed into the lungs is in a greater concentration than in the surrounding capillaries because the body tissues have taken oxygen out of the blood. The flow of oxygen in the lungs will therefore be from lungs to capillaries. Carbon dioxide will flow in the opposite direction from the capillaries to lungs since its concentration in the capillaries is higher.

OSMOSIS is the passage of water through a semipermeable membrane towards the side with the greater concentration of the dissolved substance. By this means an attempt is made to reduce the difference in concentration. The large molecules of blood proteins attract water into the capillaries. Under normal circumstances the percentage concentration of salt in the blood plasma and red blood cells is the same. If an intravenous injection of distilled water is given the salt in the plasma will be diluted. The red blood cells will take in water in an attempt to equalize the pressure, and will sometimes swell up and explode.

HOMEOSTASIS is a term used to designate the maintenance within certain normal limits of the percentage concentration of the various components of the blood, body cells, intercellular spaces, etc. For example, the percentage concentration of sugar in the blood is kept at a fairly constant level by the burning up of some and by the storing of some in the liver and muscles until required. It also includes a normal level of functioning of bodily organs so that the body temperature, blood pressure, pulse rate, rate of respiration, and other body functions, lie within normal limits.

S. 28 CELL AND TISSUE DENSITY

The density of any substance is the weight (mass) per unit volume.

Density   is expressed as:
      grams per cc.
      pounds per quart
      pounds per bushel, etc.

Examples:
      1 cc. of water weighs 1 gram,
      1 cc. of lead weighs 11.34 grams,
      1 cc. of gold weighs 19.34 grams,
Students should not confuse the term density as defined here with density referring to the density of an exposed and processed x-ray film, as in radiography.

CELLULAR, TISSUE, and ORGAN DENSITY

The density of various tissues and organs of the body varies considerably and depends upon the anatomical structure such as:

1. The composition of the cells of the organ;
2. Whether the cells are closely packed together or are widely separated;
3. The composition of the material filling in the spaces between cells - the matrix;
4. The thickness of the walls of hollow organs;
5. Whether hollow organs are empty or are filled with air or other gas, or some liquid or solid.

EXAMPLES:

Fat is very light or of low density, so much so that it will float on water. Cream may be skimmed from the top of a pan of milk after a period.
Cartilage is relatively light.
Muscle is more dense than fat or cartilage.
Bone is quite dense in comparison because of the calcium phosphate forming the matrix between cells.
The liver, being a solid organ with closely-packed cells, is dense.
The kidney, although it has a central cavity, has a thick wall and is relatively dense though hollow.
The heart has a thick wall, and its cavities are filled with blood, so it is relatively dense.
The stomach and intestines have thin walls and if empty, are light, but when filled with fluid or food are dense.
Large organs in large subjects are not more dense per unit volume, but are heavier because of their size.
The radiographic technician should become familiar with the density of various body structures because radiography is influenced by density and thickness.

MEDIA:

A medium is an agent; pl, media. The dollar is a medium of exchange, the telephone is a medium of communication, the crystal gazer is an agent by which communication may be made with some one in the other world.

In radiography a medium is an agent used to render hollow organs, often with thin walls, visible. The medium may be lighter than the organ, (translucent) or more dense than the organ, (opaque). Frequently a medium is used to fill hollow organs, such as the stomach, gall bladder, etc., so that their outline will form an image on the processed film. The organ, if empty, would not form a separate image because of the thinness of its walls and its similarity in density to adjacent tissues.

S. 29 ANATOMICAL TERMS FROM CHAPTER 2.

Divisions of the body;
  head
  neck
  trunk
  thorax
  abdomen
  pelvis
limbs or members;
  superior or upper
  inferior or lower

Body cavities;
  ventral body cavity
  thoracic
  abdominal   abdomino
  pelvic   pelvic

dorsal body cavity;
  cranial cavity
  spinal, vertebral canal

Protoplasm;
  DNA. RNA.

Organization of body
structures;
  cells
  tissues
  organs
  systems

Systems;
  skeletal + joints
  skin, integumentary
  muscular
  blood + circulatory
  or cardiovascular,
  respiratory
  digestive
  urinary
  reproductive;
    female
    male
  endocrine glands
  nervous

Cells;
  cell membrane
  nucleus
  nuclear membrane
  nucleolus
  chromosomes
  cytoplasm
  centrosome
  centriole
  mitochondria
  endoplasmic reticulum
  Golgi's apparatus
  lysosomes
  cell body
  cell processes
  somatic cells

genetic or reproductive cells
tissues;
  epithelial, epithelium
simple;
  pavement
  cuboidal
  columnar
  polyhedral
  stratified & squamous
  endothelial, endothelium
  mesothelial, mesothelium
  mucous membrane
  serous membrane
  synovial membrane
  connective tissues
  areolar or loose
  adipose or fatty
  fibrous, collagenous
  elastic
  reticular
  cartilage or
  cartilaginous
  bone or osseous

muscular tissue;-
  skeletal, or striated
  or voluntary
  visceral, or nonstriated, or
  involuntary cardiac (myocardium)
  striated,
Nervous tissue;-
  neuron or nerve cell
  axon
  dendrites, s. dendron
  cell body
  neuroglia or s. neuroglion

Density;
  scale of tissue densities
  density of hollow organs
  thickness & density
  media, s. medium
  translucent or
  lucent media
  opaque media

NOTES

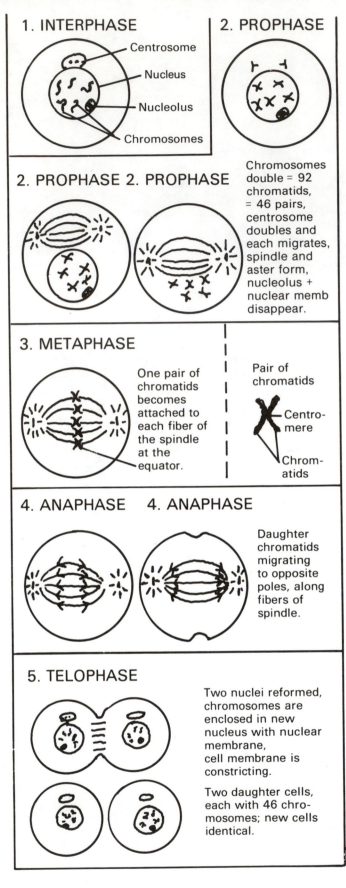

1. INTERPHASE

Centrosome
Nucleus
Nucleolus
Chromosomes

2. PROPHASE

2. PROPHASE 2. PROPHASE

Chromosomes double = 92 chromatids, = 46 pairs, centrosome doubles and each migrates, spindle and aster form, nucleolus + nuclear memb disappear.

3. METAPHASE

One pair of chromatids becomes attached to each fiber of the spindle at the equator.

Pair of chromatids

Centro-mere

Chrom-atids

4. ANAPHASE   4. ANAPHASE

Daughter chromatids migrating to opposite poles, along fibers of spindle.

5. TELOPHASE

Two nuclei reformed, chromosomes are enclosed in new nucleus with nuclear membrane, cell membrane is constricting.

Two daughter cells, each with 46 chromosomes; new cells identical.

3. CELL DIVISION

The cells of the body are divided into two groups, genetic and somatic.

Genetic cells or reproductive cells refer to the ova (eggs) and spermatoza (sperms).

Somatic cells include all the cells of the body except the genetic cells. Cell division proceeds differently in these two groups, by mitosis in the somatic, and by meiosis in genetic cells.

Growth of the body takes place by the division of existing cells, thereby multiplying the number of cells, and increasing the size of the structure. All body tissues are formed by the division, redivision and differentiation into many millions of cells of a single fertilized ovum. Two daughter cells are formed by the division of a fertilized ovum. These mature, divide, and redivide. This process is repeated over and over. When growth is complete cell division in some instances ceases.

Repair of some tissues following an injury or disease is accomplished by a similar process of division of existing cells. These replace lost tissue or fill in the gaps. For example, when a bone is fractured, or when skin is destroyed, bone or skin cells repair the damage. Some human tissues do not regenerate and the gaps are filled in by fibrous connective tissue (scar tissue). Thus when a muscle has been torn across, scar tissue bridges the gap. When a large area of skin has been destroyed a connective tissue scar will replace the original epidermis. Contraction of this scar will result in some deformity.

S. 30 **MITOTIC CELL DIVISION** - microscopic study, F. 3-1

Mitosis is cell division in somatic cells by the formation of a threadlike spindle, duplication of threadlike chromosomes of the cell nucleus, and the formation of two identical daughter cells. The cytoplasm also divides into two during this procession of division.

Phase: each stage in the process of cell division is termed a phase, although cell division is actually a single continuous process. (Phase = appearance). Four phases or stages are usually listed in the process but if the resting stage between divisions is included there are five;-
1. Interphase = resting stage or phase
2. Prophase = preliminary stage
3. Metaphase = stage of equatorial grouping
4. Anaphase = stage of separation
5. Telophase = completion of division

1. **Interphase** or resting stage is the between stage before active cell division has begun, inter = between + phase. Some of the steps listed in prophase, described below, are said to begin during the interphase.

2. **Prophase** or preliminary stage includes all the initial steps that occur in preparation for the next stage or metaphase, (pro = before + phase).
    (1) the 46 chromosomes become condensed, rodlike, and shortened, and visible by a light microscope when stained; F. 3-1
    (2) **duplication** - chromosomes become doubled and form 46 pairs of **chromatids,** (92), each one of a pair identical to its mate, each gene duplicated
    (3) paired chromatids **separate** except at one point near the center - the **centromere;**
    (4) **the centrosome divides** into two centrosomes, one of which migrates to each pole (or end) of the cell body; (centrosomes contain 2 centrioles);

    (5) **a spindle** of fibers forms between the newly formed centrosomes;
    (6) **aster formation** - other fibers radiate out from each centrosome in all directions, (aster = a star).
    (7) the nuclear membrane disappears;
    (8) the nucleolus also disappears.

3. **Metaphase** or equatorial grouping; the chromatids line up at the equator of the spindle, (meta = beyond or after + phase).
    (1) the paired chromatids become arranged at the equator of the spindle, one pair attached to a fiber of the spindle;
Note: the equator of a dividing cell is a line drawn to encircle the spindle halfway between the two poles (ends) of the spindle.

4. **Anaphase** or separation; (ana = apart)
    (1) the paired chromatids separate from each other;
    (2) one chromatid of each pair migrates along its fiber to a centrosome at each pole;
    (3) the forty-six chromatids become chromosomes or fine filaments grouped at the centrosome.

5. **Telophase** or completion of cell division; (telo = fulfilment or end of + phase).
    (1) The 46 chromosomes, threadlike filaments, form a new nucleus close to the centrosome;
    (2) the spindle and aster disappear;
    (3) a nuclear membrane forms around the 46 chromosomes at each end of the dividing cell;
    (4) the nucleolus reforms;
    (5) the cytoplasm divides; the cell membrane now becomes constricted at the equator, and pinches off forming two separate daughter cells with nuclei.
    (6) Each daughter cell is a duplication of the parent cell with 46 chromosomes and its genes.

Note: there may be some confusion as to what does occur during each phase of division, particularly during the resting stage and prophase. The important fact is that division is a continuous process.

The rather detailed description of mitosis that is included here has become necessary due to the knowledge required for radiobiology.

SOME DEFINITIONS RESTATED

Chromosomes are fine filaments, 46 in each nucleus in human somatic cells, contain DNA molecules all having many genes per chromosome, participate in cell division, becoming duplicated, form chromosomes of daughter cells.

Mitosis: cell division by spindle formation, the duplication of chromosomes, separation of daughter chromatids, (chromosomes), division of cytoplasm, formation of two identical daughter cells.

Mitotic figures: the various forms assumed by the spindle and chromosomes as seen by microscope.

Centromere: point of contact of paired daughter chromatids during cell division.

Under ordinary conditions the normal structure of the body reaches a certain size. Cell division becomes arrested except where a continuous loss of surface cells requires replacement. In some other instances, following injury or disease, tissue may regenerate by cell division. When the gap or defect has been repaired cell division stops. The new cells mature and assume the normal function for that particular type of tissue.

In cancer, instead of maturing and assuming normal function the young cells continue to divide without regard to overgrowth. Adjacent tissues are pushed aside, or invaded, or destroyed. Stained sections of cancer growths show many mitotic figures due to this repeated cell division. These immature cells may be more sensitive to radioactive agents than are mature adult cells.

S. 31 CELL DIVISION BY MIOSIS (MEIOSIS):

G. meiosis = a reduction, a lessening

MIOSIS OR MEIOSIS is a cell division in genetic or reproductive cells, ova or eggs, and spermatozoa or sperms. The number of chromosomes becomes reduced to one-half the number in the parent cell, from 46 to 23.

If in the mature ovum and the mature spermatozoon the number of chromosomes were 46 in each, the nucleus of the fertilized ovum would contain 92 chromosomes.

In order to preserve the normal number of 46 chromosomes in a fertilized ovum each parent cell goes through a process of maturing called maturation. In this process there is a reduction to one-half of the number of chromosomes. This process is called oogenesis in the ovum, and spermatogenesis in the sperm. In the very much simplified explanation given here all technical terms have been purposely omitted.

The parent cell of the ovum is formed in the ovary. This cell divides into two daughter cells, one of which (a polar body) disintegrates. The other survives and again divides forming two daughter cells, one of which, a further polar body, degenerates and disappears. The remaining cell is a mature ovum. During division the number of chromosomes has been reduced to twenty-three - haploid division or miosis. One of these chromosomes is a sex chromosome, an "x" chromosome, while twenty-two other single chromosomes persist.

The parent cell of the spermatozoon, formed in the testicle by repeated cell division, eventually produces two mature spermatozoa, each with 23 chromosomes, Unlike the polar body formed by ova, two functional sperms are developed. Each will have 23 chromosomes. One of these will have an "x" and the other a "y" sex chromosome, along with 22 other single chromosomes. An ovum fertilized by one of these will then have 46 chromosomes, or 23 pairs. Twenty-three chromosomes will have been donated by each parent cell. If an ovum is fertilized by a sperm containing an "x" sex chromosome, the embryo will be a female (x + x). If fertilization is by a sperm having a "y" sex chromosome the progeny will be a male (x + y).

S. 32 GENES

Genes are now considered to be fractions of DNA molecules of chromosomes, in many different combinations. Each gene is said to produce a single enzyme that is responsible for a single chemical reaction. Many thousands of genes are contained in a single DNA molecule. These genes determine and transmit hereditary factors to the offspring. Such traits as the color of the eyes, of hair, the shape of the head, the body build, brain capacity, etc. are thus inherited. Since each parent cell contributes 23 chromosomes the offspring will receive an equal number of genes from each parent. Because some genes are dominant the traits they represent will appear in the embryo rather than the

corresponding gene from the other parent. The suppressed genes are termed recessive. Genes for brown eyes, for instance, are dominant and an embryo having one brown-eyed parent will probably have brown eyes also.

In some instances genes become modified and mutations result. The pregnant female who contracts German measles may have some genes affected by the toxin of measles with mutations, or malformations, resulting. Radioactive agents may also affect the genes and cause mutations. Some drugs, such as thalidomide, may also cause these anomalies.

S. 33 CONGENITAL ANOMALIES

A congenital anomaly, a mutation, is a variation in structure of a part of the body from its usual form, and is present at birth. It may be due to arrested or underdevelopment, or to development in an abnormal pattern due to some abnormality of the genes.

Many anomalies are of minor importance as they do not interfere with normal functioning of the part. Others are more serious as they interfere with function, or render the part more liable to injury or disease, or may even be incompatible with life at birth or later on in life.

There are many types of anomalies although none occur very frequently. It seems to be remarkable that in an organism that passes through such a complicated development these do not happen more frequently.

A few examples are listed below and others will be described with the System of the body involved.
1. The absence of fingers, toes, or of a limb.
2. The presence of extra fingers or toes.
3. The presence of an organ on the opposite side of the body to that in which it is normally located.
4. An unusual pattern of the blood supply to an organ, limb, etc.
5. Failure of the intestine to develop an anus = imperforate anus.
6. Failure of a hollow tube to have a canal or a lumen with it, e.g. the gullet or esophagus.
7. Stenosis, the canal in a tube or hollow organ is narrower than usual.

S. 34 ANATOMICAL TERMS FROM CELL DIVISION

mitosis
mitotic cell division
mitotic figures
phases or stages
interphase or resting stage
prophase
metaphase
anaphase
telophase
chromosomes
chromatids
centrosome
spindle
aster
equator of spindle
centromere
miosis
meiosis
miotic cell division
oogenesis
spermatogenesis
genes
mutations
congenital anomalies
imperforate
atresia
stenosis

NOTES

NOTES

4. SKIN: APPENDAGES: SUBCUTANEOUS TISSUE: SECRETING GLANDS

The skin forms a covering for the body. In this chapter the parts of the skin and its modifications, or appendages, will be discussed. In addition subcutaneous tissue and secreting glands will be explained.

S. 35   THE LAYERS OF THE SKIN

1. THE EPIDERMIS

    (1) Horny layer - stratum corneum
    (2) Clear or translucent layer - s. lucidum
    (3) Granular layer - s. granulosum
    (4) Prickle cell layer - s. spinosum
    (5) Germinal or basal layer - s. germinativum

2. THE DERMIS

    (1) Papillary layer - s. papillare
    (2) Reticular layer - s. reticulare

S. 36   THE STRUCTURE OF THE SKIN F. 4-1, 2

The skin consists of two layers; an outer layer, the epidermis, and an inner layer, the dermis. Each of these is composed of layers or strata, or tiers of cells. This may prove confusing to the junior students.

1. THE EPIDERMIS or outer layer is an epithelial membrane. (Epi = upon + dermis = skin). Since it is made up of many tiers of cells placed one upon the other it is termed stratified. (Stratum = a layer). It is also called squamous since the superficial layers of cells have become flat and scaly. (Squama = scale). The term stratified squamous signifies several layers, with a scaly layer. The epidermis is formed from the ectoderm, the outer layer of the primitive embryo. In the epidermis five layers or strata can be identified. Each layer may be several tiers in thickness. The cells of any one layer are similar. From the outside these layers are named the horny layer, the translucent or clear layer, the granular layer, the prickle cell layer and the germinal or basal layer.

(1) **The horny or outer layer** is composed of several thicknesses of degenerated cells that have become mere scales with no cellular structure. The scales consist of a protein, keratin, the result of cellular degeneration

similar to that in the hair, nail, hoofs of animals, etc. (Stratum corneum).

(2) **The translucent or clear layer** lies immediately under the horny layer, and consists of tiers of cells that have lost their nuclei and cellular outlines due to the degenerative process. (Stratum lucidum).

(3) **The granular layer** lies beneath the clear layer and is composed of two or three tiers of flattened cells containing granules. These granules represent an early stage of degeneration. (stratum granulosum).

(4) **The prickle cell layer** consists of several tiers of many-sided cells. These are alive and represent mature germinal cells. (Stratum spinosum)

(5) **The germinal or basal cell** layer consists of a layer of columnar cells that forms the deepest part of the epidermis. These cells contain a pigment that determines the darkness of the skin. Colored races have pigment granules in all strata of the epidermis. (Stratum germinativum or basale).

The cells of the germinal layer divide to form daughter cells continually. These newly-formed cells push the more mature cells towards the surface. These cells pass through the various phases of degeneration described above. They eventually become scales and are rubbed off. The horny layer is very thin on most parts of the body, but is thicker on the palms of the hands and soles of the feet where callouses may form.

2. THE DERMIS, corium, or true skin lies under the epidermis. It is composed of loose (areolar) connective tissue with widely separated cells, with fibrous bands and elastic tissue fibers between. These interlace with each other between the cell bodies. Intercellular fluid fills in the spaces. Due to the fibers the dermis is flexible and elastic. The dermis is formed from the mesoderm, the middle layer of the embryo. Fat cells may be present, and blood and lymph capillaries pass freely through the dermis. Receptors for sensory nerve endings are located in the dermis but very few nerve endings penetrate into the epidermis. The epidermis is devoid of blood vessels. (G. dermis = true skin; L. corium = true skin. Distinguish from cutis = the skin).

The dermis consists of two layers or strata, the papillary and reticular layers.

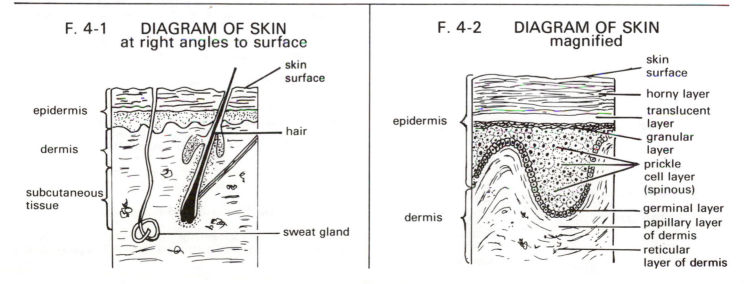

F. 4-1   DIAGRAM OF SKIN
at right angles to surface

epidermis
dermis
subcutaneous tissue

skin surface
hair
sweat gland

F. 4-2   DIAGRAM OF SKIN
magnified

epidermis
dermis

skin surface
horny layer
translucent layer
granular layer
prickle cell layer (spinous)
germinal layer
papillary layer of dermis
reticular layer of dermis

F. 4-3   THE FINGER NAIL

—dorsal view—   —lateral view—

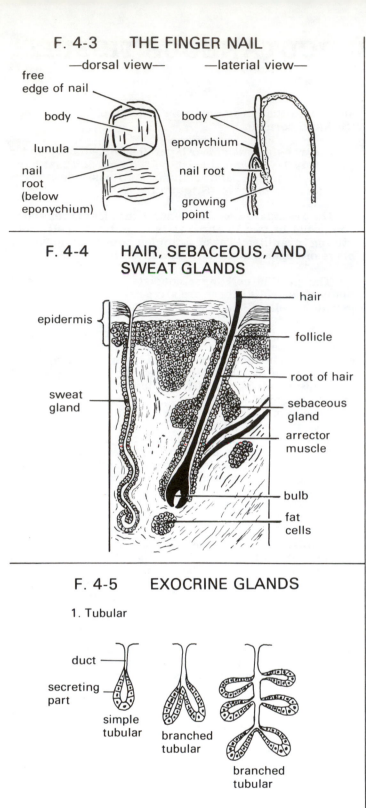

free edge of nail
body
lunula
nail root (below eponychium)
body
eponychium
nail root
growing point

F. 4-4   HAIR, SEBACEOUS, AND SWEAT GLANDS

epidermis
sweat gland
hair
follicle
root of hair
sebaceous gland
arrector muscle
bulb
fat cells

F. 4-5   EXOCRINE GLANDS

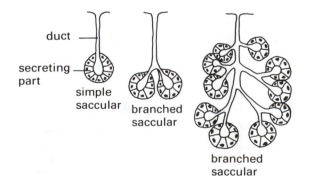

1. Tubular

duct
secreting part
simple tubular
branched tubular
branched tubular

2. Alveolar (saccular)

duct
secreting part
simple saccular
branched saccular
branched saccular

---

(1) The papillary layer lies next to the epidermis. Its outer surface is elevated into tiny projections with hollows between them, hence the name papillary. (Papilla = a nipple). The cells of the germinal layer of epidermis fit into these papillae and hollows between. This results in ridges on the skin surface - utilized in finger printing procedures.

(2) The reticular layer lies beneath the papillary layer. The cells and fibers form an interlacing network, hence the name reticular. (reticulum, a net). Blood vessels and nerves pass through this layer to reach the papillary layer.

S. 37   APPENDAGES OF THE SKIN F. 4-3, 4

1. Nails
2. Hair
3. Sebaceous glands
4. Sweat or sudoriferous glands
5. Ceruminous glands.

These structures have been described as the appendages, modifications, derivatives and accessory organs of the skin. They are composed of the epidermis, modified to perform the various functions.

S. 38   THE NAILS F. 4-3

The nails are epidermis that has been modified by becoming keratinized. The stratum lucidum, the clear layer, has thickened to form a hard protective covering for the tips of the fingers and toes. Each nail consists of a nail root body and a free edge.

The nail root is the proximal part that is covered by a fold of skin, and hidden from view. Germinal cells of the nail root form a thick layer. These cells multiply to form new nail tissue. They are converted into the hard flat nail, and are pushed by new cells towards the free edge.

The body of the nail is the exposed part, with its free edge projecting over the tip of the digit. The lunula is the white part of the nail body, shaped like a half-moon close to the nail root.

The eponychium is the outer horny layer of epidermis at the base of the nail that tends to grow out over the nail body. This may be torn - a hang nail, that sometimes becomes infected, causing pus to form under the eponychium around the base of the nail.

S. 39   THE HAIR F. 4-4

Hairs, like the nails, are modifications of the epidermis. A hair consists of a hair root, a shaft, and a hair sheath or hair follicle.

A hair follicle is a small canal opening upon the skin surface and extending down into the dermis or even the subcutaneous tissue. Each hair follicle is lined by epithelium, continuous with the epithelium of the skin. It is through this canal that a hair reaches the skin surface.

The root of a hair is that part below the skin level lying in a hair follicle.

The bulb is the inner enlarged end of each hair root. A germinal layer of epithelium within the bulb forms new cells that push the more mature cells towards the skin surface and shaft. A papilla, a nipple-like process of dermis, fits into a minute depression at the inner end of each bulb. Growth of a hair is accomplished by division of the germinal cells in the bulb. If cut off, pulled out, or destroyed, hair will regenerate as long as the germinal cells within the bulb remain intact.

S. 40   SEBACEOUS GLANDS F. 4-4

These are small secreting glands that lie beside and open into hair follicles, although some may open directly upon the skin surface. Each gland consists of a secreting part, an alveolus, that leads into a central canal (or duct). This duct leads into a follicle. The secreting cells become filled with an oily substance sebum, that is discharged into the duct and follicle. Entering a hair follicle, the secretion lubricates the hair and reaches the skin surface to lubricate the skin.

S. 41 SWEAT OR SUDORIFEROUS GLANDS F. 4-4

Sweat glands are simple tubelike glands consisting of a single canal or duct, and a coiled secreting part. The duct opens upon the skin surface and has a layer of epithelial cells surrounding its canal. These cells are continuous with the epithilium of the skin. The secreting part, coiled upon itself at the inner end, has two or three layers of cells, and is often located in the subcutaneous tissue below the dermis. The long duct passes through the dermis and epidermis to reach the skin surface. The entire body has sweat glands, but they are most numerous on the palms of the hands and soles of the feet, and in the axillae (armpits). It has been estimated that there are from 2,500 to 3,000 glands per square inch on the palms. (2.5 cm square)

Sweat glands secrete a watery solution containing sodium chloride (common salt) and minute quantities of other waste products including urea. Under normal conditions 600 cc of fluid may be lost by sweating in 24 hours. Under other conditions such as strenuous exercise, high atmospheric temperature, fever, etc., much more fluid may be expelled by the sweat glands. Sweat that accumulates on the skin surface requires body heat to evaporate it. This process causes a lowering of body temperature. Sweat glands have autonomic nerve fibers with a sweating center that controls the activity and the quantity of sweat. Sweating is a means by which the body can get rid of excessive body heat, or excessive body fluid may be excreted.

S. 42 CERUMINOUS GLANDS

These are modified sweat glands located in the external ear that secrete wax into the external acoustic (auditory) meatus.

S. 43 THE SUBCUTANEOUS TISSUE F. 4-1

Subcutaneous tissue, consisting of areolar (loose connective tissue) forms a layer between the skin and such structures as muscles, bones, or organs that lie deep to the skin. (Sub = under + cutis = skin).

It is loosely constructed with widely-spaced cells and white and elastic fibers. This looseness allows considerable freedom of movement of skin over underlying organs. Frequently there is a layer of fat cells that may be quite thick in some areas. This fat helps to insulate the body against cold. It may also provide a source of energy during starvation. Many blood vessels and nerve fibers pass through subcutaneous tissue to reach the dermis. The matrix (intracellular substance) is semiliquid.

S. 44 FUNCTIONS OF THE SKIN

1. Skin forms a protective covering to prevent injury or destruction of underlying tissues from:
    (1) Entrance of microorganisms
    (2) Harmful chemicals, acids, alkalies
    (3) Sun's rays
    (4) Extremes of temperature

2. The skin excretes water and some waste products.

3. The skin helps to regulate body temperature by the evaporation of sweat and by dilation of capillary blood vessels of the skin radiating heat.

4. The skin contains receptors of sensory nerves that pick up, and transmit to the brain, sensations of the external enviroment such as cold, heat, pain, touch and pressure. Other receptors transmit sensations of sight, hearing and smell.

S. 45 SECRETING GLANDS

Secreting glands of the body take up raw materials from blood capillaries and manufacture simple or complex substances required by the body. These must be distinguished from lymph glands or nodes that form part of the lymphatic system, and do not manufacture secretions. The name lymph nodes might be substituted for the latter to avoid confusion.

Glands are organized units of epithelial cells that extract raw materials from capillaries and intercellular spaces and manufacture new substances or secretions. There are many types of glands, each producing a specific secretion. Glands are classified as endocrine or exocrine glands.

ENDOCRINE GLANDS, or ductless glands, or glands of internal secretion have no ducts. They discharge their secretions directly into blood or lymph capillaries for distribution by the blood to all body tissues. Their secretions are called hormones. These glands will be described in detail in Chapter 20 under Endocrine Glands. (Endo = inside or within + krino = to separate).

EXOCRINE GLANDS have a secreting part and a duct. The duct is a canal through which the secretion passes into a cavity, a hollow organ, or the skin surface. The duct is formed by a single layer of epithelial cells arranged to form a hollow tube. The blind inner end of each tube consists of modified cells encircling this part of the tube. These are secreting cells and their product is discharged into the central canal or duct. The arrangement of the secreting cells varies with each type of gland. Sweat and sebaceous glands are examples of exocrine glands which have been studied. Other examples are the salivary glands, the tear glands, the glands of the digestive tract, mammary gland, etc. (Exo = out of + krino = to separate.)

CLASSIFICATION OF EXOCRINE GLANDS F. 4-5

1. **Simple tubular gland** - a single tube of cuboid cells, the inner end of which forms a secretion that is discharged through a single tube.
2. **Branched tubular gland** - a single duct with branches, with a secreting part at the end of each branch, e.g. gastric glands of stomach.
3. **Simple saccular** or alveolar gland - a single duct at the blind end of which a saclike secreting part is located.
4. **Branched saccular** or alveolar gland - a single collecting duct with branches opening into it, each branch ending in a saclike secreting end.
5. **Combined tubuloalveolar gland** - with tubular and sacular parts.

    Mucus is secreted by goblet cells.

S. 46 ANATOMICAL TERMS FROM CHAPTER 4

Skin - cutis
  epidermis
    horny layer or
     stratum corneum
    clear layer or
     stratum lucidum
    granular layer or
     stratum granulosum
    prickle cell layer or
     stratum spinosum
    germinal layer or
     stratum germinativum

dermis - corium - true
    papillary layer or
     s. papillare
    reticular layer or
     s. reticulare

Nails;
    root
    body
    lunula
    eponychium

Hair;
    hair follicle
    root
    bulb
    shaft

Sebaceous glands;
    secreting part
    duct

Sudoriferous gland
    or sweat gland
    secreting part
    duct

Ceruminous gland
    or wax gland

Exocrine glands;
Classification -
    tubular
    simple or branched
    saccular or alveolar
    simple or branched
    tubuloalveolar

Endocrine glands; or
    ductless glands or
    glands of internal secretion
    hormones

NOTES

BONES: GENERAL INFORMATION; ─ CLASSIFICATION: PARTS:
STRUCTURE: DEVELOPMENT

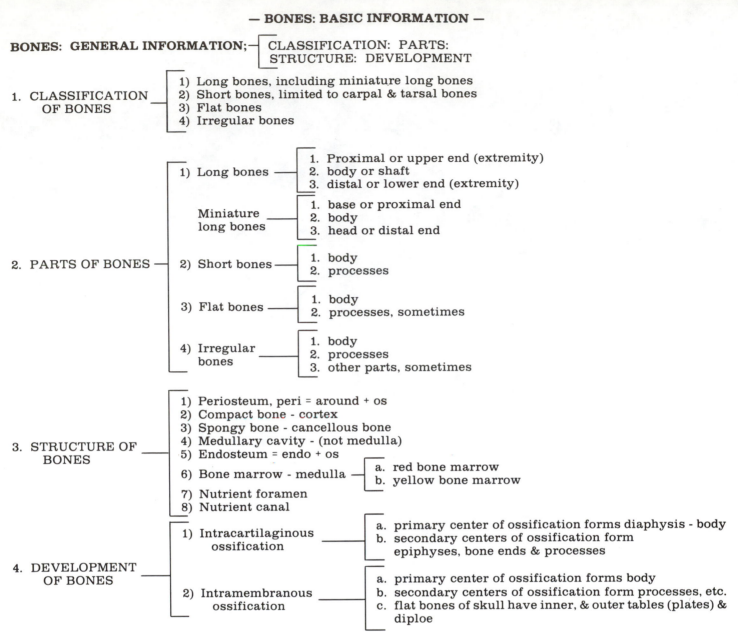

1. CLASSIFICATION OF BONES
1) Long bones, including miniature long bones
2) Short bones, limited to carpal & tarsal bones
3) Flat bones
4) Irregular bones

2. PARTS OF BONES
1) Long bones
1. Proximal or upper end (extremity)
2. body or shaft
3. distal or lower end (extremity)

Miniature long bones
1. base or proximal end
2. body
3. head or distal end

2) Short bones
1. body
2. processes

3) Flat bones
1. body
2. processes, sometimes

4) Irregular bones
1. body
2. processes
3. other parts, sometimes

3. STRUCTURE OF BONES
1) Periosteum, peri = around + os
2) Compact bone - cortex
3) Spongy bone - cancellous bone
4) Medullary cavity - (not medulla)
5) Endosteum = endo + os
6) Bone marrow - medulla
a. red bone marrow
b. yellow bone marrow
7) Nutrient foramen
8) Nutrient canal

4. DEVELOPMENT OF BONES
1) Intracartilaginous ossification
a. primary center of ossification forms diaphysis - body
b. secondary centers of ossification form epiphyses, bone ends & processes

2) Intramembranous ossification
a. primary center of ossification forms body
b. secondary centers of ossification form processes, etc.
c. flat bones of skull have inner, & outer tables (plates) & diploe

5. TERMS APPLIED TO DEVELOPING BONES

1) Diaphysis, pl. diaphyses, adj. diaphyseal — that part formed by a primary center
2) Epiphysis, pl. epiphyses, adj. epiphyseal — that part formed by a secondary center including bone ends & processes
3) Epiphyseal cartilage (line) a layer of cartilage between diaphysis & epiphysis
4) Metaphysis, pl. metaphyses, adj. metaphyseal — the part of the diaphysis next to the epiphyseal cartilage

6. DERIVATIONS OF SOME TERMS:

Os = bone; adj. osseous; ossify = to form bone
Ossification = the formation of bone; ossicle = a little bone

5. AN INTRODUCTION TO BONES AND JOINTS

1. BONES — GENERAL INFORMATION

Descriptive — terms relating to bones
Functions of bones
Classification and parts of bones
Structure of bones
Development of bones

GENERAL TERMS

Os = a bone; pl. ossa; adj. osseous
Ossicle = a little bone, e.g. ossicles of ear
Ossify = to form bone
Ossification =formation of bone
Calcification = a deposit of calcium, not the formation of bone

S. 47 BONY PROMINENCES, PROJECTIONS, DEPRESSIONS

The student should review Section 7, General anatomical terms; some of these refer to bones as well as other organs.
Parts of organs: head, neck, body, tail, lobe, lobule, segment, extremity.
Openings: aperture, foramen, hiatus, orifice, os, ostium, lumen, porus.
Depressions: fissure, fossa, hilum, sulcus, sinus.
Others: incisura (notch), meatus, ramus, septum.

Ala	= a wing; ala of sacrum, or of ilium
Condyle	= a rounded knoblike projection or knuckle; condyles of femur and tibia.
Cornu	= a horn; cornu of hyoid bone; pl. cornua
Crest	= a narrow ridge of bone the crest of the ilium (pl. crista)
Epicondyle	= a bony projection on or above a condyle; epicondyles of femur; (epi-upon).
Malleolus	= a small hammer, a rounded bony prominence; lateral malleolus of fibula, medial malleolus of tibia; pl. malleoli.
Process	= any definite or marked bony prominence; mastoid process of temporal bone.
Spine or spinous process	= a sharp slender process; the spine of the ischium; spinous processes of vertebrae.
Styloid process	= a sharp slender process; styloid process of temporal bone; styloid processes of radius and ulna.
Trochanter	= a very large rounded process; lesser and greater trochanters of the femur.
Tubercle	= a small rounded process; greater and lesser tubercles of the humerus.
Tuberosity	= a large rounded process, tuberosity of the radius or tibia.
Fovea	= a small pit or depression; fovea capitis femoris - a pit on femoral head; fovea costalis; a pit on a vertebral body or process for union with a rib (NA).
Facet	= a smooth articular surface, an older term, for smooth articular surface of vertebrae, ribs, etc.

F. 5-1   THE HUMAN SKELETON
—frontal view—

S. 48   FUNCTIONS OF BONES

1. **Protection:** bones protect many organs and other structures from injury. The skull protects the brain, ear and eye. The bones and cartilages of the thorax protect the heart and lungs. The pelvis protects the pelvic organs.

2. **Support and framework:** like the steel or wooden framework of a building, bones give support and shape to the body, and afford attachments for muscles and ligaments.

3. **Levers:** bones, together with adjacent joints form levers which permit movement in restricted and definite directions.

S. 49   CLASSIFICATION & PARTS OF BONES

CLASSIFICATION:

1. **Long bones:**	humerus, radius, ulna, femur, tibia, fibula Miniature long bones: metacarpal, metatarsal bones and phalanges of limbs
2. **Short bones:**	carpal and tarsal bones
3. **Flat bones:**	bones of vault of skull, scapula, ribs, sternum, patella
4. **Irregular bones:**	vertebrae, bones at base of skull
5. **Sesamoid bones**	

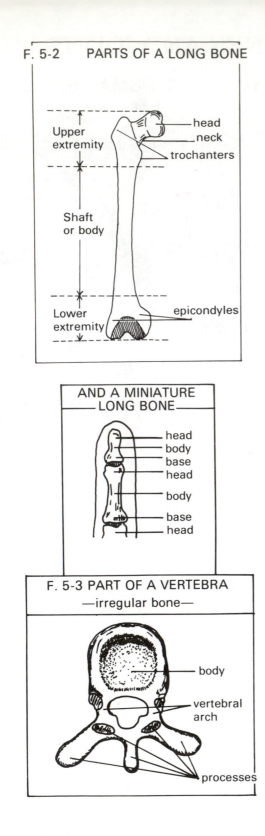

F. 5-2   PARTS OF A LONG BONE

- Upper extremity — head, neck, trochanters
- Shaft or body
- Lower extremity — epicondyles

AND A MINIATURE LONG BONE

- head
- body
- base
- head
- body
- base
- head

F. 5-3 PART OF A VERTEBRA
—irregular bone—

- body
- vertebral arch
- processes

PARTS OF BONES   F. 5-2,3

1. **Long bones:**   proximal or upper extremity (end)
   body or shaft
   distal or lower extremity (end)

   Miniature long: base or proximal extremity
   body
   head or distal extremity

2. **Short bones:**   body, sometimes processes

3. **Flat bones:**   body, other parts

4. **Irregular bones:**   bones, processes

1. **The periosteum** is a membrane that covers all bones with the exception of the articular (joint) surfaces which are covered instead by articular cartilages. The outer layer of periosteum is dense fibrous tissue while the inner layer, next to the bone, has osteoblasts or bone forming cells. (peri = around + os + bone - around a bone).

2. **Compact bone** or cortical bone is dense closely knit bone resembling ivory, made up of compact Haversian systems. It is located under the periosteum. It forms a thick layer in the bodies (shafts) of long bones. At the ends of long bones it forms a thin layer under the articular cartilages. In short and irregular bones it forms a thin outer layer under the periosteum, as well as on the articular surfaces.

3. **Spongy or cancellous bone** is porous loosely-knit bone similar in appearance to a sponge, or honeycomb, or latticework. Meshworks of slender processes and spicules of bone separate small cavities. It forms a thin layer beneath the compact bone in the bodies of long bones. At the ends of long bones and in the bodies and processes of other bones it forms all but the thin outer compact layer.

4. **The medullary cavity** or canal or marrow cavity is a central cavity extending longitudinally in the shafts of long bones. It contains bone marrow.

5. **The endosteum** is a membrane that lines the medullary cavities of long bones. (Endo = within + os).

* Articular Cartilage –covers ends of bone & joint surfaces

6. **Bone marrow** (medulla) is the tissue occupying the medullary cavities of long bones and the spaces in spongy bone. (L. medulla, G. myelos = medulla). There are two types of bone marrow:

(1) **Red bone marrow** is found in the medullary cavities of long bones and in spongy bone of children and adults. It forms red blood cells and some types of white blood cells. A marrow puncture is often done to obtain a sample of marrow. A hollow needle is pushed through the compact bone into spongy bone to get marrow tissue. As the sternum and ilium lie close to the skin surface these bones are often used for this procedure.

(2) **Yellow bone marrow** replaces red bone marrow in long bones of adults. It contains considerable fat.

7. **The nutrient foramen** is a small opening in the periosteum and opens into a nutrient canal that passes obliquely through a bone to a medullary cavity or the center of a solid bone. In long bones it is located near the center of the shaft. It carries a nerve and an artery into the bone, and provides a passage for veins and lymphatics.

8. Bones receive **their blood supply** by arteries entering through the nutrient canals as well as by small vessels that penetrate the periosteum.

9. **The flat bones** of the skull have a construction peculiar to them. Thin layers of compact bone lie adjacent to the periosteum covering the external and internal surfaces of the flat bone. These form the outer and inner tables (NA plates). A layer of spongy bone called the diploe lies between the two compact layers.
   Some of the bones of the skull have the outer and inner tables separated by an air-containing space and are referred to as pneumatic bones, with the paranasal sinuses occupying the spaces.

F. 5-4 STRUCTURE OF A LONG BONE
—the femur—

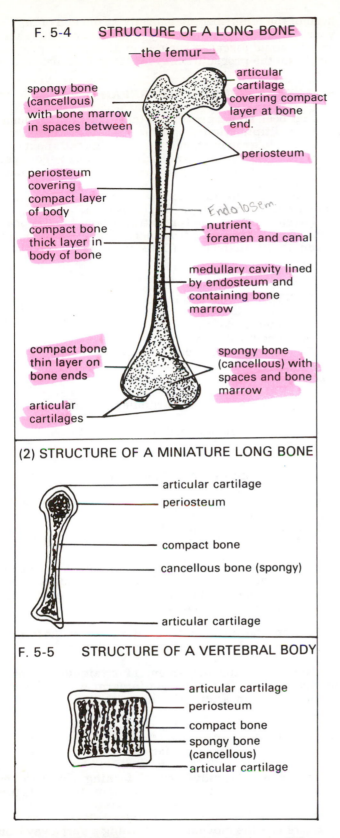

spongy bone (cancellous) with bone marrow in spaces between

articular cartilage covering compact layer at bone end.

periosteum

periosteum covering compact layer of body

Endolosem.

nutrient foramen and canal

compact bone thick layer in body of bone

medullary cavity lined by endosteum and containing bone marrow

compact bone thin layer on bone ends

spongy bone (cancellous) with spaces and bone marrow

articular cartilages

(2) STRUCTURE OF A MINIATURE LONG BONE

— articular cartilage
— periosteum

— compact bone

— cancellous bone (spongy)

— articular cartilage

F. 5-5 STRUCTURE OF A VERTEBRAL BODY

— articular cartilage
— periosteum
— compact bone
— spongy bone (cancellous)
— articular cartilage

S. 51 DEVELOPMENT OF BONES-OSSIFICATION

At an early age in the development of the human embryo the bones of the vault of the skull are formed as membranes. The other bones of the body are preformed as cartilage that is moulded into the shape of the bone to be. These membranes and cartilages are gradually replaced by bone tissue as development proceeds. The processes by which these changes are accomplished are called intracartilaginous and intramembranous ossification. (cartilaginous = adj. of cartilage or gristle; + ossification = to form bone).

S. 52 INTRACARTILAGINOUS OSSIFICATION

1. In intracartilaginous ossification the cartilage in the shafts of long bones and in the bodies of some other bones is replaced by bone while the fetus is still within the uterus. This is accomplished from a primary center of ossification. (intra = within). F. 5-6,7,8,9,10

A **Primary center of ossification** is a group of bone cells that make their appearance in the center of the bodies of long and other bones. These cells divide repeatedly and replace the cartilage of the bodies of these bones. (primary = first). **Osteoblasts** are cells that form bone.

The **diaphysis** is that part of a bone formed from a primary center of ossification and includes the body or shaft. (Dia = through + physis = growth; i.e. to grow through or along).

The bodies of long bones and of many other bones become ossified in early intrauterine life, and radiographs of the pregnant female will demonstrate them from about the twelfth week of pregnancy.

2. The ends (extremities) of most long bones and the prominences and processes of bones remain as cartilage until after birth. Ossification of these is achieved by secondary centers of ossification.

A **secondary center** of ossification is a group of bone cells that makes its appearance in the end of a bone, or in a bony prominence (or process). These bone cells divide repeatedly and replace the cartilage. They first form islands of bone within the cartilage then entirely replace it. In most bones these centers appear after birth, although about the knees they appear before birth. In girls they appear at a slightly earlier age than in boys. The age of the appearance of any one secondary center is fairly constant for each sex, but varies from bone to bone. In some bones secondary centers appear in the first year of life. Others appear at specific times during the first or second decades of life.

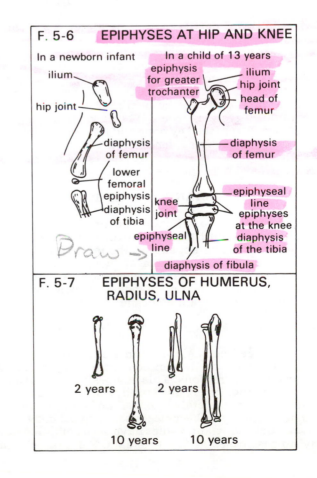

F. 5-6 EPIPHYSES AT HIP AND KNEE

In a newborn infant

ilium

hip joint

diaphysis of femur

lower femoral epiphysis

diaphysis of tibia

epiphyseal line

Draw →

In a child of 13 years

epiphysis for greater trochanter

ilium

hip joint

head of femur

diaphysis of femur

epiphyseal line epiphyses at the knee

knee joint

diaphysis of the tibia

diaphysis of fibula

F. 5-7 EPIPHYSES OF HUMERUS, RADIUS, ULNA

2 years   2 years

10 years   10 years

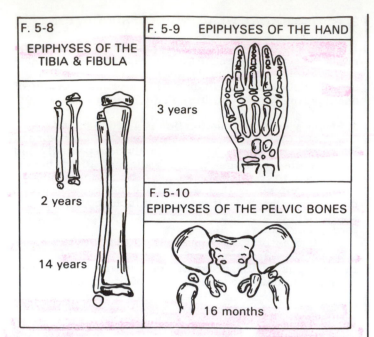

F. 5-8 EPIPHYSES OF THE TIBIA & FIBULA	F. 5-9 EPIPHYSES OF THE HAND
2 years 14 years	3 years
	F. 5-10 EPIPHYSES OF THE PELVIC BONES 16 months

The growth of secondary centers and replacement of cartilage takes place gradually over a period of months or years. Eventually the entire bone end or prominence or process becomes ossified. This ossification is completed at a definite age for each bone end or process.

In some bones more than one secondary center appears in the bone end, prominence, or process. These form as bone islands within the cartilage, gradually replacing it then fuse to form a single solid bone end or process.

The **epiphysis** of a bone is that part formed from one or more secondary centers of ossification. A long bone will have an epiphysis at each end as well as for each prominence or process. A miniature long bone will have an epiphysis at one end only as a rule. Some bones will have an epiphysis for each of its processes. (Epi = upon + physis = growth., i.e. to grow upon - at the end of).

3. **The epiphyseal cartilage** (disc or plate) is a layer of cartilage between a diaphysis and epiphysis of bone, that persists during the growing period. A bone grows in length by formation of cartilage cells in this plate. The cartilage cells next to the diaphysis or epiphysis become replaced by bone cells. Those in the center will continue to divide. The cartilage appears as a dark line in radiographs of growing bones. When growth is complete the cartilage is replaced by bone and it is impossible in a radiograph to see any evidence of it. *epiphyseal scar*

The **metaphysis** is the end of a diaphysis adjacent to an epiphyseal cartilage. (meta = beyond + physis).

S. 53   INTRAMEMBRANOUS OSSIFICATION

Intramembranous ossification occurs in the bones of the vault of the skull that form first as membranes. A center of ossification appears in the membrane of each bone before birth. By division of these bone cells the membrane is replaced by bone except for the suture, the membrane at the junction of bones. (Intra =within + membrane, i.e. within a membrane).

S. 54   BONE AGE AND CHRONOLOGICAL AGE

The bone age and the approximate age of a child or young adult may be determined from radiographs of appropriate joint regions. This is possible because the age of appearance of any one epiphysis, the age at which the cartilage is completely replaced, and the age at which the ossified epiphysis becomes fused with the body of the bone are fairly constant. Tables have been compiled from many normal cases giving this data. For example, the

exact periods

head of the femur begins to ossify at the end of the first year, and unites with the body of that bone at 18 years of age. In some diseases such as cretinism and pituitary dwarfism the procedure may be delayed or accelerated.

S. 55   BONE GROWTH AND REPAIR

It has been stated previously that bones grow in length by the multiplication of cartilage cells in the epiphyseal cartilage. In growing bones the cartilage cells next to the diaphysis and epiphysis are being replaced by bone cells with new cartilage being formed between to be replaced by bone.

Bones grow in diameter by division of osteoblastic cells from the inner layer of periosteum.

The repair of fractured or destroyed bone takes place by division of osteoblasts from periosteum. The new cells gradually fill in the gap. (Osteoblasts are bone forming cells).

II. JOINTS: GENERAL INFORMATION

Basic terms defined
Movements at joints
Classification, structure, characteristics

GENERAL TERMS

Articulation = a joint; adj. articular; (L. articulatio)
Arthron = a joint (G)
Arthrosis = a joint; adj. arthrodial; pl. arthroses
Junctura = a joint

S. 56   MOVEMENTS AT JOINTS   F. 5-11

1. **Abduction** - movement away from the median line of the body. (ab = away from + ducere = to lead, bring or draw). To abduct the upper limb would mean to move it away from the median line or plane. Abduct; abductor.

2. **Adduction** - movement towards or across the median line of the body. (ad - towards + ducere). To adduct the arm would mean to bring it towards or across the median line of the body. Adduct; adductor

3. **Flexion** - the movement of bending. To flex the forearm would mean to bend the elbow. Flex; flexor.

4. **Extension** - the movement of straightening or of stretching out. To extend the forearm would mean to stretch it out in a straight line with the arm. Extend; extensor.

Note: abduction, adduction, flexion and extension are angular movements - angulation.

5. **Inversion** - the movement of turning a part to face towards the median line. (in = in + vertere = to turn = to turn in).

6. **Eversion** - the movement of turning a part away from the median line. (e = out + vertere) evert; evertor

7. **Rotation** - the movement of turning a part in one axis. To rotate the arm would mean to simply turn it in or out without circumduction, similarly turning the head. Rotate; rotator

8. **Circumduction** - movement in a circular direction about a cone shaped axis. (circum = around + ducere). To circumduct the upper limb would mean to swing it in a circular direction, as in swinging a pail of water over the head. Circumduct; circumductor.

F. 5-11   MOVEMENTS AT JOINTS
—Frontal View—

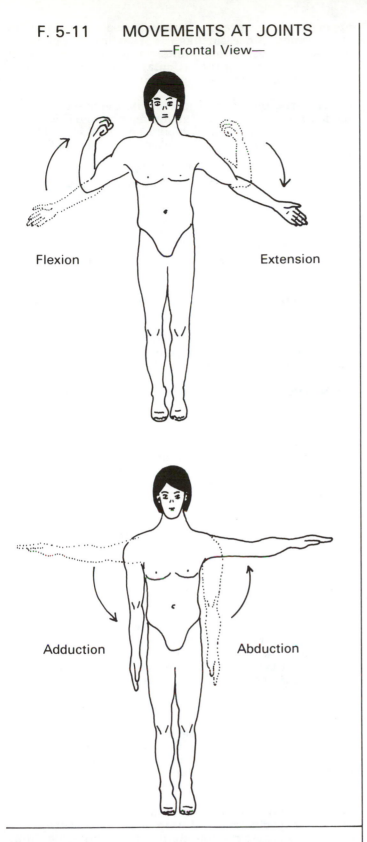

Flexion   Extension

Adduction   Abduction

9. **Supination** - the movement of turning the body or hand so that the front of the body faces upwards, i.e. lying on the back with the palm of the hand facing up. Supinate; supinator; supine.

10. **Pronation** - the movement of turning the body to face downwards, to lie face down, or turning the hand so that the palm is facing downwards. Pronate; pronator; prone

11. **Gliding** - the sliding of one bone upon another at a joint, e.g. movements at the wrist and between articular processes of adjcent vertebrae.

Additional movements

Dorsiflexion	=	bending backwards
Hyperextension	=	extension beyond the normal limit
Hyperflexion	=	flexion beyond the normal
Circumflexion	=	bending around
Forced inversion	=	forcibly inverting beyond normal
Internal rotation	=	turning inwards, in 1 axis
External rotation	=	turning outwards, in 1 axis

S. 57   CLASSIFICATION OF JOINTS

1. Fibrous joints, immovable joints, synarthrosis, or synarthrodial joints;
2. Cartilaginous joints, slightly moveable joints, amphiarthroses, or amphiarthrodial joints;
3. **Synovial** joints, freely movable joints, diarthroses, or diarthrodial joints:
1) Gliding joints
2) Hinge joints
3) Condylar joints
4) Saddle joints
5) Pivot joints
6) Ball & socket joints

   The subdivisions of fibrous and cartilaginous joints have been purposely omitted as they are of no practical importance to radiographers.

S. 58   STRUCTURE OF JOINTS   F. 5-12, 13, 14

1. **Fibrous joints** have a layer of fibrous tissue between the bone ends forming the joint. This fastens the bone ends together. As no movement is possible these are also called immovable joints. The bone ends may be irregular like the teeth of a saw, or expanded like the crowns of teeth, or may be bevelled or roughened. In later life the intervening fibrous tissue often disappears and the bone ends become united, e.g. sutures of vault of the skull, joints between facial bones, distal tibiofibular joint.

2. **Cartilaginous joints** have cartilage on the adjacent bone ends with a plate or disc of fibrocartilage uniting the two together. Ligaments pass across the joint from one bone to the other. As limited movement is possible they are called slightly movable joints, e.g. intervertebral discs between vertebrae, or the symphysis pubis - the joint between the two pubic bones.

3. **Synovial joints** have a joint cavity between the bone ends and are held together by a capsule surrounding the joint. As free movement in certain directions is possible these are called freely-movable joints. A detailed description of structure is included:

(1) Articular surfaces of bones
(2) Articular cartilages
(3) Articular capsule - synovial membrane
                           fibrous tissue layer
(4) Joint cavity
(5) Ligaments
(6) Muscles or muscle tendons

(1) **The articular surfaces** are the ends of the bones forming a joint. Each surface is smooth and consists of a thin layer of compact bone covering the spongy (cancellous) bone beneath.

(2) **The articular cartilages** cover the bone ends and adjacent bone margins. Each cartilage consists of a layer of hyaline cartilage that takes the place of periosteum. The cartilages vary in thickness from joint to joint. The cartilages are responsible for the dark space that

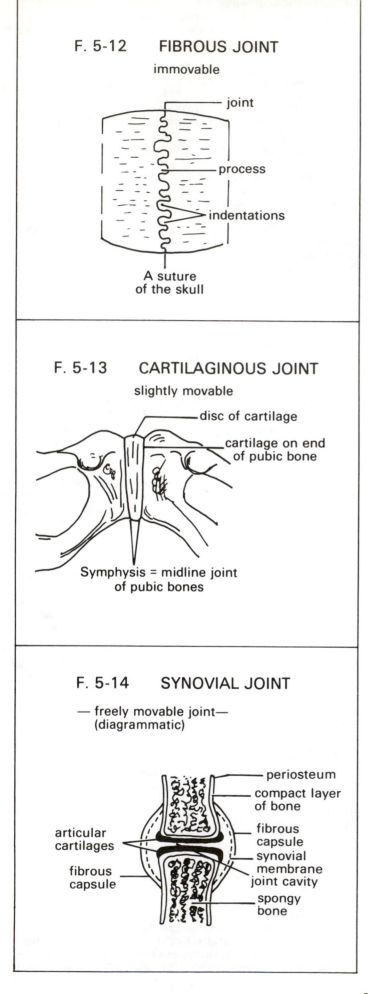

F. 5-12   FIBROUS JOINT

immovable

— joint

— process

— indentations

A suture
of the skull

F. 5-13   CARTILAGINOUS JOINT

slightly movable

— disc of cartilage

— cartilage on end
of pubic bone

Symphysis = midline joint
of pubic bones

F. 5-14   SYNOVIAL JOINT

— freely movable joint—
(diagrammatic)

— periosteum

— compact layer
of bone

— fibrous
capsule

— synovial
membrane

joint cavity

— spongy
bone

articular
cartilages

fibrous
capsule

appears to be present between the bone ends in radiographs. Since cartilage is translucent it does not obstruct x-rays. Some joints have an extra disc of cartilage separating the articular cartilages from each other, e.g. knee, temporomandibular joints.

(3) **The articular capsule** is a membrane that surrounds the joint. It forms a sleeve about the joint and is attached to the circumference of each bone beyond the limit to the articular cartilage. It is a closed sac composed of two layers:

(a) **An inner synovial membrane** that forms the lining for the joint cavity. It does not cover the articular cartilages. It secretes a fluid that lubricates the joint. (L. synovia - a joint oil).

(b) **A fibrous tissue capsule** lies outside of the synovial membrane, extending from bone to bone, completely encircling the joint. It gives support.

(4) **The joint cavity** is the potential space within the capsule.

(5) **Ligaments** composed of fibrous tissue pass from one bone across the joint to the other bone, inside or outside the capsule. These strengthen the joint.

(6) **Muscles** or muscle tendons frequently cross the joint and tend to give additional support.

A BURSA is a closed sac of synovial tissue that lies between a muscle or tendon and an adjacent bone or bony prominence. It contains a lubricating fluid that helps decrease friction as the parts move over each other. Some bursae lie close to a joint and some open into the joint cavity but many are **entirely separate**. (L. bursa = a purse, i.e. a closed sac).

CARTILAGES AT JOINTS

Hyaline cartilage forms articular cartilages at joints, costal cartilages, nasal and laryngeal cartilages, and the ringlike cartilages of the trachea and bronchi.
Fibrocartilage forms articular discs at some joints such as the intervertebral discs, semi-lunar cartilages at the knee joint, and discs at the wrist, symphysis pubis, temperomandibular, acromiclavicular and sternoclavicular joints.
Elastic cartilage forms the epiglottis, cartilages of the external ear and the auditory tube.

S. 59 **VARIETIES OF SYNOVIAL JOINTS**

1. **Gliding joints** are those that have flat or slightly-curved articular surfaces that slide over each other during movement. They have the structure of synovial joints, e.g. carpal and tarsal joints, and joints between articular processes of vertebrae. F. 5-15

2. **Hinge joints** are synovial joints that have a trochlea (pulley-shaped surface) fitting a concave surface to allow an angular motion similar to a hinge, e.g. joint between lower end of humerus and the ulna at the elbow, interphalangeal joints. F. 5-16

3. **Condylar joints**, previously called **condyloid**, have a condyle (knuckle-like process) fitting into a concave surface, allowing flexion, extension, abduction, adduction and circumduction, e.g. the metacarpophalangeal and wrist joints. F. 5-17

4. **Saddle joints** are those in which the adjacent bone ends are shaped like a western saddle, convex in one direction and concave in the other. This allows flexion, extension,

38

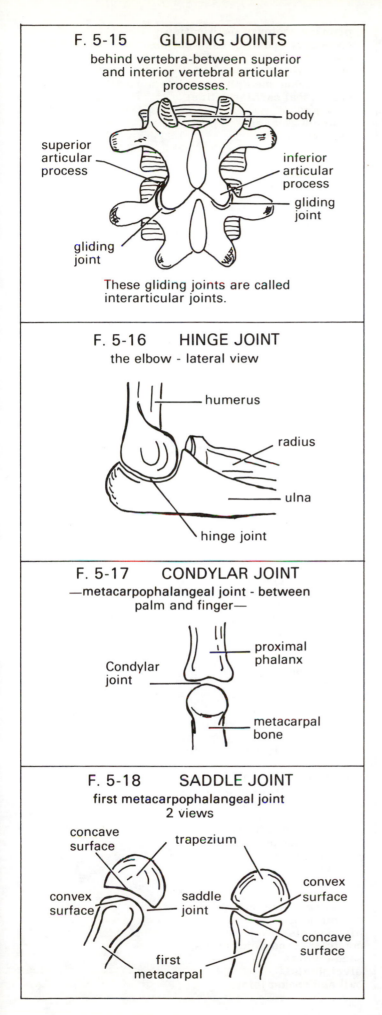

F. 5-15　GLIDING JOINTS

behind vertebra-between superior and interior vertebral articular processes.

- body
- superior articular process
- inferior articular process
- gliding joint
- gliding joint

These gliding joints are called interarticular joints.

F. 5-16　HINGE JOINT

the elbow - lateral view

- humerus
- radius
- ulna
- hinge joint

F. 5-17　CONDYLAR JOINT

—metacarpophalangeal joint - between palm and finger—

- proximal phalanx
- Condylar joint
- metacarpal bone

F. 5-18　SADDLE JOINT

first metacarpophalangeal joint
2 views

- concave surface
- trapezium
- convex surface
- saddle joint
- convex surface
- concave surface
- first metacarpal

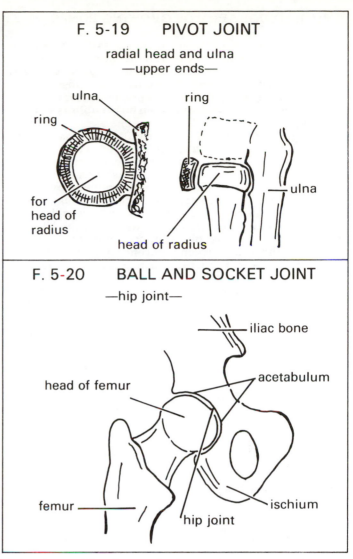

F. 5-19　PIVOT JOINT

radial head and ulna
—upper ends—

- ulna
- ring
- ring
- for head of radius
- head of radius
- ulna

F. 5-20　BALL AND SOCKET JOINT

—hip joint—

- iliac bone
- head of femur
- acetabulum
- femur
- ischium
- hip joint

abduction, adduction and circumduction, e.g. the carpometacarpal joint of the thumb. F. 5-18

5. **Pivot joints** are those in which a rounded bone end is encircled by a ring of cartilage or bone so that there is rotation or turning on one axis, e.g. the joint between the first cervical vertebra and the tooth-like dens of the second, and the proximal radioulnar joint at the elbow. F. 5-19

6. **Ball and socket joints** have a globelike end or head fitting into a cup-shaped cavity. This allows flexion, extension, abduction, adduction and circumduction, e.g. the shoulder and hip joints. F. 5-20

Note: the classification and descriptions of the three kinds of joints have been purposely kept very simple since further subclassification and structural details are confusing and unnecessary.

S. 60　**CONGENITAL ANOMALIES:**
　　　　DEVELOPING BONES

1. A secondary center of ossification may not unite with the body of a bone but may remain as a separate entity throughout life.

2. A bone may have more secondary centers of ossification than usual. Some of them may not unite with the body of the bone, and may be mistaken for a fracture.

3. A secondary center may not unite with the body of the bone to which it belongs but to an adjacent bone. This may occur in vertebrae.

RADIOGRAPHY OF GROWING BONES

In the Section on tissue and organ density it was stated that cartilage (gristle) is of low density while bone is very dense. Bones therefore absorb x-rays so that their images are white or grey. Since cartilage absorbs very little radiation, when the same exposure factors are used, their images will be dark and often not visible as separate shadows.

Radiographs of growing bones will show images of the bodies or shafts since they are ossified. The secondary centers of ossification in the epiphyses and processes will show as small or larger bone islands close to the bone ends or shafts, and will be surrounded by a clear zone. Actually, the growing bone has the shape of the adult bone with its bone ends and processes, but only those parts that are ossified are visible on film. The clear spaces are not gaps but are occupied by cartilage. Similarly, the gap between bone ends at a joint of a child or adult is due to the articular cartilages between the bone ends.

The diagnosis of pregnancy in a questionable case may be confirmed by radiography of the abdomen after about the twelfth week of pregnancy, because the fetal bones will have ossified sufficiently to be outlined in the radiographs.

S. 62 ANATOMICAL TERMS - BONES & JOINTS

Classification of bones:
  long bones &
  miniature long bones
  short bones
  flat bones
  irregular bones

Parts of long bones:
  proximal end (extremity)
  body
  distal end (extremity)
Parts of miniature long bones;
  base
  body
  head

Parts of flat bones;
  body
  other parts

Parts of irregular bones;
  body
  processes

Structure of bones:
  Perosteum
  compact bone - cortical
  spongy or cancellous bone
  medullary cavity
  endosteum
  bone marrow or medulla
    red bone marrow
    yellow bone marrow
  nutrient foramen
  nutrient canal
  Haversian systems

Structure of bones of skull;
  outer table (plate, NA)
  diploe
  inner table (plate, NA)

Development of bones:
  intramembranous ossification
  intracartilaginous ossification

primary center of ossification
secondary center of ossification
diaphysis, diaphyses, diaphyseal
epiphysis, epiphyses, epiphyseal
metaphysis, metaphyses, metaphyseal
epiphyseal cartilage (line)
bone age
chronological age
ossification
ossicle
ala
condyle
cornu, cornua
crest
epicondyle
malleolus
process
spinous process
styloid process
trochanter
tubercle
tuberosity

See also:
parts of organs, Section 7, No. 2
openings in organs, Section 7, No. 3
depressions in organs, Section 7, No. 4
additional terms, Section 7, No. 5

Joints - basic terms:
articulation
arthron
arthrosis, arthroses, arthrodial
arthrology
junctura

Movements at joints:
abduction
adduction
circumduction
flexion
extension
angular movements
inversion
eversion
rotation
supination
pronation
gliding movements
dorsiflexion
circumflexion

Classification of joints:
fibrous, immovable, synarthrodial
cartilaginous, slightly movable, amphiarthrodial
synovial, freely movable, diarthrodial

Structure of synovial joints:
articular surfaces of bone ends
articular cartilages
capsule of joint;
  synovial membrane
  fibrous tissue capsule
joint cavity
ligaments
bursa, bursae

Varieties of synovial joints:
gliding joints
hinge joints
condylar (condyloid) joints
saddle joints
pivot joints
ball and socket joints

NOTES

NOTES

NOTES

6. THE UPPER LIMB: BONES AND JOINTS

Students should learn the names and general locations of the bones and joints of the whole body before undertaking a detailed study of these structures. The list of names at the beginning of each chapter dealing with bones and joints, along with the diagrams, and access to a mounted skeleton, should enable junior students to become acquainted with this basic information. For simplicity in presenting the material the skeletal system has been divided into:
1. The upper limb, superior limb or member
2. The lower limb, inferior limb or member
3. The vertebral column
4. The thorax
5. The skull

The term "Member" is used to replace "extremity" with reference to a limb as in the Nomina Anatomica. Older personnel will probably continue to use the terms upper and lower extremities.

The axial skeleton includes the bones of:
  the skull
  the vertebral column
  the ribs and sternum
  the hyoid bone

The appendicular skeleton includes the bones of:
  the upper limb with the shoulder girdle
  the lower limb with the pelvic girdle

S. 63 BONES OF THE UPPER LIMB OR MEMBER
  F. 6-1

SHOULDER GIRDLE:
1. Scapula - shoulder blade; pl. scapulae,
   adj. scapular
2. Clavicle - collar bone; pl. clavicles,
   adj. clavicular

ARM OR BRACHIUM:
1. Humerus - arm bone; pl. humeri,
   adj. humeral

FOREARM OR ANTEBRACHIUM:
1. Radius - lateral bone of forearm;
   pl. radii, adj. radial
2. Ulna - medial bone of forearm;
   pl. ulnae, adj. ulnar

HAND: wrist, palm, digits;
1. **Wrist or carpus**; 8 bones, 2 rows of four:

   Proximal row from thumb side:
   (1) Scaphoid bone (navicular bone of wrist) or os scaphoideum
   (2) Lunate bone (semilunar bone) or os lunatum
   (3) Triquetral bone (tringular bone) or os triquetrum
   (4) Pisiform bone, or os pisiforme

   Distal row from thumb side:
   (1) Trapezium (greater multangular bone) or os trapezium
   (2) Trapezoid bone (lesser multangular bone) or os trapezoideum
   (3) Capitate bone (os magnum) or os capitatum
   (4) Hamate bone or os hamatum

2. **Palm or metacarpus**; 5 bones, from thumb side:
   (1) First metacarpal bone
   (2) Second metacarpal bone
   (3) Third metacarpal bone
   (4) Fourth metacarpal bone
   (5) Fifth metacarpal bone

3. **Digits**; five digits, fourteen phalanges,
   (s. phalanx, pl. phalanges, adj. phalangeal)
   (1) Thumb or pollex, or first digit
   (2) Index finger, or second digit
   (3) Middle finger, or third digit
   (4) Ring finger, or fourth digit
   (5) Little finger, or fifth digit
   Thumb: 2 phalanges proximal, distal
   Fingers: 3 phalanges, proximal, middle, and distal phalanges

F. 6-1 SHOULDER GIRDLE & UPPER LIMB

—Frontal View—

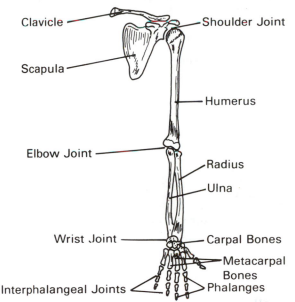

S. 64 JOINTS OF THE UPPER LIMB OR MEMBER

1. Sternoclavicular (two), sternum & clavicle
2. Acromioclavicular (two), scapula & ~~sternum~~ clavicle
3. Shoulder joint (two), scapula & humerus
4. Elbow joint (two), humerus, radius, ulna
   (1) humeroradial, (2) humeroulnar
5. Proximal radioulnar, (two), radius & ulna
6. Distal radioulnar, (two), radius & ulna.
7. Wrist or radiocarpal (NA), radius & carpals
8. Intercarpal joints, between adjacent carpals
9. Carpometacarpal, carpals & metacarpals
10. Intermetacarpal, between adjacent metacarpals
11. Metacarpophalangeal, metacarpals & phalanges
12. Interphalangeal, between adjacent phalanges
    (1) proximal, (2) distal

S. 65 IMPORTANT PARTS - BONES OF UPPER LIMB

SCAPULA: a flat bone, F. 6-2, 3, 4
1. **Body;**
      Borders: Medial or vertebral, lateral or axillary,
            superior
      Angles: superior or medial, inferior, lateral
      Spine: acromion, fossae

2. **Head:**
      glenoid cavity

3. **Neck:**
4. **Coracoid** process

CLAVICLE: a long bone F. 6-5
1. acromial extremity (end)
2. Body
3. Sternal extremity

HUMERUS: a long bone F. 6-6. 7
1. **Proximal extremity**
    head
    anatomical neck
    greater tubercle
    lesser tubercle
    intertubercular groove
    sugical neck
2. **Body** - shaft
    deltoid tubercle
3. **Condyle** or distal extremtiy
    capitulm
    trochlea
    fossae; - coronoid, radial,
              olecranon
    medial epicondyle
    lateral epicondyle
    ulnar groove

RADIUS: a long bone F. 6-8, 9, 12
1. **Proximal extremity**
    head
    neck
    radial tuberosity
2. **Body** or shaft
3. **Distal extremity**
    styloid process
    carpal articular surface
    ulnar notch

ULNA: a long bone F. 6-8, 9, 10, 11
1. **Proximal extremity**
    olecranon
    coronoid process
    trochlear notch (OT semilunar)
    radial notch
2. **Body** or shaft
3. **Distal extremity**
    head
    styloid process

HAND OR MANUS: F. 6-13
1. Wrist or carpus
2. Metacarpus
3. Digits

WRIST OR CARPUS: 8 short bones F. 6-14
1. **Proximal row:**
    (1) scaphoid bone - navicular
    (2) lunate bone - semilunar
    (3) triquetral bone - triangular
    (4) pisiform bone
2. **Distal row:**
    (1) trapezium - greater multangular
    (2) trapezoid - lesser multangular
    (3) capitate - os magnum
    (4) hamate

METACARPUS: 5 minature long bones F. 16-13
    (1) first metacarpal bone
    (2) second metacarpal bone
    (3) third metacarpal bone
    (4) fourth metacarpal bone
    (5) fifth metacarpal bone
    Parts: base or proximal extremity (end)
          body
          head or distal extremtiy (end)

DIGITS: 14 miniature long bones - phalanges
    (1) thumb, first digit, pollex
    (2) index finger, second digit
    (3) middle finger, third digit
    (4) ring finger, fourth digit
    (5) little finger, fifth digit
    Thumb, 2 phalanges, proximal & distal
    Fingers: 3 phalanges, proximal, middle, distal
    Parts: Base or proximal extremity (end)
          body
          head or distal extremity (end)

SESAMOID BONES

S. 66   DETAILED STUDY - BONES OF UPPER LIMB

The skeleton of the shoulder girdle and the upper limb includes:
    Bones of the shoulder girdle:
        scapula and clavicle
    Bone of the arm or brachium:
        humerus
    Bones of the forearm or antebrachium;
        radius and ulna
    Bones of the hand including;
        bones of the wrist - 8 carpal bones
        bones of the palm - 5 metacarpal bones
        bones of the digits - 14 phalanges

The SHOULDER GIRDLE is formed by the scapula and clavicle on each side of the body. (girdle - a support).

1. THE SCAPULA: F. 6-2, 3, 4
    Scapula - shoulder blade, pl. scapulae, adj. scapular.
    There are two scapulae, a right and a left. The scapula or shoulder blade is a flat triangular bone that lies against the upper posterolateral chest wall. The scapula has no direct connection with the bones of the thorax, but forms a joint with the lateral end of the clavicle. The clavicle in turn unites with the sternum, forming a prop or brace for the scapula. The scapula has a body, head, neck and two processes.

(1) **The body** of the scapula is flat and triangular and has three borders and three angles.
The medial border, formerly named the vertebral, is that margin closest to the median line of the body, and to the thoracic vertebrae.
The lateral border, or axillary border, is that margin farthest away from the median line of the body and adjacent to the posterior border of the armpit or axilla.
The superior border is its upper margin, It is short, and close to its lateral end is a notch, the **suprascapular notch.**
The medial angle of the scapula is at its upper medial part where the superior and medial borders meet.
The inferior angle is the lower rounded end of the scapula where the medial and lateral borders meet.
The lateral angle is at the junction of the superior and lateral borders. It is thick and blunt and forms the head of the scapula.
The spine of the scapula is a narrow ridge of bone that projects posteriorly from the dorsal surface of the body above its midpoint. It passes transversely from the medial to the lateral border. It extends posteriorly as a sort of shelf, dividing the posterior surface of the scapula into a supraspinatus fossa above, and an infraspinatus fossa below the spine. The supraspinatus and infraspinatus muscles are attached to the corresponding fossa. The spine can be palpated through the skin on the upper posterior chest.
The acromion is the flat rounded lateral end of the spine. It extends out over the shoulder joint, and can be felt through the skin at the tip of the shoulder.

F. 6-2   THE SCAPULA

—FRONTAL VIEW—

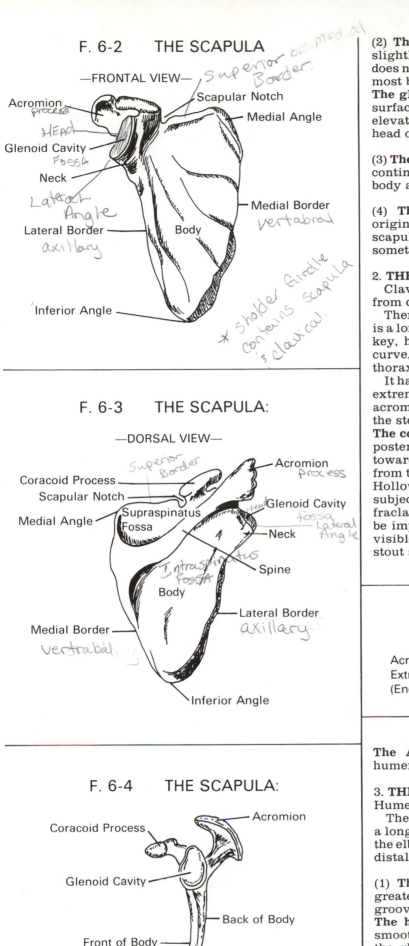

Superior or medial Border. (handwritten)

- Acromion *Process* (handwritten)
- *HEAD* (handwritten)
- Glenoid Cavity *Fossa* (handwritten)
- Neck
- *Lateral Angle* (handwritten)
- Lateral Border *axillary* (handwritten)
- Scapular Notch
- Medial Angle
- Medial Border *Vertabral* (handwritten)
- Body
- Inferior Angle

* Sholder Girdle Contains scapula & clavical. (handwritten)

F. 6-3   THE SCAPULA:

—DORSAL VIEW—

- Coracoid Process
- Scapular Notch
- Medial Angle
- Supraspinatus Fossa
- *Superior Border* (handwritten)
- Acromion *Process* (handwritten)
- *head* Glenoid Cavity *fossa Lateral Angle* (handwritten)
- Neck
- Spine
- *Intraspinatus Fossa* (handwritten)
- Body
- Medial Border *Vertrabal.* (handwritten)
- Lateral Border *axillary!* (handwritten)
- Inferior Angle

F. 6-4   THE SCAPULA:

- Coracoid Process
- Glenoid Cavity
- Front of Body
- Acromion
- Back of Body

---

(2) **The head** of the scapula is the poorly formed and slightly expanded lateral upper part of the scapula. It does not have the usual rounded form seen at the heads of most bones.

The glenoid cavity is an oval depression on the lateral surface of the head of the scapula. Its rim is slightly elevated around the concave part. It articulates with the head of the humerus to form the shoulder joint.

(3) **The neck** of the scapula is the slightly constricted part continuous with the upper part of the body between the body and head.

(4) **The coracoid process** is a beaklike **projection** originating from the anterior border of the neck of the scapula. It lies below the outer part of the clavicle and can sometimes be felt in this location.

2. THE CLAVICLE   F. 6-5

Clavicle = collar bone, pl. clavicles, adj. clavicular from clavicula.

There are two clavicles, a right and a left. The clavicle is a long flat bone somewhat resembling an old fashioned key, hence the name clavicula, a key. It has a double curve, and lies almost horizontally in front of the upper thorax. It extends from the shoulder to the sternum.

It has an acromial extremity (end), a body, and a sternal extremity (end). Its acromial end articulates with the acromion of the scapula. Its sternal end articulates with the sternum.

The conoid tubercle is a small rough prominence on the posterior margin of the inferior surface of the clavicle towards its acromial end. The conoid ligament passes from this tubercle to the coracoid process of the scapula. Hollows above and below the clavicle, visible in thin subjects are named the supraclavicular and infraclavicular fossae. They contain lymph nodes and may be important to the therapy technician. The clavicle is visible through the skin and may be palpated even in stout subjects.

F. 6-5   THE CLAVICLE

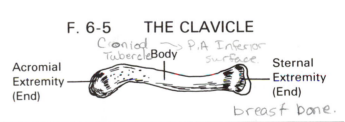

Cooniod Tubercle Body *P.A Inferior surface.* (handwritten)

- Acromial Extremity (End)
- Sternal Extremity (End)

breast bone. (handwritten)

The **ARM OR BRACHIUM** has a single bone, the humerus.

3. THE HUMERUS   F. 6-6, 7

Humerus = arm bone; pl. humeri, adj. humeral

There are two humeri, a right and a left. The humerus is a long cylindrical bone that reaches from the shoulder to the elbow. It has a proximal extremity (end), a body, and a distal extremity (end).

(1) **The proximal extremity** has a head, a neck, and greater and lesser tubercles, with an intertubercular groove (bicipital groove).

The head (L. caput humeri) is the superior expanded smooth rounded end of the humerus. It articulates with the glenoid cavity of the scapula to form the shoulder joint.

The anatomical neck is the slightly constricted obliquely directed part of the humerus between the head and remainder of the bone.

F. 6-6   THE LEFT HUMERUS:

Copy

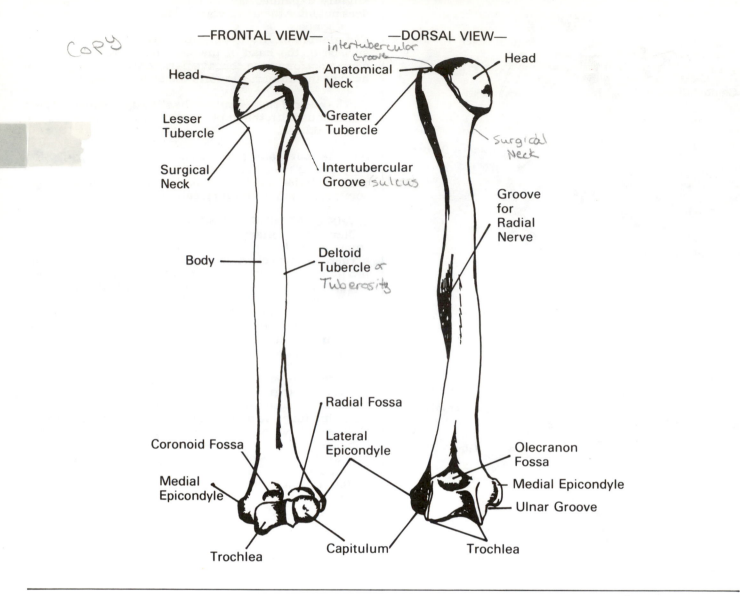

—FRONTAL VIEW—   —DORSAL VIEW—

intertubercular Groove

Head

Anatomical Neck

Head

Greater Tubercle

Lesser Tubercle

Surgical Neck

Intertubercular Groove sulcus

Surgical Neck

Groove for Radial Nerve

Body

Deltoid Tubercle or Tuberosity

Radial Fossa

Lateral Epicondyle

Coronoid Fossa

Olecranon Fossa

Medial Epicondyle

Medial Epicondyle

Ulnar Groove

Trochlea

Capitulum

Trochlea

F. 6-7   LEFT LOWER HUMERUS & ELBOW JOINT:

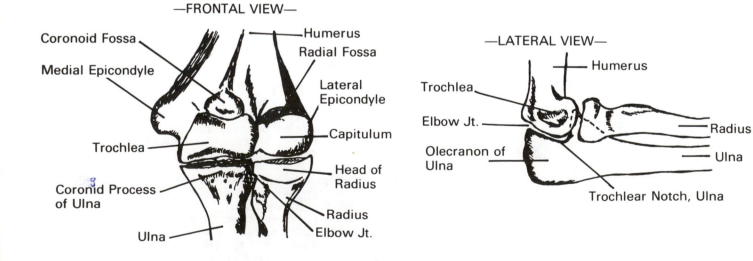

—FRONTAL VIEW—

Coronoid Fossa

Medial Epicondyle

Humerus

Radial Fossa

Lateral Epicondyle

Capitulum

Trochlea

Head of Radius

Coronoid Process of Ulna

Radius

Ulna

Elbow Jt.

—LATERAL VIEW—

Humerus

Trochlea

Elbow Jt.

Radius

Olecranon of Ulna

Ulna

Trochlear Notch, Ulna

46

The **greater tubercle** (L. tuberculum majus) is a large bony prominence on the lateral border of the humerus just below the anatomical neck. It has small depressions on its upper end that provide insertions for the spinatus muscles.

The **lesser tubercle** (L. tuberculum minus) is a smaller bony prominence on the anterior surface of the humerus just below the anatomical neck.

The **intertubercular groove** (bicipital groove) is a furrow that extends longitudinally on the anterior surface of the upper humerus between the greater and lesser tubercles. It forms a groove for the biceps tendon.

The **surgical neck** is the constricted part of the humerus below the tubercles. It has been named the surgical neck as fractures may occur here.

(2) **The body** of the humerus is the long rounded part that becomes flattened from front to back as it nears the elbow. It has a **deltoid tubercle,** a rough prominence on the anterolateral surface of the humerus close to its midpoint. The deltoid muscle inserts here. The area is sometimes not readily visible.

(3) **The distal extremity** has also been named the condyle. It has a capitulum, trochlea, medial and lateral epicondyles, three fossae (coronoid, radial, and olecranon) and an ulnar groove.

The **condyle** is the distal extremity of the humerus.

The **capitulum** (little head) is a small rounded prominence forming the lateral part of the lower articular end of the humerus. It forms a joint with the upper surface of the head of the radius. It has been named capitulum (little head) to distinguish it from the head at the upper end of the same bone.

The **trochlea** (pulley) forms the medial part of the distal articular suface of the humerus. It is shaped like a pulley or spool. It forms a joint with the trochlear (semilunar) notch of the ulna, and forms more than the medial half of the joint surface.

The **coronoid fossa** is a depression on the front of the lower humerus immediately above the trochlea. The coronoid process of the ulna fits into it when the forearm is flexed (bent).

The **radial fossa** is a small depression on the front of the lower humerus above the capitulum. The head of the radius fits into it with flexion of the forearm.

The **olecranon fossa** is a depression on the back of the lower humerus above the trochlea. The olecranon fits into it when the forearm is extended (straightened).

The **medial epicondyle** is a knucklelike rounded bony prominence on the medial border of the lower humerus above the trochlea. It is large and is readily palpable through the skin. (epi = upon + condyle, a prominence upon a condyle).

The **lateral epicondyle** is a similar but smaller bony prominence on the lateral margin of the lower humerus above the capitulum. It is also palpable.

The **ulnar groove** is a furrow on the dorsal surface of the medial epicondyle through which the ulnar nerve descends to the forearm. (the funny bone).

THE FOREARM OR ANTEBRACHIUM has two long bones, the radius and ulna, that reach from the elbow to the wrist. When the upper limb is in the anatomical position (palm of hand to the front if standing, palm directed upwards when lying in the supine position) the two bones are parallel to each other, with the radius on the lateral side.

4. RADIUS   F. 6-8, 9, 18

Radius = lateral bone of forearm, pl. radii, adj. radial. (L. a spoke, rod, ray).

There are two radii, a right and a left. The radius has the ulna on its medial side. It has a proximal extremity (end), a body, and a distal extremity (end).

(1) **The proximal extremity** of the radius **has a head, neck,** and a tuberosity.

The **head** of the radius is its upper expanded disclike end, with a slightly concave upper articular surface. The head articulates with the capitulum of the humerus.

The **neck** is the constricted part distal to the head, connecting the head to the body.

The **radial tuberosity** is a rough prominence on the anteromedial surface of the radius below its neck. The biceps tendon inserts into it.

(2) **The body** of the radius becomes gradually larger as it approaches the wrist joint and its articular surface.

(3) **The distal extremity** (end) of the radius has a styloid process, a carpal articular surface and an ulnar notch.

The **styloid process** is a large bony prominence on the lateral border and distal end of the radius. It extends down below the level of the wrist joint.

The **carpal articular surface** is a large smooth area on the distal end of the radius that articulates with the carpal bones to form the wrist joint.

The **ulnar notch** is a small depression on the medial margin of the distal end of the radius above its articular surface. The head of the ulna articulates with the radius here.

5. THE ULNA   F. 6-8, 9, 10

Ulna = medial bone of the forearm; pl. ulnae, adj. ulnar.

There are two ulnae, a right and a left. The radius lies on its lateral side. The ulna has a proximal extremity (end), a body, and a distal extremity or end.

(1) **The proximal extremity** of the ulna has an olecranon, a coronoid process, a trochlear notch, and a radial notch.

The **olecranon** is the bluntly rounded upper end of the ulna that lies posterior to the elbow joint. It has a horizontal upper surface that is directed dorsally when the elbow is flexed, and becomes palpable posterior to the joint. The anterior smooth surface is concave and helps to form the elbow joint.

The **coronoid process** extends anteriorly from the upper part of the body of the ulna. It is beaklike, and lies below and anterior to the olecranon. Its anterior flattened tip fits into the coronoid fossa of the humerus when the elbow is flexed. The brachialis muscle is attached to it.

The **trochlear notch** (semilunar notch) is a concave half-moon shaped hollow on the anterior surface of the upper ulna. It is formed by the anterior curved surface of the olecranon and the upper curved surface of the coronoid process. Its smooth articular surface forms a joint with the trochlea of the humerus.

The **radial notch** is a depression on the lateral surface of the upper ulna below the trochlear notch. The medial margin of the head of the radius fits into this notch to form the proximal radioulnar joint.

(2) **The body** of the ulna, large at its superior end, becomes smaller as it approaches the wrist.

(3) **The distal extremity** has a head, and a styloid process.

The **head** of the ulna is its lower expanded end and can be felt through the skin. It is also visible as a prominence on the posteromedial border of the wrist.

The **styloid process** of the ulna is a small, pointed prominence that extends distally from the posteromedial border of the ulnar head. It is palpable. The student should note that the radial styloid process is much larger than the ulnar styloid, and extends down to a lower level beyond the wrist joint.

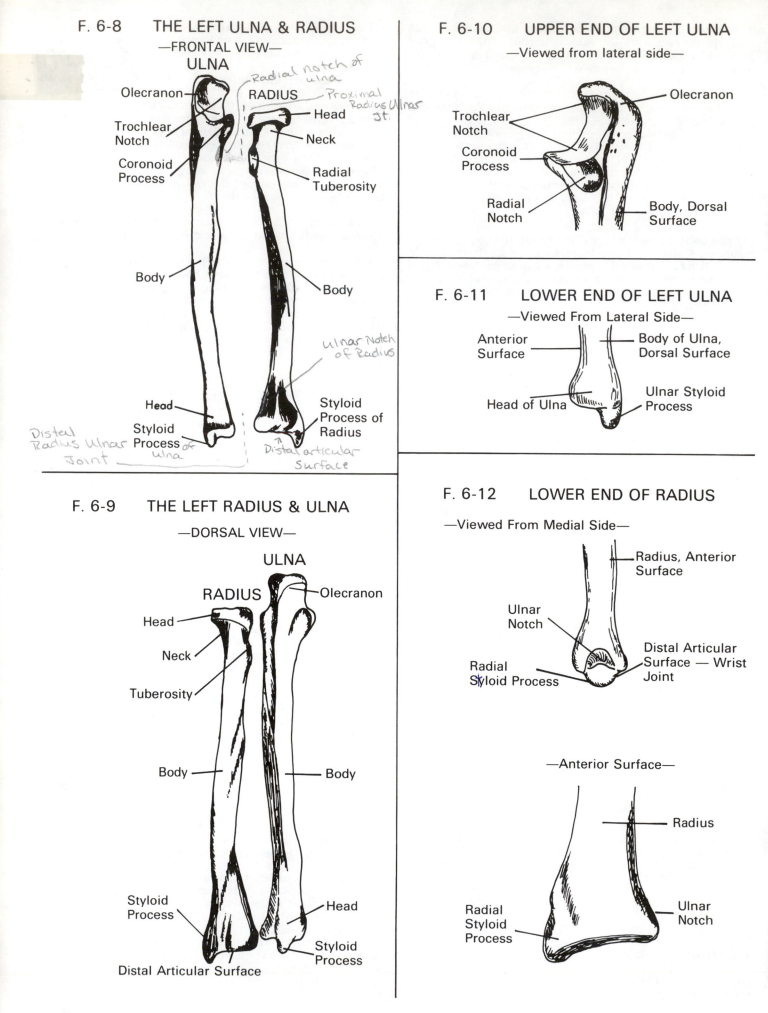

F. 6-8   THE LEFT ULNA & RADIUS
—FRONTAL VIEW—
ULNA
Olecranon
Radial Notch of ulna
RADIUS
Proximal Radius Ulnar Jt.
Trochlear Notch
Head
Coronoid Process
Neck
Radial Tuberosity
Body
Body
Ulnar Notch of Radius
Head
Styloid Process of Radius
Distal Radius Ulnar Joint
Styloid Process of ulna
Distal articular Surface

F. 6-9   THE LEFT RADIUS & ULNA
—DORSAL VIEW—
ULNA
RADIUS
Olecranon
Head
Neck
Tuberosity
Body
Body
Styloid Process
Head
Styloid Process
Distal Articular Surface

F. 6-10   UPPER END OF LEFT ULNA
—Viewed from lateral side—
Olecranon
Trochlear Notch
Coronoid Process
Radial Notch
Body, Dorsal Surface

F. 6-11   LOWER END OF LEFT ULNA
—Viewed From Lateral Side—
Anterior Surface
Body of Ulna, Dorsal Surface
Head of Ulna
Ulnar Styloid Process

F. 6-12   LOWER END OF RADIUS
—Viewed From Medial Side—
Radius, Anterior Surface
Ulnar Notch
Distal Articular Surface — Wrist Joint
Radial Styloid Process
—Anterior Surface—
Radius
Radial Styloid Process
Ulnar Notch

48

F. 6-13   LEFT HAND

—Viewed from the back DORSAL SURFACE –

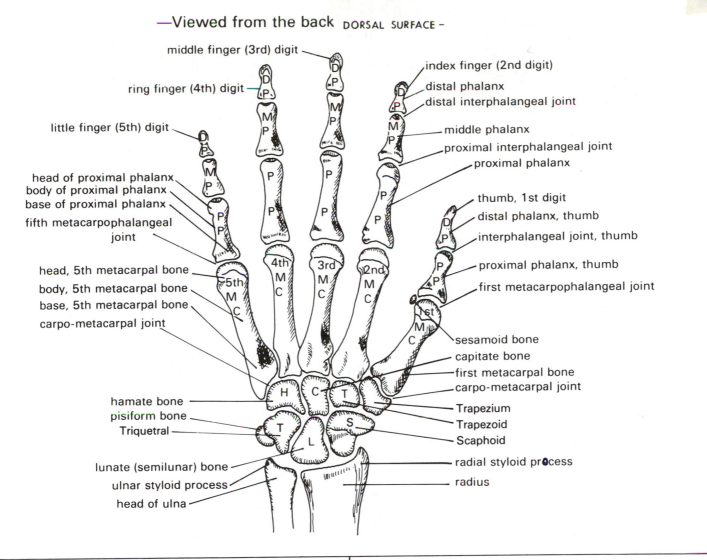

- middle finger (3rd) digit
- ring finger (4th) digit
- little finger (5th) digit
- head of proximal phalanx
- body of proximal phalanx
- base of proximal phalanx
- fifth metacarpophalangeal joint
- head, 5th metacarpal bone
- body, 5th metacarpal bone
- base, 5th metacarpal bone
- carpo-metacarpal joint
- hamate bone
- pisiform bone
- Triquetral
- lunate (semilunar) bone
- ulnar styloid process
- head of ulna

- index finger (2nd digit)
- distal phalanx
- distal interphalangeal joint
- middle phalanx
- proximal interphalangeal joint
- proximal phalanx
- thumb, 1st digit
- distal phalanx, thumb
- interphalangeal joint, thumb
- proximal phalanx, thumb
- first metacarpophalangeal joint
- sesamoid bone
- capitate bone
- first metacarpal bone
- carpo-metacarpal joint
- Trapezium
- Trapezoid
- Scaphoid
- radial styloid process
- radius

THE HAND OR MANUS   F. 6-13

The skeleton of the hand includes all of the bones of the wrist (carpal bones), the bones of the palm (metacarpal bones) and the bones of the digits (phalanges).

6. THE WRIST OR CARPUS   F. 6-13, 14

There are eight carpal bones, classified as short bones. Four of them form a **proximal row** distal to the lower end of the radius. The other four form a **distal row** proximal to the metacarpal bones. It is customary to name the proximal row first, beginning at the thumb side or radial border of the wrist. The distal row is then named in the same order from the thumb side. Some of these bones have two names in common use. The student should learn both names as the older medical personnel will continue to use the old names.

Proximal row:
(1) scaphoid bone
(2) lunate bone
(3) triquetral bone
(4) pisiform bone

Distal row:
(1) trapezium
(2) trapezoid bone
(3) capitate bone
(4) hamate bone

The scaphoid bone, navicular bone, or os scaphoideum, is boat shaped and lies on the radial side of the wrist. It has a tubercle on its anterolateral margin. (G. skaphe = boat + oid = like, so boat shaped). The term navicular has now been restricted by (NA) to refer to a bone of the foot.

The lunate bone, semilunar bone or os lunatum, lies between the scaphoid and triquetral bones. It is crescent shaped, resembling a half-moon, hence the terms lunate - moon, and semilunar - half-moon.

The triquetral bone, triangular bone os triquetrum, lies on the ulnar side of the wrist and is somewhat triangular in shape. (triquetrus = triangle).

The pisiform bone, or os pisiforme lies anterior to the triquetral bone, not medial to it. It forms a prominence on the anteromedial part of the wrist and is readily palpable. (L. pisum = a pea + form = shape, hence pea shaped).

The trapezium, or greater multangular bone, lies on the radial or thumb side of the wrist in the distal row of carpal bones. (G. trapezium = a table, a four sided figure).

The trapezoid bone, or lesser multangular bone lies between the trapezium and the capitate bone. (oid = like + trapezium, i.e. like trapezium).

The capitate bone, os magnum, or os capitatum, lies between the trapezoid and hamate bones. Its proximal end is rounded, hence the name capitate from caput = head. It is the largest bone of the wrist, hence the older name os magnum.

The hamate bone lies on the ulnar border of the wrist. It has a hooklike process - the hamulus. (L. hamus = a hook)

7. THE METACARPUS F. 6-13

Five metacarpal bones form the bony framework of the palm of the hand. They are named from the thumb or radial side as the first, second, third, fourth and fifth metacarpal bones. They are classified as miniature long bones, not as short bones. (meta = beyond + carpus = wrist, i.e. beyond the wrist).
Each metacarpal bone has:
 a base or proximal or carpal extremity (end)
 a body
 a head, or distal extremity (end)

8. THE DIGITS F. 6-13, 15

Fourteen phalanges form the digits. The thumb has two phalanges - a proximal and a distal. Each of the fingers has three phalanges - a proximal, a middle, and a distal phalanx.
Each phalanx has:
 a base or proximal extremity
 a body
 a head or distal extremity
The digits are named from the thumb or radial side as the first, second, third, fourth and fifth digits. The digits are also named the thumb, or pollex, the index finger, the middle finger, the ring finger and the little finger.
Sometimes they are incorrectly named the first finger, the second finger, the third finger, the fourth finger and fifth finger.

9. THE SESAMOID BONES F. 6-13

The sesamoid bones are small oval or rounded masses of bone tissue that develop in tendons. They are composed of bone cells, and are not just deposits of calcium. In the hand they are located on the palmar surfaces of the metacarpophalangeal joints in the flexor tendons of the digits. The thumb frequently has two of them, and the index and other digits may also have them. Sesamoids may also be present on the plantar surfaces of the metatarsophalangeal joints of the toes, with two frequently located beneath the great toe. They may also be present in other tendons of the extremities. The patella, or knee cap, is actually a sesamoid bone in the tendon of the quadriceps tendon above the patellar ligament.

S. 67 PARTS OF BONES FORMING THE JOINTS OF THE UPPER LIMB

Sternoclavicular joint; a gliding joint,
sternal end of clavicle
clavicular notch of sternum
Acromioclavicular joint; a gliding joint
acromial end of clavicle
acromion of scapula
Shoulder joint; a ball and socket joint,
head of humerus
glenoid cavity of head of scapula
Elbow joint; a hinge joint,
1. Humeroulnar; trochlea of humerus
                 trochlear notch of ulna
2. Humeroradial; capitulum of humerus
                  head of radius
Proximal radioulnar joint; a pivot joint
head of radius (circumference)
radial notch of ulna
Distal radioulnar joint;
head of ulna
ulnar notch of radius
Wrist joint; a condylar joint
distal articular surface of radius
scaphoid, lunate, triquetral bones - wrist
Intercarpal joints; gliding joints,
between adjacent carpal bones

F. 6-14 LEFT WRIST
—FRONTAL VIEW—

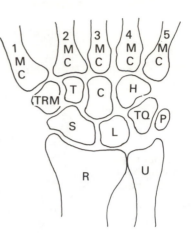

—LATERAL VIEW—

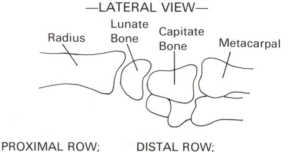

PROXIMAL ROW;		DISTAL ROW;			
S	= Scaphoid	TRM	= Trapezium	R	= Radius
L	= Lunate	T	= Trapezoid	U	= Ulna
TQ	= Triquetral	C	= Capitate	MC	= Metacarpal
P	= Pisiform	H	= Hamate		

F. 6-15 MIDDLE FINGER
BONES & JOINTS

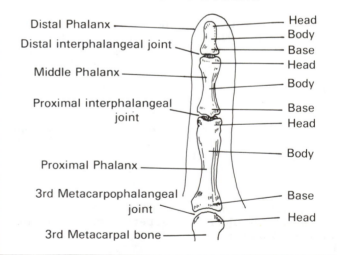

Carpometacarpal joint;
at thumb = saddle joint.
others = gliding joints.
distal row of carpal bones
bases of adjacent metacarpal bones
Intermetacarpal joints; gliding joints
between bases of adjacent metacarpals
Metacarpophalangeal joints; condylar joints
distal end of a metacarpal
base proximal phalanx of digit

Proximal interphalangeal joints; hinge joints
head of proximal phalanx of digit
base of middle phalanx
thumb = 2 phalanges, one interphalangeal jt.
Distal interphalangeal joints; hinge joints
head of middle phalanx
base of a distal phalanx   of fingers

S. 68 DETAILED STUDY: JOINTS OF UPPER LIMB

Students should recall the structure of the types of
joints as described in the preceding chapter:
Synovial joints are freely movable in definite directions.
They have a joint cavity, a **capsule,** consisting of a
synovial membrane, and an outer fibrous tissue layer
surrounding the joint, reinforced by ligaments and
sometimes by muscles or tendons, e.g. shoulder joint.
Cartilaginous joints have restricted movement, with
cartilage between the bone ends, and with ligaments
extending across the joint from one bone to the other to
hold the bone ends in position.
Fibrous joints have fibrous tissue between the bone ends
with no movement permitted. The fibrous tissue may
disappear and bone ends may unite.

1. **Sternoclavicular joints:** Rt. & Lt.   F. 6-16
- located at the upper lateral margin of the manubrium on
each side, and palpable here.
- formed by sternal end of clavicle articulating with the
clavicular notch and adjacent margin of the sternum.
- a synovial gliding joint, with some movement in
practically all directions.
-an articular disc lies between the adjacent bone ends, so
there are two synovial cavities.

2. **Acromioclavicular joints:** Rt. & Lt.   F. 6-17
- located at the lateral end of the clavicle at the tip of the
shoulder.
-formed by the acromial end of the clavicle and the medial
border of the acromion of the scapula. Sometimes an
articular disc is present.
- a gliding joint with a sliding movement.

3. **Shoulder joints:** Rt. & Lt.   F. 6 - 17
   (NA) humeral joints.
- located beneath the acromion, at the shoulder, and
covered by the deltoid muscle so not palpable.
- formed by the head of the humerus articulating with the
glenoid cavity of the scapula.
- a synovial ball and socket joint the glenoid cavity of the
humerus deepened by a ring of cartilage attached to the
bony rim of the glenoid.
- flexion, (forward movement) of the humerus, extension,
(backward movement), abduction, adduction, rotation,
and circumduction are all possible.
- capsule and ligaments at the shoulder may be loose, but
the bone ends are held together by the muscles and
tendons that cross the joint from the trunk to the
humerus. Dislocation is more frequent at the shoulder
than at the hip joint.

4. **Elbow joints:** Rt. & Lt.   F. 6-18
- (NA) cubital joints; AS. elnboga, L. cubitus
- located at the lower end of the humerus
- formed by the condyle (NA) of the humerus and the
proximal ends of the radius and ulna,
- trochlea of the humerus fits into trochlear notch
(semilunar notch) of the ulna = the humeroulnar joint.
- capitulum of the humerus articulates with the head of
the radius i.e. the humeroradial joint.
- a synovial hinge joint, with flexion and extension, and
with the radioulnar joints pronation and supination of
the hand. The synovial membrane encloses the elbow
joint and also the proximal radioulnar joint.
Humeroradial joint is a hinge and gliding joint.

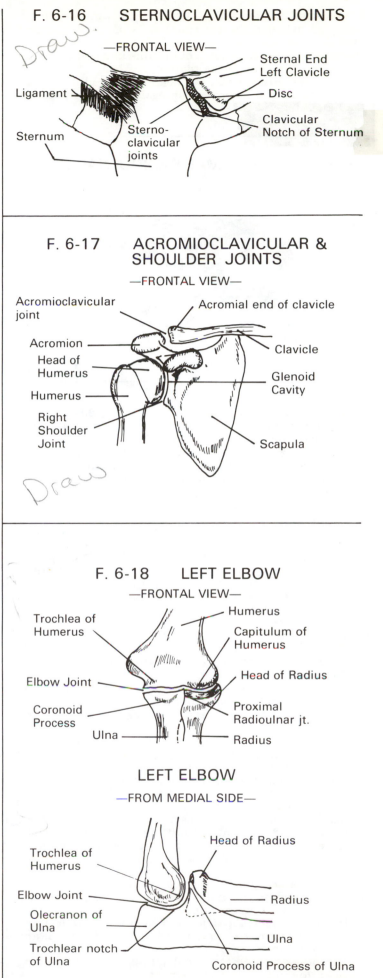

F. 6-16   STERNOCLAVICULAR JOINTS
—FRONTAL VIEW—
Draw.
Sternal End
Left Clavicle
Ligament
Disc
Sternum
Sterno-
clavicular
joints
Clavicular
Notch of Sternum

F. 6-17   ACROMIOCLAVICULAR &
SHOULDER JOINTS
—FRONTAL VIEW—
Acromioclavicular
joint
Acromial end of clavicle
Acromion
Head of
Humerus
Clavicle
Glenoid
Cavity
Humerus
Right
Shoulder
Joint
Scapula
Draw

F. 6-18   LEFT ELBOW
—FRONTAL VIEW—
Trochlea of
Humerus
Humerus
Capitulum of
Humerus
Elbow Joint
Head of Radius
Coronoid
Process
Proximal
Radioulnar jt.
Ulna
Radius

LEFT ELBOW
—FROM MEDIAL SIDE—
Head of Radius
Trochlea of
Humerus
Elbow Joint
Radius
Olecranon of
Ulna
Ulna
Trochlear notch
of Ulna
Coronoid Process of Ulna

5. Proximal radioulnar joints: Rt. & Lt.   F. 6-18, 19
- located between the proximal ends of the radius and ulna.
- formed by the radial notch of the ulna and a fibrous ring encircling the head of the radius.
- a synovial pivot joint, the head of the radius rotates within the ring to produce supination and pronation of the hand.

F. 6-19   PROXIMAL RADIOULNAR JOINT

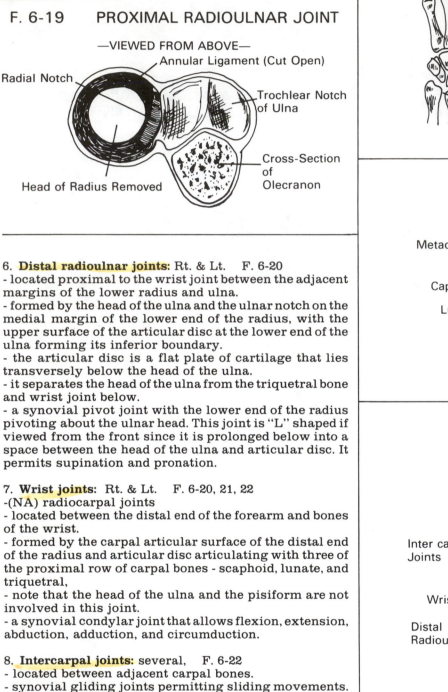

—VIEWED FROM ABOVE—

Radial Notch
Annular Ligament (Cut Open)
Trochlear Notch of Ulna
Cross-Section of Olecranon
Head of Radius Removed

6. Distal radioulnar joints: Rt. & Lt.   F. 6-20
- located proximal to the wrist joint between the adjacent margins of the lower radius and ulna.
- formed by the head of the ulna and the ulnar notch on the medial margin of the lower end of the radius, with the upper surface of the articular disc at the lower end of the ulna forming its inferior boundary.
- the articular disc is a flat plate of cartilage that lies transversely below the head of the ulna.
- it separates the head of the ulna from the triquetral bone and wrist joint below.
- a synovial pivot joint with the lower end of the radius pivoting about the ulnar head. This joint is "L" shaped if viewed from the front since it is prolonged below into a space between the head of the ulna and articular disc. It permits supination and pronation.

7. Wrist joints: Rt. & Lt.   F. 6-20, 21, 22
-(NA) radiocarpal joints
- located between the distal end of the forearm and bones of the wrist.
- formed by the carpal articular surface of the distal end of the radius and articular disc articulating with three of the proximal row of carpal bones - scaphoid, lunate, and triquetral,
- note that the head of the ulna and the pisiform are not involved in this joint.
- a synovial condylar joint that allows flexion, extension, abduction, adduction, and circumduction.

8. Intercarpal joints: several,   F. 6-22
- located between adjacent carpal bones.
- synovial gliding joints permitting sliding movements.
   The **midcarpal joint** is an intercarpal joint between the proximal and distal rows of carpal bones.

9. Carpometacarpal joints: five,   F. 6-20
- located between the distal row of carpals and the bases of adjacent metacarpal bones.
- **first carpometacarpal joint:** a synovial saddle joint, allowing flexion, extension, abduction, adduction, and circumduction.
- four other carpometacarpal joints; synovial gliding joints allowing limited movement.

F. 6-20   JOINTS OF WRIST, PALM & DIGITS

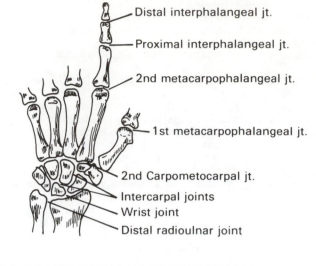

Distal interphalangeal jt.
Proximal interphalangeal jt.
2nd metacarpophalangeal jt.
1st metacarpophalangeal jt.
2nd Carpometocarpal jt.
Intercarpal joints
Wrist joint
Distal radioulnar joint

F. 6-21   WRIST JOINT
—LATERAL VIEW—

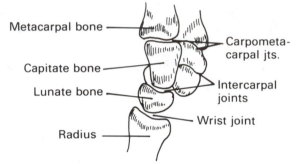

Metacarpal bone
Capitate bone
Lunate bone
Radius
Carpometacarpal jts.
Intercarpal joints
Wrist joint

F. 6-22   LEFT WRIST—
—DORSAL VIEW—

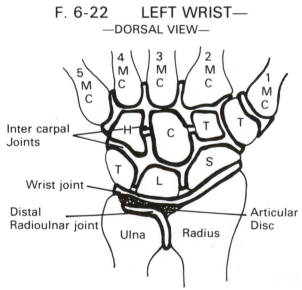

5 MC   4 MC   3 MC   2 MC   1 MC
Inter carpal Joints
H   C   T   T
T   L   S
Wrist joint
Distal Radioulnar joint
Ulna   Radius
Articular Disc

10. Metacarpophalangeal joints: F. 6-20, 22
-five in each hand.
- located at the knuckles, between the metacarpal bones and bases of proximal phalanges.
- synovial condylar joints, permitting flexion, extension, abduction, adduction, and circumduction.

11. Interphalangeal joints: F. 6-20
- nine in each hand.
- located between the head of one phalanx and the base of the phalanx distal to it.
- **thumb** with only two phalanges has only one interphalangeal joint.
- other digits have two interphalangeal joints - a proximal and a distal.
- **proximal** interphalangeal joint located between head of a proximal and base of a middle phalanx.
- **distal** interphalangeal joint located between head of a middle phalanx and base of a distal phalanx
- synovial hinge joints, allowing flexion and extension.

The carpal tunnel is formed anterior to the wrist by a transverse carpal ligament that is attached to the scaphoid and trapezium laterally and to the pisiform and hook of the hamate medially. This ligament, together with the anterior curved surfaces of the carpal bones, forms a tunnel for the tendons, blood vessels, and nerves to pass from the forearm into the hand.

S. 69   SOME CONGENITAL ANOMALIES OF THE UPPER LIMB

One or both upper limbs may be absent - abrachium (a = without + brachium = without an arm)
One or several fingers or the whole hand may be absent. There may be an extra digit present - a supernumerary digit.
Sprengel's deformity - elevation and deformity of the scapula.
Madelung's deformity - a curvature of the body of the radius, so that the wrist joint lies obliquely.
Fusion - a joining together of the upper ends of the radius and ulna, with inability to supinate or pronate the hand.

S. 70   LANDMARKS AND PROMINENCES OF THE UPPER LIMB

Shoulder:
The spine and inferior angle of the scapula can be palpated (felt).
The tip of the acromion overhangs the shoulder joint and can be palpated there.
The coracoid process of the scapula can sometimes be palpated below the clavicle near to and below the lateral end of the clavicle.
The clavicle can be palpated throughout its entire length, and in thin subjects is also visible.
The deltoid tubercle cannot be palpated but may be located on the lateral border of the midhumerus as a dimple of the skin there.

Elbow:
Three bony prominences are readily palpable at the elbow, the epicondyles and the olecranon.
The medial epicondyle of the humerus may be visualized and palpated at the medial margin of the elbow about 1 cm or 0.5 inches above the level of the elbow joint.
The lateral epicondyle of the humerus is much smaller, but may also be palpated at the lateral margin of the elbow, slightly above the joint level, particularly if the forearm is flexed.
The olecranon forms the tip of the elbow and can be palpated on the posterior surface of the elbow, particularly when flexed.
The head of the radius is not usually visible but may be palpated as a bony ridge rotating under a finger placed on the posterior and lateral part of the elbow, as the hand is being pronated and supinated.

Wrist:
The styloid process of the radius is large and can be palpated on the lateral margin of the wrist. The head of

the ulna forms a rounded prominence on the posteromedial surface of the wrist, and can be seen and palpated.
The styloid process of the ulna although small can be palpated distal to the ulnar head.
The pisiform bone is the only carpal bone that is palpable as a separate prominence. It lies in front of the triquetral bone and can be palpated on the anterior surface and medial border of the wrist.

Metacarpals and digits:
The metacarpophalangeal joints, or knuckles, are quite prominent on the dorsum of the hand with the digits flexed, and are visible and palpable.
The digits and their joints present no problem radiologically.

S. 71   ANATOMICAL PECULIARITIES AND RADIOGRAPHY

For this study a mounted skeleton is required. If one is not available a set of bones of the upper limb and shoulder girdle, together with radiographs of each part, should be utilized.
The student should imagine that his/her eye is the anode of an x-ray tube, and view each structure as it would appear to the x-ray tube.
In any description of the bones and joints of the shoulder and upper limb it is always assumed the subject is in the anatomical position. This applies to films taken in the upright, supine, or prone positions. For lateral views the parts should be positioned so as to be at right angles to the anteroposterior or posteroanterior anatomical positions.

The scapula lies obliquely against the upper chest wall. Because of this the body must be rotated from the anatomical position so that it can be viewed "face on", or anteroposteriorly. The opposite shoulder must therefore be brought forward with the subject upright, or elevated from the table top if supine. To view the scapula laterally or in the "face on" position the subject must be turned away from the anatomical position.

The shoulder joint also lies obliquely to the viewing eye with the subject in the anatomical position. To see through this joint adjustments similar to those required for the scapula must be made.

Calcifications about the shoulder frequently lie above or posterior to the greater tubercle of the humerus, in or about the tendons here. They may not be visible through the humerus. Sometimes they may be demonstrated by internally and externally rotating the humerus.

The humerus in frontal views presents no problem but for lateral views it may be necessary to place the forearm across the chest or to place the hand upon the iliac crest. Following an injury the limb is often supported in a sling. No attempt should be made to move the limb into the anatomical position but the entire body including the limb may be rotated to view the humerus from the front or laterally.

The elbow joint and forearm in the anatomical position will have the radius and ulna lying parallel, whereas with the palm of the hand down the radius crosses the ulna. Further, if the subject lies upon the table top or sits upon a low stool beside the table the arm, forearm and hand will be at the same level, and in contact with the film. If it is unwise to extend the forearm because of an injury anteroposterior views may be obtained, one with the forearm resting upon the film, the other with the arm upon the film and the partly flexed forearm supported by a pillow. To see into the elbow joint for lateral views the shoulder should be placed at the same level as the limb. This also applies to anteroposterior views of the humerus.

The **wrist** because of the overlapping of the carpal bones with each other renders radiography more exacting. While there is some overlapping of adjacent bones in any position, multiple views in several oblique positions should make it possible to see each of these bones. Abducting and adducting the hand should also help.

The **metacarpal bones** lie close together and have limited movement so that in lateral views there will be overlapping. By oblique positioning, additional information may be obtained.

The **digits** are more mobile. To obtain views of a single digit it is a simple matter to have the other digits held out of the way. For lateral views of all the digits with minimum overlapping the fingers may be placed in varying degrees of flexion.

To obtain a list of the anatomical terms that have been introduced in this chapter the student should refer to S. 63, S. 64, and S. 65 at the beginning of this study.

7. THE LOWER LIMB: BONES AND JOINTS

S. 72 BONES OF THE LOWER LIMB OR MEMBER

PELVIC GIRDLE

1. Right hip bone or os coxae or coxal or innominate bone
2. Left hip bone - as above for right
Three fused bones:
    ilium, pl. ilia, adj. iliac
    ischium, pl. ischia, adj. ischial
    pubis, pl. pubes. adj. pubic

PELVIS
1. Right hip bone
2. Left hip bone
3. Sacrum
4. Coccyx

THIGH
1. Femur or thigh bone
    pl. femora, adj. femoral
2. Patella or knee cap
    pl. patellae, adj. patellar

LEG OR CRUS
1. Tibia or shin bone
    pl. tibiae, adj. tibial
2. Fibula or calf bone
    pl. fibulae, adj. fibular

FOOT: tarsus, metatarsus, digits
1. **Tarsus** ~~or instep~~: seven bones

    Posterior tarsal bones: three bones
    (1) Talus or astragalus
    (2) Calcaneus or os calcis, heel bone
    (3) Navicular bone or tarsal scaphoid (OT)

    Anterior tarsal bones: four bones from medial side
    (1) Medial cuneiform or first cuneiform
    (2) Intermediate cuneiform or second cuneiform
    (3) Lateral cuneiform or third cuneiform
    (4) Cuboid bone

2. **Metatarsus:** five bones, from medial side
    (1) First metatarsal bone
    (2) Second metatarsal bone
    (3) Third metatarsal bone
    (4) Fourth metatarsal bone
    (5) Fifth metatarsal bone

3. **Digits or toes:** five toes, fourteen phalanges
    s. phalanx, pl. phalanges, adj. phalangeal
    (1) Great toe, hallux or first digit
    (2) Second toe or second digit
    (3) Third toe or third digit
    (4) Fourth toe or fourth digit
    (5) Fifth toe or fifth digit
    Great toe - two phalanges proximal and distal phalanges
    Other toes - three phalanges proximal, middle, distal phalanges

4. **Sesamoid Bones**

S. 73 JOINTS OF THE LOWER LIMB OR MEMBER

1. Sacroiliac (two) sacrum and ilium
2. Symphysis pubis (one) between pubic bones
3. Hip joints (two) hip bone and femur
4. Knee joints (two) femur and tibia
5. Proximal tibiofibular (two) upper tibia and fibula

F. 7-1   BONES OF PELVIS & LOWER LIMB

— FRONTAL VIEW —

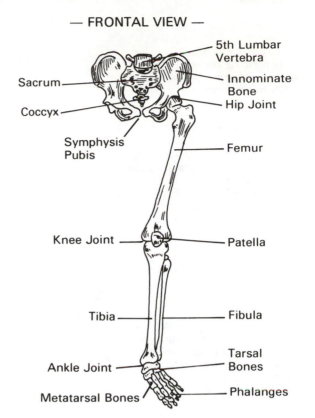

5th Lumbar Vertebra
Sacrum
Innominate Bone
Coccyx
Hip Joint
Symphysis Pubis
Femur
Knee Joint
Patella
Tibia
Fibula
Ankle Joint
Tarsal Bones
Metatarsal Bones
Phalanges

6. Distal tibiofibular (two) lower tibia and fibula
7. Ankle or talocrural (NA), (two) talus, tibia, fibula
8. Intertarsal joints (several) between adjacent tarsal bones
9. Tarsometatarsal joints (five) each foot between tarsals and metatarsals
10. Intermetatarsal joints (four) each foot between adjacent metatarsals
11. Metatarsophalangeal joints (five) each foot between metatarsals and proximal phalanges
12. Interphalangeal joints (nine) each foot (1) proximal, (2) distal

S. 74   IMPORTANT PARTS - BONES OF LOWER LIMB

HIP BONE: OS COXAE, INNOMINATE BONE;   F. 7-2, 3
    acetabulum
    acetabular notch
    acetabular fossa
    obturator foramen
    3 fused bones:
      ilium
      ischium
      pubis

ILIUM   F. 7-3
1. **Body**
2. **Wing** or ala
    crest
    iliac fossa
    auricular surface
    four spines:
      anterior superior spine
      anterior inferior spine
      posterior superior spine
      posterior inferior spine

ISCHIUM F. 7-3
1. **Body**
   spine of ischium, ischial spine
   tuberosity of ischium or ischial tuberosity
2. **Ramus**

PUBIS F. 7-3
1. Body
2. Superior ramus
3. Inferior ramus
4. Pubic arch
5. Pubic tubercle
6. Pubic crest

WHOLE PELVIS. F. 7-2
1. Linea terminalis or arcuate line i. e. terminal line
2. Superior aperture = inlet
3. Inferior aperture = outlet
4. True pelvis
5. False pelvis
6. Greater sciatic notch
7. Lesser sciatic notch
8. Sacrum
9. Coccyx

FEMUR a long bone F. 7-4, 5, 6
1. **Proximal extremity** or end
   head & fovea capitis femoris
   neck
   greater trochanter
   lesser trochanter
   intertrochanteric crest
   intertrochanteric line
2. **Body** or shaft
   linea aspera
   popliteal surface
3. **Distal extremity** or end
   medial condyle
   lateral condyle
   intercondylar fossa
   patellar surface
   medial epicondyle
   lateral epicondyle
   adductor tubercle
   popliteal groove

PATELLA a flat bone F. 7-7
1. Base or proximal margin
2. Apex or pointed end
3. Articular surface

TIBIA a long bone F. 7-8, 9
1. **Proximal extremity** or end
   medial condyle
   lateral condyle
   intercondylar eminence
   tibial tuberosity
   fibular articular surface
2. **Body** of tibia
3. **Distal extremity** or end
   inferior articular surface
   medial malleolus
   fibular notch

FIBULA a long bone F. 7-8, 9
1. **Proximal extremity** or end
   head, apex, styloid process
2. **Body** of fibula
3. **Distal extremity** or end
   lateral malleolus

TARSAL BONES seven short bones F. 7-10
1. **Posterior tarsal bones** — three
   (1) talus, body, neck, head, trochlea
   (2) Calcaneus — tuberosity, sustentaculum tali

(3) Navicular bone — tuberosity
2. **Anterior tarsal bones** — four
   (1) Medial cuneiform or first
   (2) Intermediate cuneiform or second
   (3) Lateral cuneiform or third
   (4) Cuboid bone

METATARSAL BONES five miniature long bones
F. 7-10
   First, second, third, fourth, fifth
   Each has a base, or proximal extremity
      a body
      a head or distal extremity
   Fifth metatarsal has a tuberosity

Digits five toes, 14 miniature long bones
fourteen phalanges: F. 7-10
   Great toe or hallux or first digit proximal and distal
   phalanges only
   Other four toes or 2nd, 3rd, 4th, 5th digits each with
   proximal, middle, and distal phalanges
   Each has a base or proximal extremity
      a body
      a head or distal extremity

S. 75 **DETAILED STUDY OF BONES OF LOWER LIMB**

The skeleton of the lower limb includes:
Bones of the pelvic girdle, hip bones
Bones of thigh, femur, patella
Bones of the leg, tibia, fibula
Bones of the foot, tarsal, metatarsal bones, phalanges, and sesamoid bones

The pelvic girdle, consisting of the right and left hip bones, connects the lower limbs and trunk.
The pelvis has the right and left hip bones and the sacrum and coccyx.

1. **THE HIP BONE** F. 7-2, 3
   Hip bone = os coxae (bone of hip), or coxal bone, or innominate bone.

   There are two hip bones, a right and a left. The hip bone is a large irregular bone that connects the lower limb to the trunk.
The acetabulum is a deep circular cup shaped depression on the lateral surface of the hip bone below its middle. This circular depression forms a socket for the head of the femur at the hip joint. (L. acetabulum = a vinegar cup, acetum = vinegar).
The acetabular notch is a gap on the inferior margin of the rim of the acetabulum.
The acetabular fossa is the flat central part of the acetabulum. The head of the femur rests upon the rim of this cup, not on the nonarticular fossa.
The obturator foramen is a large opening in the inferior part of the hip bone, between the pubis and ischium.
   The hip bone consists of three bones that are fused together - the ilium, ischium and pubis. In early life they are distinctly separate bones with cartilage between their adjacent margins. They are formed from separate centers of ossification. They meet at the acetabulum, and the cartilage between them is "Y" shaped. At the eighteenth to the twentieth year the cartilage will be replaced by bone tissue to form a single bone.

2. **THE ILIUM** F. 7-2, 3
   Ilium or flank bone; pl. ilia, adj. iliac. The ilium forms the upper two-fifths of the acetabulum, and all of the hip bone above it. The ilium has a body and a wing or ala. The name of this bone must not be confused with the name of the distal part of the small intestine, the ileum. Note that ilium has "ium", not "eum".

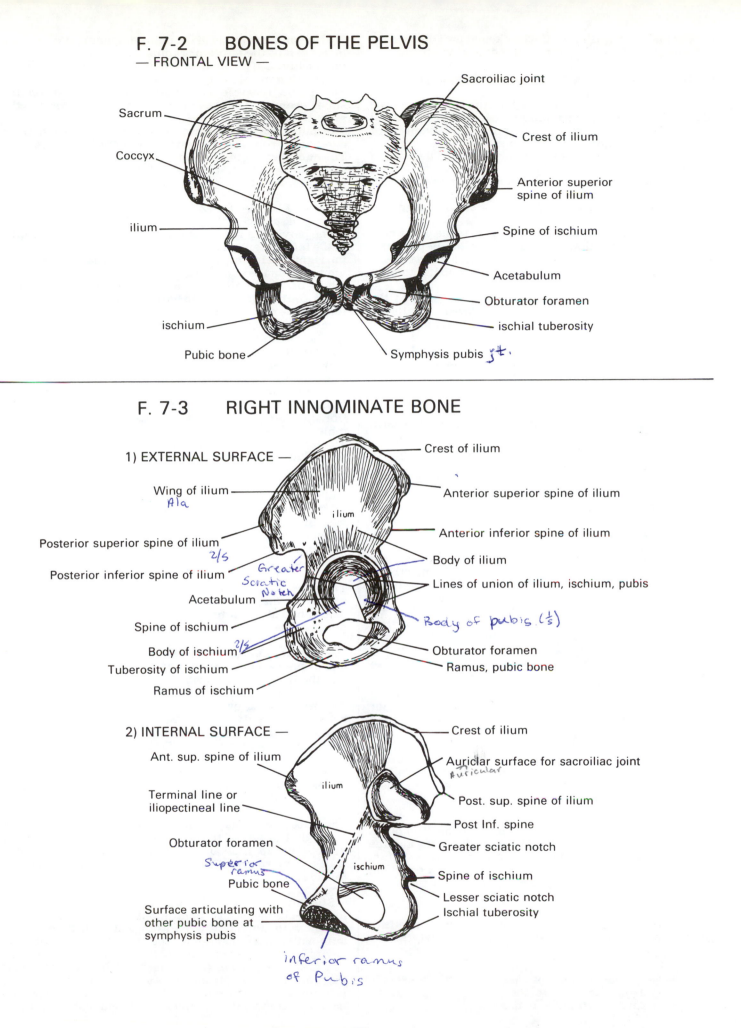

F. 7-2   BONES OF THE PELVIS
— FRONTAL VIEW —

Sacroiliac joint

Sacrum

Crest of ilium

Coccyx

Anterior superior
spine of ilium

ilium

Spine of ischium

Acetabulum

Obturator foramen

ischium

ischial tuberosity

Pubic bone

Symphysis pubis *jt.*

F. 7-3   RIGHT INNOMINATE BONE

1) EXTERNAL SURFACE —

Crest of ilium

Wing of ilium
Ala

Anterior superior spine of ilium

ilium

Anterior inferior spine of ilium

Posterior superior spine of ilium
2/5

Body of ilium

Posterior inferior spine of ilium
Greater Sciatic Notch

Lines of union of ilium, ischium, pubis

Acetabulum

Body of pubis (1/5)

Spine of ischium

Body of ischium *2/5*

Obturator foramen

Tuberosity of ischium

Ramus, pubic bone

Ramus of ischium

2) INTERNAL SURFACE —

Crest of ilium

Ant. sup. spine of ilium

Auriclar surface for sacroiliac joint
Auricular

ilium

Terminal line or
iliopectineal line

Post. sup. spine of ilium

Post Inf. spine

Obturator foramen

ischium

Greater sciatic notch

Superior ramus

Spine of ischium

Pubic bone

Lesser sciatic notch

Surface articulating with
other pubic bone at
symphysis pubis

Ischial tuberosity

inferior ramus of Pubis

57

(1) The body of the ilium is the thickened part close to the acetabulum, of which it forms the upper two-fifths.

(2) The ala or wing is the thin flattened upper part of the ilium above the body.

The crest of the ilium is the upper curved border of the ala. It can be palpated throughout its length on the lateral wall of the abdomen. It is a very important landmark for radiological technicians, and is used for centering.

The iliac fossa is the concave inner surface of the wing of the ilium.

The auricular surface is a rough, ear-shaped part posterior to the iliac fossa that articulates with the upper lateral part of the sacrum. (Auricula = ear, i.e., ear or ear shaped).

The anterior superior spine of the ilium is the prominent anterior end of the iliac crest; it can be easily palpated.

The anterior inferior spine lies on the anterior margin of the ala about one inch below the anterior superior spine. It cannot be palpated.

The posterior superior spine is a prominence at the posterior end of the crest. While it is not readily palpable a dimple in the skin marks its position. This same dimple marks the midpoint of the sacroiliac joint.

The posterior inferior spine lies at a lower level on the posterior margin of the ilium about one inch below the posterior superior spine; it is not palpable.

3. THE ISCHIUM   F. 7-2, 3

Ischium, pl. ischia, adj. ischial; also forms the term sciatic.

The ischium forms the lower two-fifths of the acetabulum and that part of the hip bone below and dorsal to the acetabulum. The ischium has a body and a ramus.

(1) The body, in addition to forming part of the acetabulum, forms the inferior and posterior part of the ischium. This lower part was formerly named the superior ramus of the ischium.

The spine of the ischium or ischial spine is a pointed process that extends dorsally and medially from the body of the ischium. The extent to which it is directed medially is important in the female as the fetal head must pass between the two ischia during childbirth.

The tuberosity of the ischium, or ischial tuberosity, is a large rounded process on the posterior surface of the lower end of the body of the ischium. It supports the weight of the body when the subject is sitting down. It can be palpated on the lateral side of the anus.

(2) The ramus of the ischium, formerly named the inferior or ascending ramus, extends anteriorly, medially and upwards from the body. It unites with the inferior ramus of the pubis. In a child a layer of cartilage separates these rami. (See a radiograph of a child's pelvis). This cartilage becomes replaced by bone when growth is complete. The two rami form the lower margin of the obturator foramen.

4. THE PUBIS   F. 7-2, 3

s. pubis, pl. pubes, adj. pubic; pubic bone. The pubis forms the anterior one-fifth of the acetabulum and the anterior part of the hip bone in front of the acetabulum and obturator foramen. The pubic bone has a body and two rami, and with the opposite pubic bone forms the pubic arch.

(1) The body of the pubic bone forms the anterior one-fifth of the acetabulum and is continued anteriorly as the superior ramus.

(2) The superior ramus of the pubic bone extends anteriorly from the body to the symphysis pubis, the joint between the two pubic bones.

(3) The inferior ramus of the pubis extends down from the lower end of the superior ramus, and joins the ramus of the ischium. The student should note the change in description of this bone.

The pubic tubercle is a very small process on the superior margin of the pubic bone slightly lateral to the symphysis pubis.

The pubic crest is a small ridge on the upper margin extending from the symphysis to the pubic tubercle.

The pubic arch is the curved surface formed by the medial margins of both inferior rami of the pubic bones and the lower border of the symphysis pubis.

5. THE PELVIS   F. 7-2

Pelvis, pl. pelves, adj. pelvic; L. pelvis = a basin.

The pelvis is a hollow curved cylinder with bony walls, at the inferior end of the vertebral column. It is formed by the right and left hip bones, the sacrum and coccyx. When viewed from the side, in the median plane it resembles an elbow or a bent pipe. The anterior vertical diameter is short, the posterior curved border is much longer since it follows the curve of the sacrum. At the lower end the ring is incomplete. The gaps between the pubic arch anteriorly the ischial tuberosities laterally and the coccyx posteriorly are completed by ligaments stretched between them to complete the ring on each side.

The floor of the pelvis is formed by a hammock of muscles slung between the pelvic bones and ligaments. In the pelvic floor there are three openings in the female (for the urethra, vagina and anus) and only two in the male.

The linea terminalis (terminal line) is a ridge of bone that begins at the upper margin of the symphysis pubis. It runs along the upper margin of the pubis. It then crosses the medial surface of the ilium at the junction of the body and wing of the ilium, passing obliquely to the sacroiliac joint. It then crosses the prominent anterior border of the base of the sacrum, where it meets a similar ridge on the opposite hip bone. The complete ring is visible on the pelvic, or inner, surface of the pelvis. This ring is also named the ileopectineal, or arcuate, line.

The superior aperture (opening) of the pelvis, pelvic rim or inlet, is marked by the terminal line. It is a somewhat heart shaped opening into the true pelvis, and forms the plane of the inlet.

The inferior aperture or outlet is an opening at the lower end of the pelvis composed of the incomplete bony ring and the ligaments described above.

The true pelvis, or minor pelvis, is the cavity located between the pelvic inlet (superior aperture) and the pelvic outlet (inferior aperture). It is surrounded by the bones of the pelvis which afford some protection for the pelvic organs. The female minor pelvis is larger and more circular than that of the male. (See details below). It forms the birth canal and the fetus must pass through it during childbirth.

The false pelvis, major pelvis, lies above the plane of the inlet, that is above the true pelvis. The lower lumbar vertebrae and iliac bones form its dorsal and lateral walls. Anteriorly there are no bony structures; the muscles of the anterior abdominal wall support the organs here.

Students should realize that the false and true pelves form a single cavity continuous with the abdominal cavity with no actual separation between them. Some authorities have adopted the term abdominopelvic cavity, with abdominal and pelvic parts.

The greater sciatic notch is a deep notch on the posterior margin of the hip bone. It lies between the posterior inferior spine of the ilium above, and the spine of the ischium below. A ligament stretched between these

— FRONTAL VIEW —

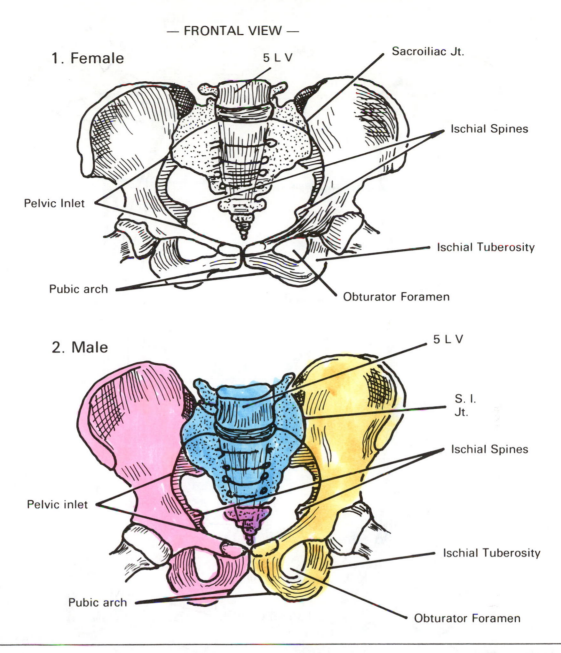

1. Female

5 L V

Sacroiliac Jt.

Ischial Spines

Pelvic Inlet

Ischial Tuberosity

Pubic arch

Obturator Foramen

2. Male

5 L V

S. I.
Jt.

Ischial Spines

Pelvic inlet

Ischial Tuberosity

Pubic arch

Obturator Foramen

prominences converts it into a foramen. (Sciatic = a corruption of ischiadikos - the ischium).
The lesser sciatic notch lies between the spine of the ischium above, and the ischial tuberosity below. It is below the greater notch and is much smaller.

COMPARISON OF FEMALE AND MALE PELVES
F. 7-2A

Because the fetus must pass through the birth canal, the female pelvis, during childbirth, there are several structural differences between the female and male pelves.

The hips in the female appear to be more prominent, the iliac crests are flared outward more, and they are less curved.
The anterior spines of the ilia are further apart.
The superior aperture or inlet is larger and more rounded, while in the males it is more heart shaped.
The acetabula and ischial tuberosities are further apart and flare outwards.

The ischial spines are not directed as far medially as in the male.
The female inferior aperture or outlet is also larger.
The pubic arch is broader, and the joint at the symphysis pubis is shorter than in the male.

6. **THE FEMUR**   F. 7-4, 5, 6
Femur - thigh bone, pl. femora, adj. femoral.
   There are two femora, a right and a left. The femur is a long bone, and the longest bone in the body, extending from the hip to the knee. It is slightly curved, its convexity being located anteriorly. The femur has a proximal extremity, a body, and a distal extremity.

(1) **The proximal extremity** has a head, neck, and greater and lesser trochanters with intertrochanteric line and crest.
The head, or caput femoris, is the upper expanded rounded end that articulates with the acetabulum to form the hip joint. The head is not palpable but its center lies opposite the upper margin of the greater trochanter.
The pit of the femoral head (fovea capitis femoris) is a small but definite hole at the center of the head. The

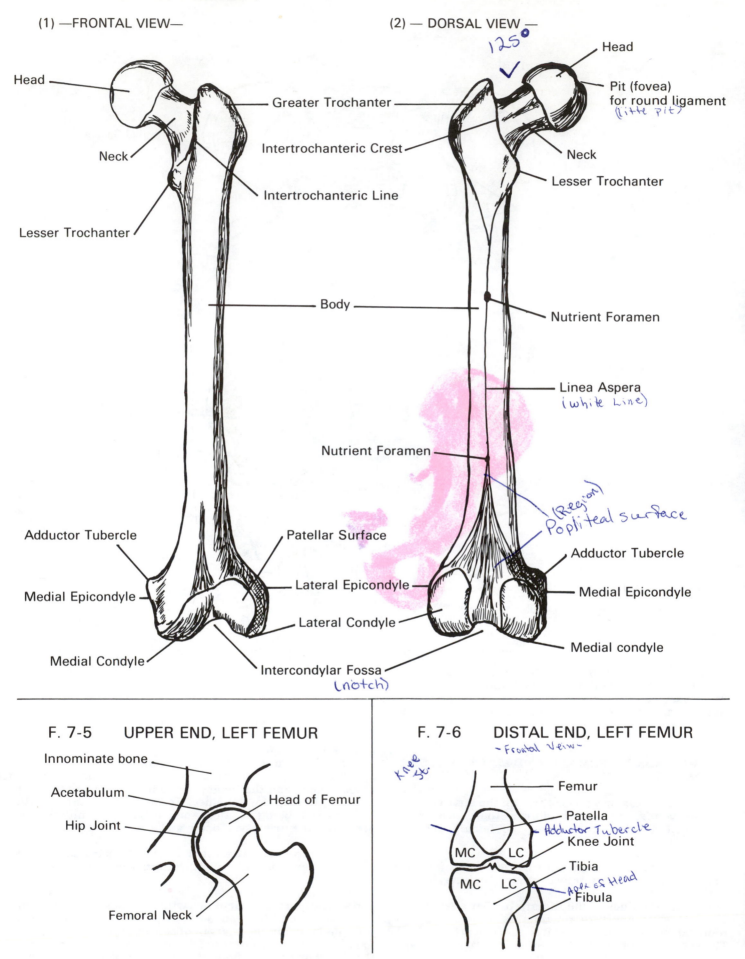

F. 7-4   THE LEFT FEMUR

(1) —FRONTAL VIEW—

Head

Neck

Lesser Trochanter

Greater Trochanter

Intertrochanteric Crest

Intertrochanteric Line

Body

Adductor Tubercle

Medial Epicondyle

Medial Condyle

Patellar Surface

Lateral Epicondyle

Lateral Condyle

Intercondylar Fossa
(notch)

(2) — DORSAL VIEW —

125°

Head

Pit (fovea)
for round ligament
(little pit)

Neck

Lesser Trochanter

Nutrient Foramen

Linea Aspera
(white Line)

Nutrient Foramen

(Region)
Popliteal surface

Adductor Tubercle

Medial Epicondyle

Medial condyle

F. 7-5   UPPER END, LEFT FEMUR

Innominate bone

Acetabulum

Hip Joint

Head of Femur

Femoral Neck

F. 7-6   DISTAL END, LEFT FEMUR

~Frontal View~

Knee Jt.

Femur

Patella

Adductor Tubercle

Knee Joint

Tibia

Apex of Head

Fibula

MC   LC

MC   LC

60

ligament of the femoral head, or ligamentum teres, is attached to this depression and to the acetabular notch of the hip bone.

The **neck** is the constricted part connecting the head to the body of the femur. It is directed obliquely posteriorly and laterally and is not palpable.

The **greater trochanter** is a large prominence on the lateral surface of the upper femur. It can be palpated through the skin and is a useful landmark.

The **lesser trochanter** is a much smaller rounded process that extends medially from the posteromedial margin of the upper femur at the junction of the neck and shaft. It lies at a lower level than the greater trochanter and is not palpable.

The **intertrochanteric crest** is a ridge of bone that passes obliquely across the back of the upper femur between the greater and lesser trochanters.

The **intertrochanteric line** is a ridge that extends obliquely across the anterior surface of the upper femur from the greater trochanter to the medial border of the femur in front of the lesser trochanter.

Note: the crest is posterior; the line is anterior.

(2) The **body** or shaft of the femur is the long cylindrical part that becomes broadened as it approaches the knee.

The **linea aspera** is a double bony ridge that passes longitudinally down the posterior surface of the body of the femur. It is formed by the convergence of three smaller ridges. It divides below into two ridges that are directed to the margins of the femur at the condyles. These two form the margins of the popliteal surface.

The **popliteal surface** of the femur is the flat area on the dorsal surface of the lower femur between the divided ridges of the linea aspera.

(3) The **distal extremity** of the femur has medial and lateral condyles, medial and lateral epicondyles, an intercondylar fossa, an adductor tubercle, and a patellar surface.

The **medial condyle** is a rounded knoblike process that forms the medial part of the distal end of the femur. Its smooth surface articulates with the medial condyle of the tibia at the knee joint.

The **lateral condyle** is a similar rounded process forming the lateral part of the distal end of the femur. It articulates with the lateral condyle of the tibia forming part of the knee joint. The two condyles are separated from each other.

The **intercondylar fossa** is a deep notch located between the posterior parts of the femoral condyles.

The **patellar surface** is the smooth anterior surface of the distal part of the femur between the anterior parts of the femoral condyles. It forms a joint with the patella.

The **medial epicondyle** is a large prominence on the medial surface of the lower femur just above the medial condyle. It is palpable. (Epi = upon +condyle - upon a condyle).

The **lateral epicondyle** is a smaller rounded process on the lateral surface of the lower femur above the lateral condyle. It is palpable.

The **adductor tubercle** is a small process on the medial surface of the medial epicondyle.

7. THE PATELLA   F. 7-7

Patella - knee cap; pl. patellae, adj. patellar.

There are two patellae, a right and a left. The patella is a sesamoid bone lying within the tendon of the quadriceps muscle tendon. It is a flat rounded bone and lies anterior to the knee joint. It is palpable, and has a base and apex.

The **base** is the proximal slightly rounded upper border.

The **apex** is the lower pointed end. The patellar ligament lies between it and the tibial tuberosity.

The **posterior surface** is smooth and glides over the smooth patellar surface of the femur as the knee is flexed or extended.

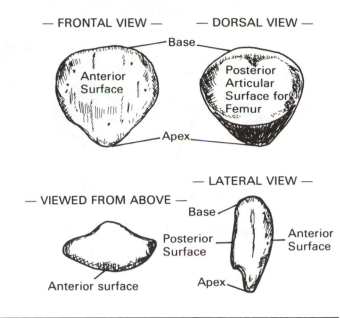

THE LEG OR CRUS   F. 7-8

Each leg has two bones, a tibia and a fibula, that extend from the knee to the ankle. These bones lie paralled to each other, with the fibula on the lateral side of the tibia. Movements of pronation and supination are not possible in the leg; the two bones retain the same relative position to each other during movements.

8. THE TIBIA   F. 7-8, 9

Tibia - shin bone; pl. tibiae, adj. tibial.

There are two tibiae, a right and a left. The tibia is a large sturdy bone and lies medial to the fibula. It is expanded at its upper and lower ends, at the knee and ankle joints. Its anterior border is narrow and sharp except at the lower end and lies close to the skin surface. The tibia has a proximal extremity, a body, and a distal extremity.

(1) The **proximal extremity** of the tibia has medial and lateral condyles, an intercondylar eminence, a tuberosity, and a fibular articular surface.

The **medial condyle** is the medial part of the upper expanded end of the tibia. Its smooth, slightly concave proximal surface articulates with the medial condyle of the femur at the knee joint. It is palpable.

The **lateral condyle** is the lateral part of the upper expanded end of the tibia. Its smooth superior surface articulates with the lateral femoral condyle at the knee joint.

The **intercondylar eminence** is a small, double-pointed process on the proximal end of the tibia, located between the articular surfaces of the medial and lateral condyles of the tibia, slightly posterior to its midpoint. Two small tubercles can be identified here. This process extends up into the intercondylar fossa of the femur.

The **tibial tuberosity** is a rounded prominence on the anterior surface of the tibia below the condyles. It can be palpated through the skin. The patellar ligament is inserted into it.

The **fibular articular surface** (fibular facet) is a small smooth rounded area on the posterolateral border of the lateral tibial condyle. It articulates with the head of the fibula.

(2) The **body** of the tibia is the long sturdy part that becomes flattened as it approaches the ankle.

THE LEFT TIBIA AND FIBULA:

(1) — FRONTAL VIEW —
(2) — DORSAL VIEW —

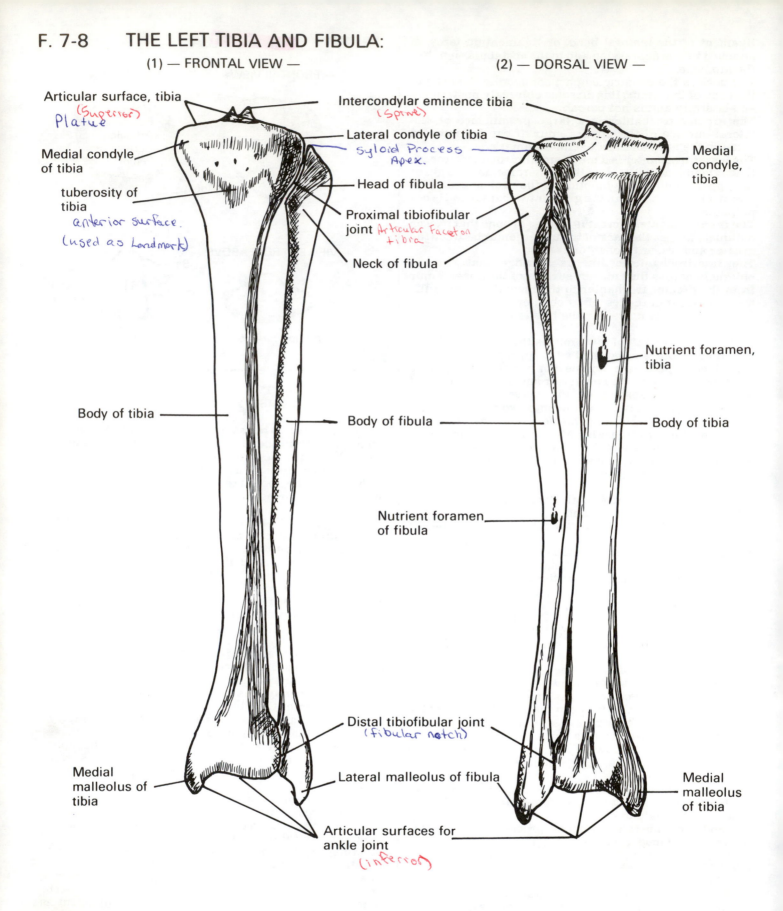

Articular surface, tibia
(Superior)
Platue

Medial condyle
of tibia

tuberosity of
tibia
anterior surface.
(used as Landmark)

Intercondylar eminence tibia
(Spine)

Lateral condyle of tibia
Syloid Process
Apex.

Head of fibula

Proximal tibiofibular
joint *Articular Facet on*
tibia

Neck of fibula

Medial
condyle,
tibia

Body of tibia

Body of fibula

Body of tibia

Nutrient foramen,
tibia

Nutrient foramen
of fibula

Distal tibiofibular joint
(fibular notch)

Medial
malleolus of
tibia

Lateral malleolus of fibula

Medial
malleolus
of tibia

Articular surfaces for
ankle joint
(inferior)

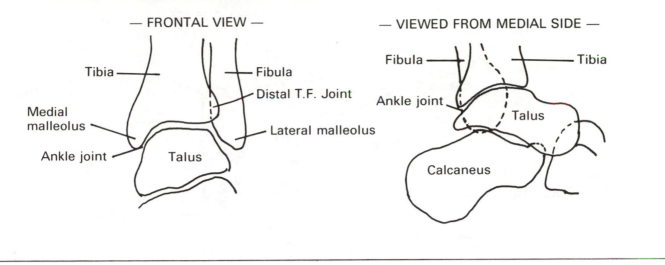

— FRONTAL VIEW —

— VIEWED FROM MEDIAL SIDE —

(3) **The distal extremity** has an inferior articular surface, a medial malleolus, and a fibular notch.
The inferior articular surface is a four sided smooth surface on the distal end of the tibia that articulates with the talus, forming the ankle joint.
The medial malleolus projects down from the medial margin of the distal end of the tibia. It reaches about one-half inch below the level of the ankle joint. It forms a large prominence on the medial border of the ankle that is visible and palpable. (Malleus = a hammer, + olus = small, a small hammer).
The fibular notch is a concave depression on the lateral surface of the tibia just above the ankle joint. It articulates with the medial border of the lower fibula to form the distal tibiofibular joint.

9. THE FIBULA   F. 7-8, 9
Fibula - calf bone, pl. fibulae, adj. fibular.
(L. fibula, that which fastens, the pin of a brooch, or the tongue of a buckle).
There are two fibulae, a right and a left. The fibula is a long slender bone lying on the lateral side of the tibia and parallel to it. It does not form any part of the knee joint but helps form the ankle joint. The fibula has a head, body, and a lateral malleolus.

(1) **The head** of the fibula is its upper expanded end that articulates with the fibular articular surface of the tibia, forming the proximal tibiofibular joint.
The apex or styloid process of the fibula is the pointed upper end of the head.

(2) **The body** is long and very slender.

(3) **The lateral malleolus** is the distal expanded end of the fibula. It extends down on the lateral margin of the ankle joint for about three-quarters of an inch. Its tip is at a lower level than that of the medial malleolus. Along with the medial malleolus it forms the ankle mortice. Above the ankle joint it articulates with the fibular notch of the tibia forming the distal tibiofibular joint.

THE FOOT OR PES   F. 7-10
The skeleton of the foot includes the tarsal bones, the metatarsal bones, and the phalanges of the toes.

10. THE TARSUS ~~OR INSTEP~~   F. 7-10

There are seven tarsal bones, classified as short bones. They form the posterior part of the foot. To facilitate their study they may be divided into posterior and anterior groups instead of proximal and distal rows as with the carpal bones. Each of the anterior group forms a joint with a metatarsal bone.

Posterior tarsal bones
  (1) talus
  (2) calcaneus
  (3) navicular
Anterior tarsal bones
  (1) medial cuneiform
  (2) intermediate cuneiform
  (3) lateral cuneiform
  (4) cuboid bone

The talus or astragalus is sometimes referred to as the ankle bone. It lies between the distal end of the tibia and the calcaneus, transmitting the body weight to the calcaneus.
The body is the large posterior part.
The trochlea (pulley) is the upper smooth convex surface that articulates with the lower end of the tibia. The medial and lateral surfaces of the body below the trochlea are smooth and articulate with the medial and lateral malleoli forming the ankle mortice.
The neck is a constricted part lying between the body and head.
The head is the rounded anterior end of the talus. Its convex anterior surface articulates with the posterior surface of the navicular bone.
The inferior surface of the talus has three articular surfaces that form joints with smooth surfaces on the calcaneus. The middle one lies above the sustentaculum tali.

The calcaneus, or os calcis, or heel bone.
pl. calcanei, adj. calcaneal. This bone is the largest tarsal bone and forms the heel. It lies below the talus. Its upper surface has three smooth areas that form joints with the talus. Its anterior end forms a joint with the posterior surface of the cuboid bone. The calcaneus has a tuberosity and a sustentaculum tali.

F. 7-10   BONES OF THE LEFT FOOT
— VIEWED FROM ABOVE — DORSAL ASPECT

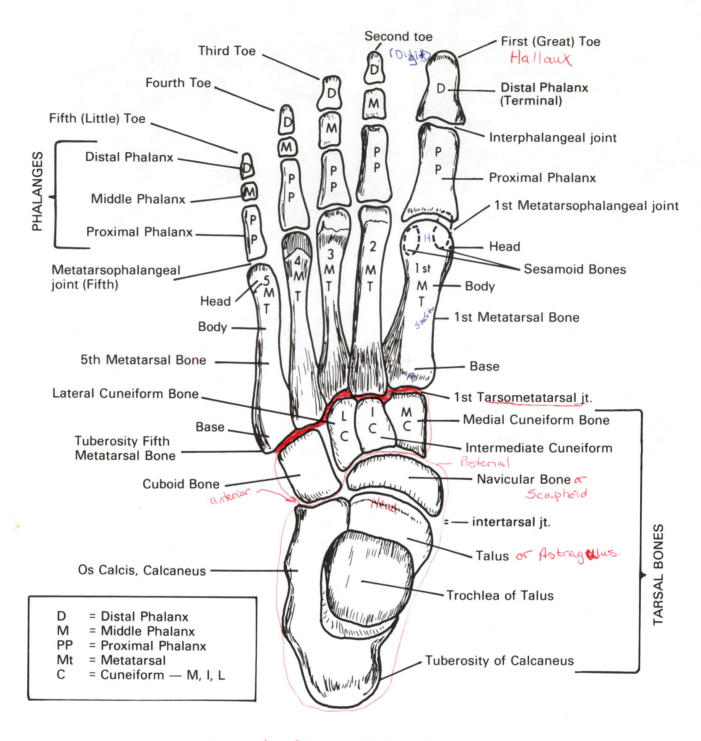

Second toe *(Digit)*

Third Toe

Fourth Toe

Fifth (Little) Toe

PHALANGES

Distal Phalanx

Middle Phalanx

Proximal Phalanx

Metatarsophalangeal joint (Fifth)

Head

Body

5th Metatarsal Bone

Lateral Cuneiform Bone

Base

Tuberosity Fifth Metatarsal Bone

Cuboid Bone

anterior

Os Calcis, Calcaneus

First (Great) Toe — *Hallaux*

Distal Phalanx (Terminal)

Interphalangeal joint

Proximal Phalanx

1st Metatarsophalangeal joint

Head

Sesamoid Bones

Body

1st Metatarsal Bone

Base

1st Tarsometatarsal jt.

Medial Cuneiform Bone

Intermediate Cuneiform

Posterior

Navicular Bone *or Scaphoid*

= intertarsal jt.

Head

Talus *or Astragalus.*

Trochlea of Talus

Tuberosity of Calcaneus

TARSAL BONES

D	= Distal Phalanx
M	= Middle Phalanx
PP	= Proximal Phalanx
Mt	= Metatarsal
C	= Cuneiform — M, I, L

Come to Cuba Next Christmas.

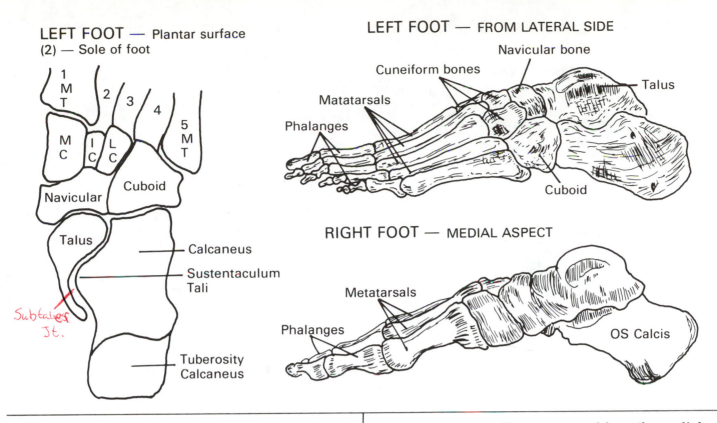

LEFT FOOT — Plantar surface
(2) — Sole of foot

1 M T
2
3
4
5 M T
M C
I C
L C
Navicular
Cuboid
Talus
Calcaneus
Sustentaculum Tali
Subtalar Jt.
Tuberosity Calcaneus

LEFT FOOT — FROM LATERAL SIDE

Navicular bone
Cuneiform bones
Matatarsals
Phalanges
Talus
Cuboid

RIGHT FOOT — MEDIAL ASPECT

Metatarsals
Phalanges
OS Calcis

The calcaneal tuberosity is its enlarged posterior end — visible and palpable. The Achilles tendon is attached to the posterior surface.

The sustentaculum tali is a small but definite shelf of bone that projects medially from the medial surface of the calcaneus immediately below the talus, which it helps to support.
(L. sustentaculum = a support or prop).

The navicular bone (tarsal scaphoid) is boat shaped, hence the name navicular (navis = ship). It lies anterior to the talus and behind the three cuneiform bones, on the medial side of the ankle. It has a tuberosity.
The tuberosity of the navicular bone is its prominent medial border. It is palpable below and anterior to the medial malleolus. The navicular articulates behind with the talus and in front with the three cuneiform bones.

The medial cuneiform bone (first cuneiform) is located on the medial border of the foot anterior to the navicular and posterior to the first metatarsal bone. The cuneiform bones are wedge shaped. (L. cuneus = a wedge + form = shape).

The intermediate cuneiform bone (second cuneiform) is located lateral to the medial cuneiform, posterior to the second metatarsal bone and anterior to the navicular.

The lateral cuneiform bone (third cuneiform) lies lateral to the intermediate cuneiform, posterior to the third metatarsal, and in front of the navicular bone.
All three cuneiform bones articulate with the navicular bone and with a metatarsal bone.
The cuboid bone is roughly cube shaped. It lies on the lateral side of the foot, and has the calcaneus behind it, with the fourth and fifth metatarsals in front.
The student should note that the calcaneus and cuboid form a joint (C & C) while the talus and navicular form a joint (T & N).

11. THE METATARSUS   F. 7-10

Five metatarsal bones lie side by side in front of the tarsals and behind the toes. (meta = beyond + tarsus = beyond the tarsals). They are named from the medial or great toe side of the foot as first, second, third, fourth, and fifth metatarsals. They are classified as miniature long bones. Each of them has —
  a base or proximal end
  a body
  a head or distal end.
The fifth metatarsal bone has a tuberosity, a prominence on the lateral margin of the base. As this is protruding it may become fractured off.
The bases of the metatarsals form joints with the anterior tarsal bones. Their heads articulate with the bases of the proximal phalanges of the toes.

12. THE DIGITS OR TOES   F. 7-10

Fourteen phalanges form the five digits or toes; s. phalanx, pl. phalanges, adj. phalangeal. The digits or toes are named from the medial or great toe side as the great or first toe, the second toe, third toe, fourth toe, and the fifth toe, or as the first digit, second digit, third digit, fourth digit, and fifth digit.
The great toe has two phalanges, a proximal and a distal. Each of the other four toes has three phalanges, a proximal, middle and distal. Hallux is an alternate name for the great toe.
Each phalanx has —
  a base or proximal extremity or end
  a body
  a head or distal extremity or end.

13. THE SESAMOID BONES   F. 7-10

The lower limb has the largest sesamoid bone of the body, the patella. One is frequently present posterior to and above the lateral condyle of the femur, the fabella. Sesamoids are often present in tendons on the plantar surfaces of the metatarsophalangeal joint of the great toe, second toe, and sometimes the other toes. See — Bones of upper limb, No. 9.

S. 76   PARTS OF BONES FORMING JOINTS OF THE LOWER LIMB
This information has now been included in the next section S. 77 rather than separately.

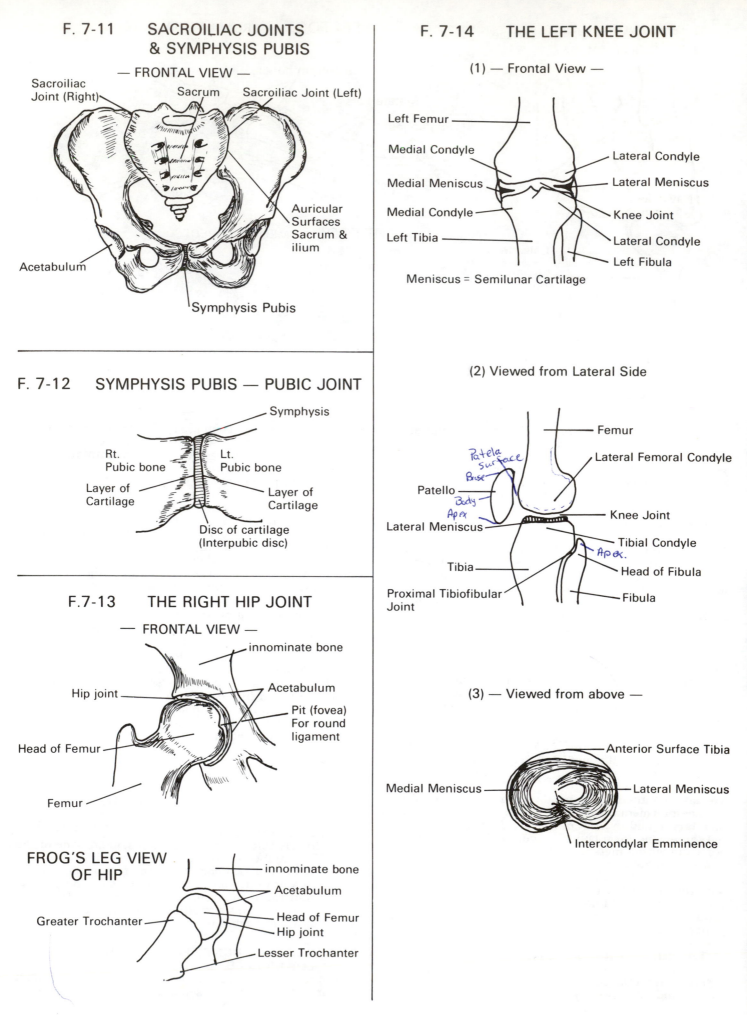

F. 7-11 SACROILIAC JOINTS & SYMPHYSIS PUBIS

— FRONTAL VIEW —

Sacroiliac Joint (Right)
Sacrum
Sacroiliac Joint (Left)
Auricular Surfaces Sacrum & ilium
Acetabulum
Symphysis Pubis

F. 7-12 SYMPHYSIS PUBIS — PUBIC JOINT

Symphysis
Rt. Pubic bone
Lt. Pubic bone
Layer of Cartilage
Layer of Cartilage
Disc of cartilage (Interpubic disc)

F.7-13 THE RIGHT HIP JOINT

— FRONTAL VIEW —

innominate bone
Acetabulum
Hip joint
Pit (fovea) For round ligament
Head of Femur
Femur

FROG'S LEG VIEW OF HIP

innominate bone
Acetabulum
Greater Trochanter
Head of Femur
Hip joint
Lesser Trochanter

F. 7-14 THE LEFT KNEE JOINT

(1) — Frontal View —

Left Femur
Medial Condyle
Lateral Condyle
Medial Meniscus
Lateral Meniscus
Medial Condyle
Knee Joint
Left Tibia
Lateral Condyle
Left Fibula

Meniscus = Semilunar Cartilage

(2) Viewed from Lateral Side

Femur
Patela surface Base
Lateral Femoral Condyle
Patello Body Apex
Lateral Meniscus
Knee Joint
Tibial Condyle
Apex.
Tibia
Head of Fibula
Proximal Tibiofibular Joint
Fibula

(3) — Viewed from above —

Anterior Surface Tibia
Medial Meniscus
Lateral Meniscus
Intercondylar Emminence

66

S. 77 DETAILED STUDY - JOINTS OF THE LOWER LIMB

Students should review S. 68 in which the structures of the 3 types of joints are listed. See also S. 73 that lists the various joints.

1. Sacroiliac joints   F. 7-11
Two, right and left.
Located in posterior part of the pelvis.
Formed by the auricular surfaces of the sacrum and ilium.
A synovial gliding joint also classified by some authorities as a cartilaginous joint, with a limited gliding movement.

2. Pubic symphysis   F. 7-12
NA — symphysis pubica, a single joint.
Located at the midline between the two pubic bones.
Formed by the adjacent medial margins of the pubic bones.
An articular cartilage covers each bone end, and a further cartilaginous disc, an interpubic disc, lies between these two articular cartilages.
A cartilaginous joint with slight movement.

3. Hip joints   F. 7-13
Coxal joints, a right and a left.
Located where the pelvis meets the thigh.
Formed by the head of the femur articulating with the cup shaped acetabulum of the hip bone.
A synovial ball and socket joint.
The round ligament (ligamentum teres) or the ligament of the femoral head is attached to the pit on the femoral head (fovea capitis femoris) above and to the acetabular notch of the hip bone below. This is a band of fibrous tissue.

4. The knee joints: F. 7-14
Genu = knee, two, a right and a left.
Formed by the medial and lateral condyles of the femur articulating with the medial and lateral condyles of the tibia, and the patellar surface of the femur articulating with the patella.
A synovial hinge and double condylar and a gliding joint.
Flexion and extension are permitted and when the knee is flexed internal and external rotation may occur.
The menisci or semilunar cartilages are two flat half-moon shaped discs of cartilage within the knee joint. One, the lateral meniscus, is located between the lateral femoral and tibial condyles. The other, the medial meniscus, lies between the medial condyles of the femur and tibia. They act as cushions to lessen jarring at the knee joint. The outer convex margin is attached to the capsule of the knee joint. The inner margin is free except at each end where it is attached to the intercondylar space of the tibia. The femoral condyles move upon these cartilages. They are sometimes torn during physical exercise. (L. meniscus = a crescent, pl. menisci).
The cruciate ligaments are two bands of fibrous tissue that pass up from the intercondylar part of the tibia, one to each margin of the intercondylar fossa. They cross each other like the letter "X" hence the name cruciate. (Crux = a cross).
The synovial membrane of the knee joint is prolonged upwards on the anterior surface of the femur behind the patella and quadriceps tendon to form the suprapatellar bursa. It opens into the knee joint. Twelve other bursae are located about the knee; some of them open into the knee joint while others are separate from it.
The tendon of the quadriceps femoris muscle, (a large muscle on the front of the thigh) is attached to the base (upper margin) of the patella, and also forms a covering that encloses the patella.

The patellar ligament extends from the apex (lower end) of the patella to the tibial tuberosity. These structures are intimately associated with the front of the knee and give support to it.

5. Superior tibiofibular joints   F. 7-7, 14
Two, a right and a left.
Located between the adjacent upper margins of the tibia and fibula.
Formed by a fibular articular surface on the upper posterior margin of the lateral tibial condyle articulating with the articular surface on the medial margin of the head of the fibula.
A synovial gliding joint, limited movement.

6. Inferior tibiofibular joints: F. 7-15
Two, a right and a left.
Located between the inferior margins of the tibia and fibula above the ankle joint.
Formed by the medial surface of the lower part of the fibula articulating with the fibular notch of the tibia, on its lateral side.
A cartilaginous joint.

7. Ankle joint   F. 7-9, 15
NA talocrural joints, (talus = ankle + crus = the leg); two, a right and a left.
Located at the distal ends of the tibia and fibula where they join the foot.
Formed by the inferior articular surface of the tibia articulating with the trochlea of the talus, and the medial and lateral malleoli with the medial and lateral surfaces of the talus.
A synovial hinge joint permits plantar flexion and dorsiflexion. These terms require an explanation. The foot lies at right angles to the leg. The upper surface of the foot has been named its dorsal surface, and its inferior surface has been named the plantar surface. Compare these with the dorsal and palmar surfaces of the hand. In plantar flexion the anterior part of the foot is bent downwards, while in dorsiflexion the anterior foot is raised towards the front of the leg. (Dorsi here = upper).

8. Intertarsal joints   F. 7-16; (several)
Located between adjacent tarsal bones, and named from the bones involved as calcaneocuboid, talonavicular, etc. The joints mentioned here are the most movable, the one between the anterior end of the calcaneus and cuboid, and that between the anterior end of the talus and the navicular.
Synovial gliding joints permitting slight movement, with some inversion and eversion of the foot, mostly at the two joints listed above.

9. Tarsometatarsal joints   F. 7-16
Five joints in each foot.
Located between the tarsal and metatarsal bones of the foot.
Formed by the medial cuneiform articulating with the base of the first metatarsal, the intermediate cuneiform with the base of the second metatarsal, the lateral cuneiform with the base of the third metatarsal, and the cuboid with the bases of the fourth and fifth metatarsals.
Synovial gliding joints, permitting slight movement.

10. Intermetatarsal joint   F. 7-16
Four joints in each foot.
Located between the bases of the adjacent metatarsal bones.
Gliding joints with slight movement.

11. Metatarsophalangeal joints   F. 7-16
Five in each foot.
Located between the metatarsal bones and the toes (digits).

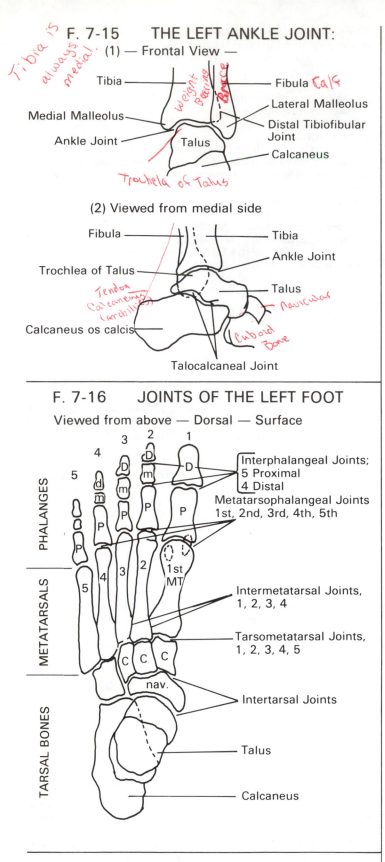

F. 7-15　THE LEFT ANKLE JOINT:
(1) — Frontal View —

Tibia is always medial.

- Tibia
- Medial Malleolus
- Ankle Joint
- Talus
- Weight Bearing
- Brace
- Fibula　*Calf*
- Lateral Malleolus
- Distal Tibiofibular Joint
- Calcaneus

Trochlea of Talus

(2) Viewed from medial side

- Fibula
- Trochlea of Talus
- *Tendon Calcaneus (archillies)*
- Calcaneus os calcis
- Tibia
- Ankle Joint
- Talus
- *navicular*
- *Cuboid Bone*
- Talocalcaneal Joint

F. 7-16　JOINTS OF THE LEFT FOOT
Viewed from above — Dorsal — Surface

PHALANGES
METATARSALS
TARSAL BONES

- Interphalangeal Joints;
  5 Proximal
  4 Distal
- Metatarsophalangeal Joints
  1st, 2nd, 3rd, 4th, 5th
- Intermetatarsal Joints, 1, 2, 3, 4
- Tarsometatarsal Joints, 1, 2, 3, 4, 5
- Intertarsal Joints
- Talus
- Calcaneus

Formed by the heads of the metatarsal bones articulating with the bases, i.e. proximal ends, of the corresponding proximal phalanges of the toes.

Synovial condylar joints, permitting abuction, abduction, flexion, extension, and some circumduction.

12. **Interphalangeal joints**　F. 7-16

Nine in each foot.

Located between the phalanges of each toe.

Formed by the head of one phalanx articulating with the base of the phalanx adjacent to it.

Synovial hinge joints, permitting flexion and extension.

The great toe with only two phalanges has only one interphalangeal joint.

The other four toes have two each, a proximal and a distal interphalangeal joint.

S. 78　THE ARCHES OF THE FOOT　F. 7-10

The plantar surface or sole of the foot is curved in two directions to form the longitudinal and transverse arches.

The **longitudinal arch** is visible when the foot is viewed on the medial border. The arch is formed by the tarsal and metatarsal bones. When the subject is standing the calcaneus and the heads of the metatarsals are in contact with the floor. The arch is maintained by strong ligaments that bind adjacent bones together.

The **transverse arch** is a curve from the medial side of the plantar surface to the lateral margin of the foot. In flat feet the arches disappear and the whole foot rests upon the floor.

S. 79　CONGENITAL ANOMALTIES OF THE PELVIS AND LOWER LIMB

Congenital dislocation of one or both hips may occur. The femoral head lies outside of and above the acetabulum. The socket of the acetabulum is very shallow or absent and lies almost vertically. The first sign observed may be a limp when the infant begins to walk.

Absence of one or both patellae: this bone may fail to develop at one or both knees.

Bipartite patella: the patella may present as two separate bones. Sometimes the upper lateral part of it develops from a separate center of ossification that does not unite with the remainder of the bone.

Os trigonum: a small triangular bone posterior to the talus that has formed from a separate epiphysis to that of the talus, and has failed to unite with it. To the inexperienced this may look like a fracture of the talus.

Accessory navicular bone: the posteromedial part of the navicular bone may be separate from the main part of this bone.

Many other anomalies may occur. They are frequently bilateral (present in both limbs) so it is a routine practice to take radiographs of both limbs in children.

S. 80　LANDMARKS AND BONY PROMINENCES OF THE PELVIS AND LOWER LIMB

The pelvis:　F. 7-2

The crest of the ilium can be palpated as a curved bony ridge in the lateral wall of the abdomen. It usually lies at the level of the navel but is a much more reliable landmark as the location of the navel varies from one subject to another.

The anterior superior spine of the ilium forms a very definite prominence at the anterior end of the iliac crest. It can be readily palpated and is a useful landmark.

The posterior superior spine of the ilium corresponds to a dimple on the skin on the medial part of the buttock posteriorly.

The tuberosity of the ischium is a definite prominence palpable at the perineum and lateral to the anus.

The sacroiliac joint only a small part of the sacroiliac joint can be felt at the dimple marking the posterior superior spine of the ilium.

The symphysis pubis can be palpated as a depression in the midline anteriorly between the two pubic bones.

The pubic arch can be felt below the symphysis as a curved bony border between the pubic bones.

The greater trochanter of the femur lies close to the skin surface and can be palpated as a large prominence on the lateral surface of the hip.

The head of the femur lies deeply and cannot be palpated. If a line be drawn from the anterior superior spine of the ilium to the upper margin of the symphysis pubis, and if this line be bisected, the hip joint will lie at the point of bisection.

The knee: F. 7-14

The patella can be seen and palpated on the anterior surface of the knee above the joint.

The tibial tuberosity is palpable as a definite prominence on the front of the upper tibia. The patellar ligament can be traced from the apex (lower end) of the patella to this tuberosity.

The knee joint may be located by placing the thumb on one side and the index finger on the other side of the patellar ligament below the patella. A ridge of bone can be felt here, passing transversely across the knee. The ridge is the upper margin of the tibial condyles, and the joint lies above it. This landmark is more readily palpated with the knee flexed. The knee joint lies opposite the lower end of the patella.

The condyles of the femur are palpable as two rounded knoblike prominences, one on either side of the patella above the ridge of the tibial condyles, when the knee is flexed.

The medial epicondyle of the femur can be felt as a large prominence on the medial border of the knee above the knee joint.

The lateral epicondyle is palpable on the lateral border of the knee above the knee joint.

The head of the fibula can be palpated on the lateral margin of the knee below the knee joint and towards the posterior surface.

The ankle: F. 7-15

The medial malleolus is the large prominence on the medial surface of the ankle, visible and palpable here.

The lateral malleolus of the fibula is the prominence on the lateral surface of the ankle. It is visible and palpable.

The ankle joint lies about 2 cm or three-quarters of an inch above the tip of the lateral malleolus; the tips of the malleoli must not be taken for the level of this joint.

The calcaneal tuberosity forms the heel and is visible and palpable.

The tuberosity of the navicular bone can be palpated on the medial border of the foot just below and in front of the medial malleolus.

The tuberosity at the base of the fifth metatarsal bone is a definite prominence on the lateral border of the foot well below and in front of the lateral malleolus.

The metatarsal bones and phalanges should not cause any problems in radiography.

The metatarsophalangeal joints do not lie opposite the webs between the toes but proximal to them.

S. 81   ANATOMICAL PECULIARITIES AND RADIOGRAPHY

A mounted skeleton and processed films are necessary for this study. As in the study of the upper limb the student's eye must represent the anode of an x-ray tube and view each part as it would appear to this x-ray tube. Place the mounted skeleton on the table top or suspend it upright.

The sacroiliac joints are directed obliquely from front to back so that their posterior borders lie closer to the median line of the body than to the anterior. To look through a sacroiliac joint on one side this hip must be elevated from the table top or, if the subject is upright, must be turned obliquely towards the eye. To look into the opposite sacroiliac joint the subject must be turned into the opposite oblique.

The hip or innominate bones curve around the pelvis. To view one of them "face on" the opposite hip must be raised from the table top, and steadied by flexing the knee and thigh of the affected side.

The necks of the femora lie obliquely and are directed laterally and posteriorly. If the feet are turned out in their natural positions, the necks are foreshortened when viewed from the front. To offset this the medial margins of the feet must be touching each other, or, better still, have the great toes touching and the heels separated. (Review the anatomical position). If the skeleton is placed in the lateral position the femoral heads, necks and bodies will be superimposed when viewed from the front. If instead the skeleton is placed in the supine position, and one leg and thigh flexed, then rotated laterally that femoral head and neck become visible in the lateral position.

If a subject has suffered a recent injury then no attempt must be made to move the injured limb. Utilizing the skeleton in the supine position the uninjured leg may be flexed at the hip and knee and a view of the injured femoral head and neck may be obtained by directing the x-ray beam laterally under the flexed uninjured leg towards the injured limb.

The patella, when viewed from the front or back, will be superimposed upon the lower part of the femur. By placing the skeleton face down and flexing the leg at the knee the patella becomes visible without superimposition.

With the lower limb in the lateral position the patella becomes visible without any overlapping.

The knee joint is visible in the anteroposterior or lateral positions without overlapping provided the examining eye is at the level of the joint.

The ankle joint, as stated previously, lies about 2 cm or .75 inches above the tip of the lateral malleolus. The eye must therefore be centered this distance above the lateral malleolus. In addition the medial margin of the foot must be placed at right angles to the table top, and not turned out in the natural position.

The calcaneus is readily visible from the side. With the skeleton supine this bone is obscured by the other tarsal bones. If instead the skeleton is placed face down, and the examining eye is directed at an angle towards the foot, most of the os calcis is visible without superimposition.

The metatarsal bones overlap each other when viewed from above or laterally. This is partly due to the arching of the foot. If the skeleton is placed with the medial margin of the foot touching the table top and the lateral border elevated there will be less overlapping of bones.

The toes are normally partly flexed in the living subject. They must be straightened out by pressing them against the top of the table. Lateral views of each toe are possible if the toes are flexed to different degrees to overcome overlapping.

In applying the knowledge of anatomy to radiography the peculiarities of each area must be taken into consideration, and views devised to minimize the degree of overlapping of bones. Following recent injury, however, the parts must be moved as little as possible. Two views at right angles to each other can usually be obtained by rotating the whole limb or whole body. NEVER rotate or straighten the injured part of a limb.

Note: as the bones, joints and parts are listed at the beginning of this chapter they will not be listed again. Students should refer to S. 72, S. 73, and S. 74 for these terms.

NOTES

8. THE VERTEBRAL COLUMN: BONES AND JOINTS

S. 82 BONES OF THE VERTEBRAL COLUMN   F. 8-1

Vertebrae — 33 bones
  s. vertebra, pl. vertebrae, adj. vertebral
Cervical vertebrae — 7 bones
Thoracic vertebrae — 12 bones
Lumbar vertebrae — 5 bones
Sacrum — 5 fused bones
  pl. sacra, adj. sacral
Coccyx — 4 or 3 or 5 incomplete bones
  adj. coccygeal

S. 83 JOINTS OF THE VERTEBRAL COLUMN

Intervertebral joints or discs
  between adjacent vertebrae
Interarticular joints
  between articular processes
Atlantooccipital joints, between
  occipital bone and atlas
Atlantoaxial joints — three
  Median — one, Lateral — two
Lumbosacral joint — one — between
  5th lumbar vertebra & sacrum
Sacrococcygeal joint — one
  between sacrum and coccyx
Costovertebral joints — between
  a rib and a vertebra
Costotransverse joints — between
  tubercle of a rib & transverse
  process of vertebra

F. 8-1   THE VERTEBRAL COLUMN:
— Frontal View —   — Lateral view —

Seven cervical vertebrae

Twelve thoracic vertebrae

Five lumbar vertebrae

Sacrum 5 fused

Coccyx (3-4)

Cervical curvature (lordotic)

Thoracic curvature (kyphotic)

Lumbar curvature (lordotic)

Pelvic curvature (kyphotic)

**S. 84 PARTS AND PROMINENCES OF THE
         VERTEBRAE**
True or movable vertebrae
False or fixed vertebrae

DIVISIONS OF THE VERTEBRAL COLUMN   F. 8-1
  1. Cervical vertebrae, seven in the neck
  2. Thoracic vertebrae, twelve in the thorax
  3. Lumbar vertebrae, five behind the abdomen
  4. Sacrum, five fused vertebrae, in pelvis
  5. Coccyx, possibly 3 or 4 or 5 vertebrae in the pelvis

NORMAL CURVATURES OF VERTEBRAL COLUMN
      F. 8-1
  1. Cervical curvature, lordotic type
  2. Thoracic curvature, kyphotic type
  3. Lumbar curvature, lordotic type
  4. Pelvic curvature, kyphotic type

STRUCTURE OF VERTEBRAE   F. 8-2 (1) (2) (3)
1. Body
2. Arch: two pedicles, a right and a left
    two laminae (plates), Rt. & Lt.
3. Seven processes:
    two transverse, right & left
    two superior articular, Rt. & Lt.
    two inferior articular, Rt. & Lt.
    one spinous process
4. Vertebral arch, one
5. Vertebral canal, one
6. Vertebral notches, four
    two superior
    two inferior
7. Intervertebral foramina, two, Rt. & Lt.

**INDENTIFYING CHARACTERISTICS OF SOME
VERTEBRAE**
  1. **Cervical:** F. 8-3
     forked spinous processes
     foramina transversaria, two, Rt. & Lt.
     Atlas: first cervical vertebra
     no body
     two arches, anterior, posterior
     Axis: second cervical vertebra
     a dens or odontoid process
     Vertebra prominens: 7th cervical
     large, long spinous process
  2. **Thoracic:** F. 8-6
     costal pits or foveae
         on vertebral body, 2 or 4 right & left
         on transverse processes, 2 right & left
  3. **Lumbar:** F. 8-2
     large size
     no pits
     no foramina transversaria
  4. **Sacrum:** F. 8-7 (1) (2) (3)
     five fused bodies,
     lateral parts fused,
     2 alae, a base, an apex,
     promontory,
     superior articular processes, Rt. & Lt.,
     sacral canal, hiatus & two cornua,
     sacral foramina; 8 anterior, 8 posterior.
  5. **Coccyx:** 4 or 3, or 5 incomplete, 2 cornua F. 8-8

**S. 85 DETAILED STUDY OF BONES OF
        VERTEBRAL COLUMN**

All animals including humans with vertebral columns are named vertebrates. The vertebral column or back bone extends from the base of the skull to the tail bone and includes this bone.

The vertebral column is made up of thirty-three bones named vertebrae. Twenty-four of them remain as separate bones during life and have been named true or movable vertebrae. Five become fused to form the sacrum, the other four form the coccyx. These nine are named false vertebrae.

S. 86 DIVISIONS OF THE VERTEBRAL COLUMN
F. 8-1

The vertebral column is composed of five parts or divisions, named according to the location:
1. Cervical vertebrae, seven in the neck
2. Thoracic vertebrae, twelve in the thorax
3. Lumbar vertebrae, five, posterior to abdomen
4. Sacrum, five fused vertebrae in the pelvis
5. Coccyx, 4 or 3, or 5 incomplete, in pelvis
   The vertebrae in each division are further named as
first cervical, second cervical, etc.,
first thoracic, second thoracic, etc.,
first lumbar, second lumbar, etc., sacrum or
first sacral segment, second sacral segment, etc.,
coccyx or first, second, etc., coccygeal segments.

S. 87 CURVATURES OF VERTEBRAL COLUMN
F. 8-1

Each division of the vertebral column presents a characteristic curve when viewed from the side or lateral aspect.
The cervical curvature is a lordotic curve with an anterior convexity. The midcervical vertebrae therefore project more anteriorly than the upper or lower cervical vertebrae.
The thoracic curvature is a kyphotic curve with its concave surface directed anteriorly, and the convexity posteriorly. The midthoracic vertebrae project more posteriorly than the upper or lower thoracic.
The lumbar curvature is a lordotic curve like the cervical with convex curve anterior. The mid-lumbar bodies project more anteriorly than the upper or lower lumbar.
The pelvic curvature is a lordotic curve with its concave surface anterior, similar to the thoracic curve. The sacrum and coccyx form the pelvic curvature.

S. 88 STRUCTURE OF A VERTEBRA F. 8-2

Vertebra = a bone of the vertebral column. pl. vertebrae, **adj.** vertebral (L. vertere = to turn).
A vertebra has a body, an arch, seven processes and a vertebral foramen. In the present study a lumbar vertebra has been used to show the parts. F. 8-2 (1), (2), (3)

1. **The body** of a vertebra is the solid anterior part shaped like a cylinder, with the posterior surface flattened.

2. **The arch** of a vertebra is the curved posterior part enclosing an opening. The arch has two pedicles and two laminae that, together with the posterior surface of the body, form a vertebral foramen.
The pedicles are two short rounded processes that extend posteriorly, one from each lateral margin of the dorsal surface of a vertebral body (pedicle = a little foot, a stock, stem, root).
The laminae are two flattened plates of bone, one passing towards the midline from each pedicle. They unite posteriorly in the midline to form the spinous process, and to complete the vertebral arch. (L. lamina = a plate, pl. laminae).
The vertebral foramen is an opening formed by the vertebral arch with its pedicles and laminae, and the posterior surface of a vertebral body.
The vertebral canal is a tubular passage extending from the foramen magnum of the skull to the lower sacrum. It is formed by the vertebral foramina of the vertebrae and the ligaments that join successive vertebral arches together. The spinal cord extends down through this canal.

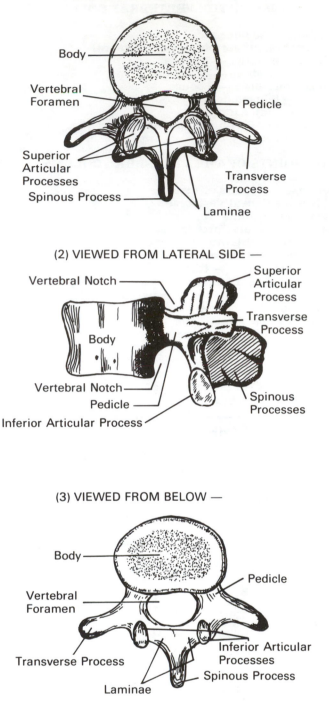

F. 8-2 A TYPICAL VERTEBRA — LUMBAR VERTEBRA
(1) VIEWED FROM ABOVE —
Body
Vertebral Foramen
Pedicle
Superior Articular Processes
Transverse Process
Spinous Process
Laminae

(2) VIEWED FROM LATERAL SIDE —
Vertebral Notch
Superior Articular Process
Transverse Process
Body
Vertebral Notch
Pedicle
Inferior Articular Process
Spinous Processes

(3) VIEWED FROM BELOW —
Body
Pedicle
Vertebral Foramen
Inferior Articular Processes
Transverse Process
Spinous Process
Laminae

The **vertebral notch** is formed by either the upper or lower curved surface of a pedicle; there is a superior and an inferior notch on each side at each level, posterior to the vertebral body.
The **intervertebral foramen** is an opening on each side of the arch, formed by an inferior notch of one vertebra and the superior notch of the vertebra below. A spinal nerve leaves the spinal cord through each of these openings.

3. **Seven processes:**

The transverse processes, right and left, extend laterally from the junction of a pedicle with a lamina on each side.
The superior articular processes, right and left, extend up towards the head from the junction of each pedicle and lamina. Their smooth articular surfaces are directed somewhat posteriorly.

The inferior acticular processes, right and left, project caudally (downwards) from the junction of each pedicle and lamina. Their smooth articular surfaces are directed anteriorly.

The inferior articular processes of one vertebra form joints with the superior articular processes of the vertebra below. As the inferior articular processes lie posterior to the corresponding superior processes a forward displacement of the upper of the two vertebrae cannot occur unless there are defects in the inferior articular processes of the upper vertebra. If present these might be congenital or due to some injury, and will result in spondylolisthesis.

The single spinous process extends posteriorly from the dorsal margin of each vertebral arch, at the junction of the right and left laminae.

The bodies of vertebrae are composed of cancellous or spongy bone with a thin covering of compact bone. The arches and processes have a core of cancellous bone with an outer layer of compact bone.

S. 89 CHARACTERISTICS THAT IDENTIFY SOME OF THE VERTEBRAE

The vertebrae become progressively larger from the cervical to the lumbar divisions. There are certain features that are common to all vertebrae. Those of each division have other features in common. The vertebrae of each division may have some differences in structure as well.

1. **The cervical vertebrae** (F. 8-3) are small, with spinous processes usually forked or bifid or double. There is an opening in each of the transverse processes — a foramen transversarium or transverse foramen through which the vertebral artery passes to reach the skull. The superior and inferior articular processes on each side of each vertebra are united to form a pillar or column of bone behind the transverse process. The articular facets, or foveae on the columns are inclined slightly posteriorly from the horizontal. They do not lie vertically as do those of other vertebrae. Three cervical vertebrae have alternate names and special features.

The atlas or first cervical vertebra (F. 8-4) has no body and no spinous process. Instead it has two lateral masses of bone and two arches. These together surround the vertebral foramen.

The lateral masses are solid masses of bone that lie one on either side of the foramen and support the articular facets (foveae). These masses are joined anteriorly by an anterior bony arch, and posteriorly by a posterior arch of bone.

The anterior arch forms the front wall of the foramen by uniting the two lateral masses here.

The posterior arch is a curved structure that joins the posterior parts of the lateral masses together, completing a bony ring.

The vertebral foramen is large and is formed in front by the anterior arch, laterally by the lateral mass, and behind by the posterior arch. A transverse ligament passes across from one lateral mass to the other dividing the foramen into two parts. The anterior part is occupied by the dens (odontoid process) of the second cervical vertebra, while the posterior part contains the spinal cord.

Superior and inferior articular facets (foveae) are located on the superior and inferior surfaces of each lateral mass. The superior one articulates with a condyle of the occipital bone on the same side. The inferior ones articulate with the superior articular processes of the second cervical vertebra below. (Atlas from Greek mythology). Atlas carried the world on his shoulders; here the atlas carries the head.

F. 8-4 THE ATLAS — FIRST CERVICAL VERTEBRA — FROM ABOVE —

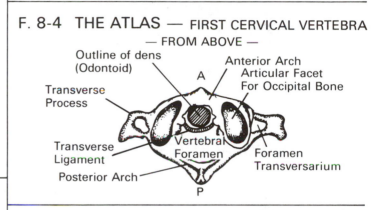

Outline of dens (Odontoid)
Transverse Process
Transverse Ligament
Posterior Arch
Vertebral Foramen
A
Anterior Arch
Articular Facet For Occipital Bone
Foramen Transversarium
P

The axis or second cervical vertebra (F. 8-5) has a tooth shaped process extending upwards from its body. This dens or odontoid process fits into the anterior arch of the atlas above. It forms a pivot joint with this arch, allowing rotation of the head and atlas upon the axis. (Axis — a straight line or an axle about which turning occurs). (L. dens = a tooth or toothlike process).

F. 8-3 A TYPICAL CERVICAL VERTEBRA

(1) VIEWED FROM ABOVE — Superior Aspect

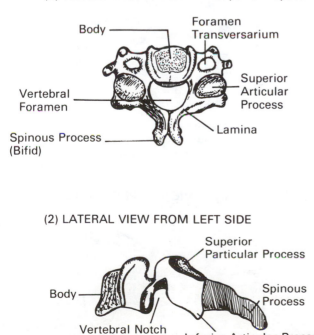

Body
Foramen Transversarium
Vertebral Foramen
Superior Articular Process
Spinous Process (Bifid)
Lamina

(2) LATERAL VIEW FROM LEFT SIDE

Superior Particular Process
Body
Vertebral Notch
Spinous Process
Inferior Articular Process

F. 8-5 THE AXIS — SECOND CERVICAL VERTEBRA
(1) VIEWED FROM ABOVE & BEHIND

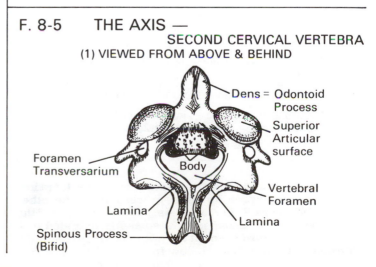

Dens = Odontoid Process
Superior Articular surface
Foramen Transversarium
Body
Lamina
Vertebral Foramen
Lamina
Spinous Process (Bifid)

F. 8-5   THE AXIS — continued

(2) LATERAL VIEW FROM LEFT SIDE

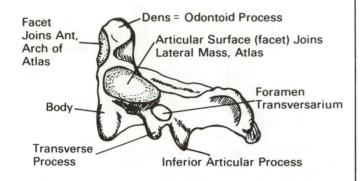

Facet Joins Ant, Arch of Atlas

Dens = Odontoid Process

Articular Surface (facet) Joins Lateral Mass, Atlas

Body

Foramen Transversarium

Transverse Process

Inferior Articular Process

F. 8-6   A THORACIC VERTEBRA

(1) VIEWED FROM ABOVE

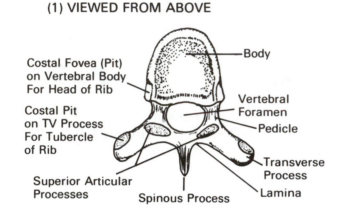

Costal Fovea (Pit) on Vertebral Body For Head of Rib

Body

Costal Pit on TV Process For Tubercle of Rib

Vertebral Foramen

Pedicle

Superior Articular Processes

Transverse Process

Spinous Process

Lamina

F. 8-6   A THORACIC VERTEBRA — continued

(2) LATERAL VIEW — FROM LEFT SIDE

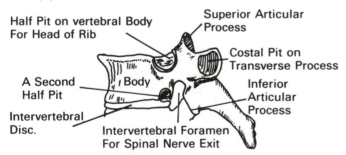

Half Pit on vertebral Body For Head of Rib

Superior Articular Process

Costal Pit on Transverse Process

A Second Half Pit

Body

Inferior Articular Process

Intervertebral Disc.

Intervertebral Foramen For Spinal Nerve Exit

Note;—

1st thoracic vertebra = 1 & ½ costal pits or each side

2nd to 8th thoracic = two half pits on each side

10th to 12th thoracic = single pit each side
  9th thoracic = variable, 1, or 2 half pits

The vertebra prominens or seventh cervical vertebra has a longer and larger spinous process than the other cervical vertebra. It is readily palpable at the base of the neck posteriorly. Because it is so easily palpated it is used as a landmark to count the spinous processes of the thoracic vertebrae lying below it.

---

2. **The thoracic vertebrae** (F. 8-6) are larger then the cervical. Their spinous processes are directed almost vertically caudally (towards the tail end of the body). The tip of one of these may lie opposite the inferior margin of the vertebra below the one to which it is attached. The thoracic vertebrae have costal pits named foveae. (s. fovea).

A costal pit (NA) fovea costalis or facet, is a smooth round or semicircular smooth depression located on a vertebral body or its transverse process.

(1) **Costal pit of a thoracic vertebra,** or a facet, lies on the lateral surface of a thoracic vertebra close to its posterior margin, to articulate with the head of a rib. As some vertebrae articulate with two ribs there are two semilunar pits, one on the upper and the other on the lower margin, rather than one circular pit at the center of the body. The tenth, eleventh, and twelfth vertebrae have a single pit on each side as they form joints with one rib only.

(2) **Costal pit on a transverse process:** this rounded pit lies on the anterior surface of a transverse process of thoracic vertebrae. If forms a joint with a tubercle of a rib.

3. **The lumbar vertebrae** (F. 8-2) are larger than the thoracic, and become progressively larger towards the sacrum. Their spinous processes are directed almost horizontally and posteriorly. Their transverse processes have no foramina and no costal pits.

4. **The sacrum** (F. 8-7) pl. sacra, adj. sacral, is made up of five vertebrae that have become united to form a single bone. In the fetus and young child the sacral segments are separated by cartilages. The sacrum has a central part or body, two lateral parts, an arch, and a sacral canal.

(1) **The body** of the sacrum or central part is the large curved area that is formed from the fused bodies of the five sacral segments. The transverse ridges on the pelvic (anterior) surfaces at each level denote the points of union of these segments.

F. 8-7   THE SACRUM

(1) FRONTAL VIEW —

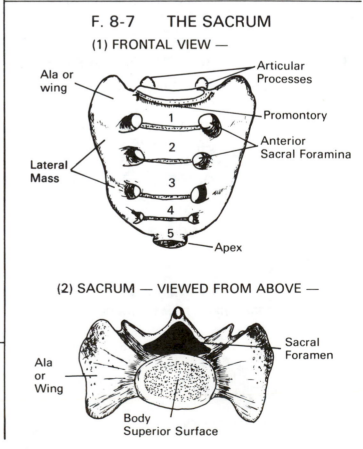

Ala or wing

Articular Processes

Promontory

Anterior Sacral Foramina

Lateral Mass

1
2
3
4
5

Apex

(2) SACRUM — VIEWED FROM ABOVE —

Ala or Wing

Sacral Foramen

Body Superior Surface

(3) SACRUM — LATERAL VIEW — RIGHT SIDE

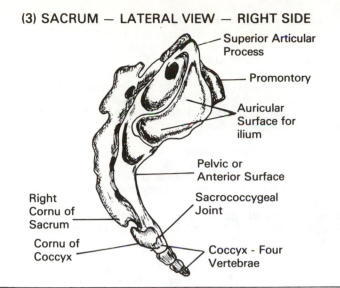

- Superior Articular Process
- Promontory
- Auricular Surface for ilium
- Pelvic or Anterior Surface
- Sacrococcygeal Joint
- Right Cornu of Sacrum
- Cornu of Coccyx
- Coccyx - Four Vertebrae

(2) The lateral parts (lateral masses) of the sacrum lie one on either side of the fused bodies, lateral to the sacral foramina. Each lateral part is made up of the fused transverse processes and costal elements of the five fused vertebrae. (NA. pars lateralis).

The ala is the winglike upper part of each lateral mass, viewed from above.

The base of the sacrum is its broad upper end consisting of the upper surface of the body of the first sacral segment and the alae on either side.

The promontory of the sacrum is the prominent anterior margin of the upper sacral segment.

The superior articular processes extend up from the posterolateral margins of the sacrum on each side. Their articular surfaces are directed posteriorly and form joints with the inferior articular processes of the fifth lumbar vertebra.

The apex of the sacrum is its small inferior end which articulates with the first segment of the coccyx.

The sacral canal is a tubelike passage that extends through the sacrum from top to bottom. The vertebral foramina of the original vertebrae have formed this channel with the fused bodies in front and the fused arches behind. The sacral and coccygeal nerves pass down in this canal.

The sacral hiatus is an opening in the sacral canal posteriorly at its lower end, causing a gap between the laminae of the fifth sacral segment.

The sacral cornua are two small bony projections that extend caudally from the lower sacrum posteriorly, one on each side of the sacral hiatus. They form joints with similar cornua of the coccyx. (L. cornu = a horn).

Four dorsal sacral foramina open into the canal on either side. They lie lateral to the fused bodies and transmit the dorsal branches of the sacral nerves from the sacral canal.

Four pelvic (anterior) sacral foramina open into the sacral canal on each side from the pelvic surface of the sacrum. They transmit anterior branches of the sacral nerves to the pelvis.

Two auricular surfaces are located one on each lateral surface of the sacrum. These rough areas join the auricular surfaces of the iliac bones to form the sacroiliac joints. (Auricle = ear, or earlike).

5. The coccyx: adj. coccygeal. F. 8-8
The coccyx is usually formed by four (or occasionally three, or five) incompletely developed vertebrae. The segments often remain as separate bones. The upper end of the coccyx forms a joint with the apex of the sacrum.
The cornua are two small bony processes that extend up from the posterior surface of the coccyx to meet similar cornua of the sacrum.

F. 8-8 COCCYX

FRONTAL —VIEW—

- Cornua (Two)
- Transverse Process

F. 8-9 THE INTERVERTEBRAL DISC-JOINT

(1) LATERAL VIEW FROM LEFT SIDE —

- Anterior Part
- Posterior Part
- Vertebral Body
- Intervertebral Disc.
- Intervertebral Foramen
- Vertebral Body
- Intervertebral Disc.

(2) MIDSAGITTAL SECTION FROM LEFT SIDE

- Vertebral Body
- Lamina
- Sp. Process
- Interv. Disc.
- Intervertebral For.
- Interv. Disc.
- Verterbral Body
- Lamina
- Lig. Flava
- Anterior Longitudinal Lig.
- Posterior Longitudinal Ligament

(3) DISC MAGNIFIED —

- Bone
- Disc
- Cartilage on vertebra
- Anulus fibrosus
- Cartilage on vertebra
- Anulus fibrosus
- Nucleus pulposus

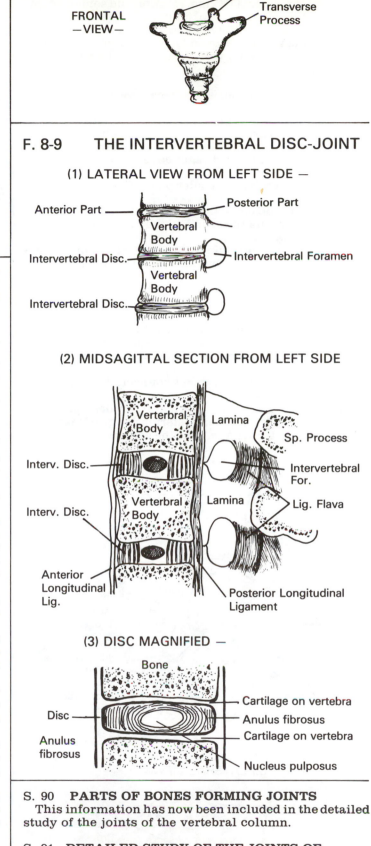

S. 90 PARTS OF BONES FORMING JOINTS
This information has now been included in the detailed study of the joints of the vertebral column.

S. 91 DETAILED STUDY OF THE JOINTS OF THE VERTEBRAL COLUMN

1. Intervertebral joints: F. 8-9
(NA, intervertebral discs)
Located between the flat articular surfaces of bodies of adjacent vertebrae from the second cervical vertebra to and including the lumbosacral joint between the fifth lumbar body and the sacrum.

These are cartilaginous joints permitting slight flexion, extension, lateral flexion, rotation and some circumduction. Movement at one joint is limited, but since several joints are involved considerable movement is possible. Most of the movement occurs in the cervical and lumbar divisions, with minimal movement in the thoracic part.

Formed by:

(1) the articular surfaces of adjacent vertebrae covered by compact bone;

(2) thin articular cartilages covering the articular surface of each vertebra, composed of hyaline cartilage;

(3) **an intervertebral disc,** a flat circular plate lying between the articular cartilages covering each bone, and consisting of:

an anulus fibrosus — an outer pad of fibrous tissue and fibrocartilage;

a nucleus pulposus — the pulpy center of a disc in the center of an anulus fibrosus and surrounded by it.

(4) **ligamenta**

- an anterior longitudinal ligament of fibrous tissue extending lengthwise along the anterior surfaces of the vertebrae;

- a posterior longitudinal ligament passes lengthwise between adjacent vertebrae along their posterior surfaces;

- the ligamenta flava pass lenghtwise between the laminae of adjacent vertebrae. (s. ligamentum flavum);

- interspinal ligaments connect adjacent spinous processes;

- the supraspinatus ligament connects the tips of the spinous processes.

- as stated earlier the vertebral arches, posterior surfaces of the vertebrae, posterior longitudinal ligament, and ligamenta flava, form the vertebral canal down which the spinal cord passes.

2. **Interarticular joints:** F. 8-10

(NA) Zygaphyseal joints; right and left;

Located between articular processes of the adjacent vertebrae on each side.

Formed by the inferior articular process of one vertebra articulating with the superior articular process of the vertebrae below on the same side of the body.

These are synovial gliding joints, the articular surfaces sliding over each other.

3. **Atlantooccipital joints:**

Two, a right and a left.

Located between the atlas and occipital bone.

Formed by an occipital condyle of the occipital bone articulating with the superior articular surface on the lateral mass of the atlas on the same side.

Synovial condylar joints, permitting some flexion, extension, and slight lateral flexion.

4. **Atlantoaxial joints:** F. 8-11

One median and two lateral, right and left located between the atlas and axis.

Median atlantoaxial joint:

Located between the dens (odontoid process) of the axis and a ring formed by the anterior arch of the altas and its transverse ligament.

A synovial pivot joint with two synovial cavities, one anterior, one posterior to the dens, permitting rotation of atlas and head upon the axis.

Lateral atlantoaxial joints: F. 8-11

Two, a right and a left

Located between the inferior articular surface (fovea) of the lateral mass of the atlas, and the superior articular process of the axis on each side.

Synovial gliding joints.

F. 8-10 INTERARTICULAR OR ZYGAPOPHYSEAL JOINTS

(1) DORSAL VIEW

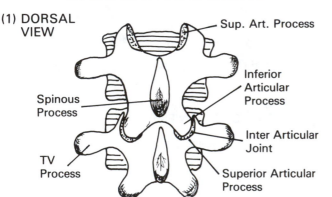

(2) LATERAL VIEW

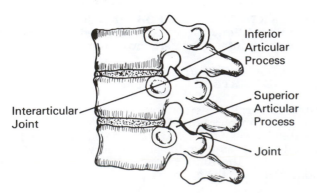

F. 8-11 ATLANTOAXIAL JOINTS

MEDIAN & LATERAL BETWEEN ATLAS & AXIS

(1) RADIOGRAPHIC APPEARANCE — Front Front —

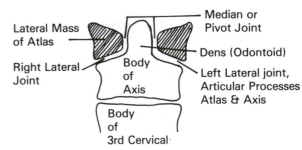

MEDIAN ATLANTOAXIAL JOINT

(2) FROM ABOVE UPPER SURFACE OF ATLAS

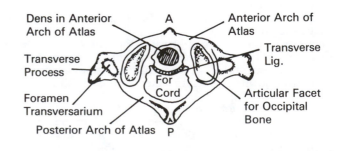

5. **Accessory cervical joints:** several,

Located between the lateral margins of articular surfaces of lower cervical vertebrae, on each side.

F. 8-12 LUMBOSACRAL & SACROCOCCYGEAL JOINTS

MIDSAGITTAL SECTION - SACRUM COCCYX

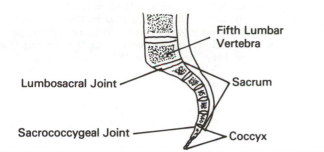

Fifth Lumbar Vertebra

Lumbosacral Joint

Sacrum

Sacrococcygeal Joint

Coccyx

6. Lumbosacral joint: F. 8-12
a single joint.
Located between fifth lumbar vertebra and the upper end of sacrum.
Formed by the inferior articular surface of the fifth lumbar vertebra and the superior articular surface of the sacrum.
A cartilaginous joint with an intervertebral disc, similar to other intervertebral ones.
Interarticular joints, similar to No. 2 above are also present between the inferior articular processes of the fifth lumbar vertebra and articular processes of upper end of sacrum; gliding joints, right and left.

7. Sacroccygeal joint: F. 8-12
Located between sacrum and coccyx.
Formed by distal end of sacrum forming a joint with the first segment of the coccyx. Cartilaginous joint with a disc.
Intercoccygeal joints with discs are also present between the segments of the coccyx. The discs disappear in later life and the segments unite.

S. 92 CONGENITAL ANOMALIES OF VERTEBRAE

Fusion of vertebrae: two or more vertebrae may be united with no joint between them.
Hemivertebra: the right or left half of a vertebra may be absent, or one half of a vertebra may fuse with the vertebra above or below it, leaving the other half as a separate bone.
Spina bifida: the right and left laminae may fail to unite with each other at the median line posteriorly. The vertebral arch will be incomplete with a gap in the midline. Several vertebrae may be involved, particularly in the lower lumbar and upper sacral areas. The spinal cord may bulge or herniate out through this gap. A swelling, called a meningocele, will be present in the midline of the back low down.
Articular defects: defects that may be present between the superior and inferior articular processes of a vertebrae at the isthmus where these join. This is usually bilateral, and occurs most frequently behind the fifth lumbar body. A gap is visible in the bone on radiographs. This condition is named spondylolisthesis.

Absence of sacrum: the lower part of the sacrum may fail to develop; if so the coccyx will also be absent.
Sacralization: the fifth lumbar vertebra may be partly or completely fused with the upper sacrum. Occasionally the first sacral segment may remain a separate bone, resulting in six lumbar vertebrae.

S. 93 LANDMARKS, PROMINENCES, OF VERTEBRAE

Spinous processes: the tips of the spinous processes of the lower cervical, thoracic and lumbar vertebrae. They form a vertical chain of knoblike prominences that are easily palpable in the midline posteriorly.

Spinous process of seventh cervical vertebra: the spinous process of this vertebra is large and is readily palpable at the base of the neck as a prominence. It is used in locating the position of other vertebrae. This is done by counting the spinous processes of the vertebrae down to the desired level.
The spinous processes of the thoracic vertebrae lie at a lower level than the corresponding vertebral bodies.

The vertebral bodies are not palpable except for the lumbar, which can sometimes be palpated through the anterior abdomen.

Sacrum and coccyx: these are palpable through the crease between the buttocks posteriorly, and can also be felt by a finger inserted into the rectum.
Locating a lumbar vertebra on a processed film; when the lumbosacral joint is not included on the radiograph the twelfth thoracic vertebra may be determined as that to which the last rib is attached. The lumbar vertebrae can then be counted down from this twelfth one.

SURFACE MARKINGS ON ANTERIOR BODY SURFACE CORRESPONDING TO VARIOUS VERTEBRAE

C = cervical, T = thoracic, L = lumbar
Hard palate - 1st cervical vertebra
Hyoid bone - 2nd or 3rd cervical vertebrae
Thyroid cartilage - upper margin C4
Cricoid cartilage - cervical 6
Jugular notch - between T2 and T3
Sternal angle - joint between T4 and T5
Xiphisternal junction - joint between T9 and 10
Transpyloric plane - L1
Subcostal plane - L3
Umbilicus - joint between L3 and L4
Iliac crests - L5
Anterior superior spine - 2nd sacral segment.

S. 94 ANATOMICAL PECULIARITIES AND RADIOGRAPHY F. 8-13

In order to verify the statements made here the student should place a mounted skeleton upon a table top or study a mounted skeleton upright.
Curvatures of the vertebral column: the cervical and lumbar lordotic curves, and the thoracic and pelvic kyphotic curves, should be noted. Because of these curves it is difficult to see the vertebral bodies and joints without the overlapping of one upon the other. This may be done by attempting to straighten the curves, or by moving the eye up or down in order to see through each joint. In radiography, therefore, careful centering using small x-ray films is necessary.
In the cervical area flexing the head upon the chest will obliterate the curve, but the facial bones will then obscure the lower cervical bodies. If the mouth be opened the upper cervical vertebrae will become visible.
The thoracic vertebrae have limited movement and it may be impossible to get rid of the kyphotic curve. The examining eye or x-ray tube must therefore be moved up or down for each part and only small areas should be radiographed at a time.
The lumbar vertebrae are quite movable. If a hand be placed upon the table top under the lumbar vertebrae, with the subject supine, the midlumbar vertebrae will not be touching the table. If the knees and thighs are flexed this curve will disappear and the vertebrae will touch the hand. This maneuver is frequently used in the radiography of the lumbar division. In the lateral position there will be no problem.

F. 8-13   VERTEBRAL COLUMN

—Lateral View Outlining Normal Curves—

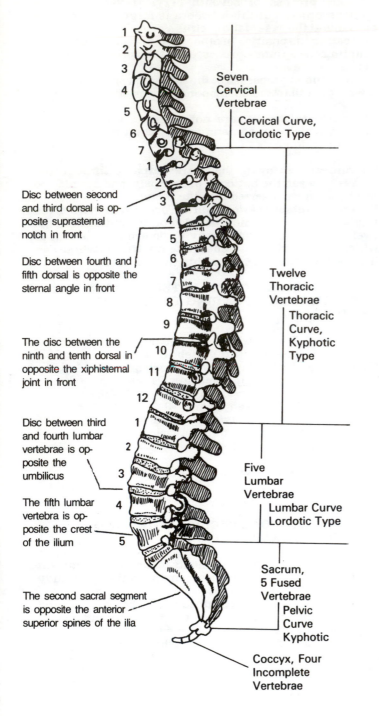

Disc between second and third dorsal is opposite suprasternal notch in front

Disc between fourth and fifth dorsal is opposite the sternal angle in front

The disc between the ninth and tenth dorsal in opposite the xiphisternal joint in front

Disc between third and fourth lumbar vertebrae is opposite the umbilicus

The fifth lumbar vertebra is opposite the crest of the ilium

The second sacral segment is opposite the anterior superior spines of the ilia

Seven Cervical Vertebrae

Cervical Curve, Lordotic Type

Twelve Thoracic Vertebrae

Thoracic Curve, Kyphotic Type

Five Lumbar Vertebrae

Lumbar Curve Lordotic Type

Sacrum, 5 Fused Vertebrae

Pelvic Curve Kyphotic

Coccyx, Four Incomplete Vertebrae

The **interarticular joints** lie obliquely to the sagittal plane so that it is not possible to see through them with the subject supine. The skeleton must be turned into a oblique position in order to view them satisfactorily. The lumbar division should be examined in this exercise.

The **sacrum and coccyx** form the pelvic curve and appear to be foreshortened when viewed with the eye at their level. It is therefore necessary to look upwards from slightly below them or, in radiography, to angulate the tube towards the head.

The **lumbosacral joint** is directed obliguely and cannot be seen without overlapping of bones in the supine position. The eye must be directed upwards towards the head.

Scoliosis of the vertebral column is a curvature in a lateral direction to the right or left. It does not occur normally. It becomes visible with the subject supine.

S. 95 ANATOMICAL TERMS USED IN STUDYING THE VERTEBRAL COLUMN

anulus fibrosus
apophyseal joint
zygapophyseal joint
arch of a vertebra
articular cartilage
 of a vertebra
atlas
atlantooccipital joint
axis
atlantoaxial joint
body of a vertebra
cervical vertebra
coccyx
coccygeal
congenital anomaly
cornu, cornua
cornu of coccyx
cornu of sacrum
dens or
odontoid process
dorsal vertebra
facet of vertebra
facet of rib
foramen, foramina

foramen transversarium
fovea = a pit
hemivertebra
inferior articular
 process

interarticular isthmus
interarticular joints
intervertebral foramen
kyphosis
kyphotic
lamina, laminae
lordosis
lordotic
lumbar vertebra
lumbosacral joint
nucleus pulposus
pedicle of vertebra
sacrum, sacra, sacral
sacrococcygeal joint
sacroiliac joint
sacrovertebral joint
scoliosis
scoliotic
spina bifida
spinous process of a
 vertebra
superior articular
 process
thoracic vertebra
vertebra, vertebrae
vertebral
vertebral foramen
vertebral foramina
vertebral notch
vertebra prominens

NOTES

NOTES

9. THE THORAX: BONES AND JOINTS

S. 96 BONES OF THE THORAX

Thoracic vertebrae — 12 F. 9-1
See vertebral column S. 88, 89

Sternum — breast bone — one only
pl. sterna, adj. sternal

Ribs or costae — 12 pairs, right, left
s. costa, adj. costal

Costal cartilages — 12 pairs
right and left

S. 97 JOINTS OF THE THORAX

Costovertebral — between a rib and the body of a vertebra
Costotransverse — between a rib and a transverse process of a vertebra
Sternocostal or **costosternal** — between a rib cartilage and the sternum
Costochondral — between a rib and its costal cartilage
Interchondral — between the adjacent anterior margins of costal cartilages of sixth to tenth ribs
Sternoclavicular — between the sternum and the clavicle
Manubriosternal — between the manubrium and the body of the sternum
Xiphisternal — between the body of the sternum and its xiphoid process

S. 98 PARTS AND PROMINENCES OF BONES OF THORAX

Thoracic vertebrae See S. 88, 89

Sternum — a flat bone
1. Manubrium:
   jugular notch (suprasternal notch)
   clavicular notches
   costal notches, complete and half, Rt. Lt.
2. Body or gladiolus;
   costal notches — 4 and 2 half notches, Rt. Lt.
3. Xiphoid process — one half notch, Rt. Lt.

Ribs (costae) 12 pairs - right and left:
1. dorsal extremity or vertebral extremity (end) with a head neck and tubercle
2. body or shaft with an angle
3. ventral extremity or sternal extremity

Costal cartilages — 12 pairs - right and left:
seven upper ribs articulating with sternum separately, eighth, ninth, tenth, join cartilage of rib above, the eleventh and twelfth have free anterior ends so are floating

True and false ribs:
true ribs — upper seven pairs
false ribs — lower five pairs

Alternate classification:
vertebrosternal — upper seven pairs
vertebrochondral ribs — eighth, ninth, tenth pairs
vertebral ribs — eleventh and twelfth pairs

S. 99 DETAILED STUDY OF THE BONES OF THE THORAX

The thorax, the upper part of the trunk, is a cage composed of bones and cartilages, which affords protection for several very vital organs: the heart with its trunk vessels and the respiratory organs. The skeleton of the thorax includes:
twelve thoracic vertebrae
the sternum
twelve pairs of ribs
twelve pairs of costal cartilages

F. 9-1   THE THORAX: STERNUM: RIBS: COSTAL CARTILAGES:

—FRONTAL VIEW—

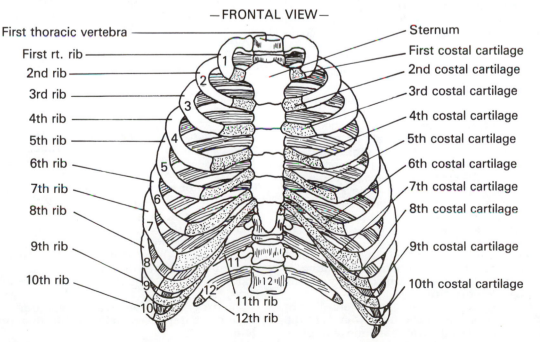

First thoracic vertebra — Sternum
First rt. rib — First costal cartilage
2nd rib — 2nd costal cartilage
3rd rib — 3rd costal cartilage
4th rib — 4th costal cartilage
5th rib — 5th costal cartilage
6th rib — 6th costal cartilage
7th rib — 7th costal cartilage
8th rib — 8th costal cartilage
9th rib — 9th costal cartilage
10th rib — 10th costal cartilage
11th rib
12th rib

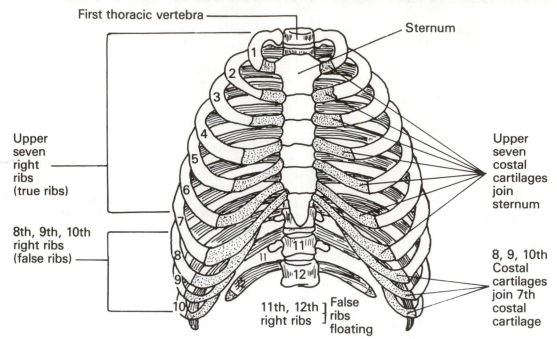

First thoracic vertebra

Sternum

Upper seven right ribs (true ribs)

8th, 9th, 10th right ribs (false ribs)

Upper seven costal cartilages join sternum

8, 9, 10th Costal cartilages join 7th costal cartilage

11th, 12th right ribs] False ribs floating

The thoracic vertebrae have been studied in the preceding chapter. The student should review these paragraphs dealing with the costal pits (fovea) on the vertebral bodies, and their transverse processes.

1. The sternum: F. 9-1, 2,3
Sternum — breast bone; pl. sterna, adj. sternal

The sternum is a long flat bone that lies vertically in the midline of the anterior chest wall. Because this bone is shaped somewhat like a sword or dagger, its parts have been given names similar to the parts of a sword — a handle, blade and point. The sternum has a manubrium, a body, and a xiphoid process.
(1) **The manubrium** of the sternum is its upper part and represents the handle of the sword. The manubrium is separated from the body of the sternum by cartilage. In later life these two parts may fuse. The manubrium has a jugular notch, clavicular notches and costal notches. (L. manubrium = a handle)   The jugular notch formerly named the suprasternal notch (OT) is a depression on the upper surface of the manubrium of the sternum.
The clavicular notches, a right and a left, are placed obliquely on either side of the jugular notch. They form joints with the medial ends of the clavicles — the sternoclavicular joints.
(2) **The body** or gladiolus of the sternum extends from the manubrium to the xiphoid process below and is the largest part of the sternum. It consists of four segments formed from separate centers of ossification. They unite during adolescence to form a single bone. The body has costal notches. (L. gladiolus = sword, body of sword).
The sternal angle (angle of Louis or Ludwig) is marked by a prominent transverse ridge at the junction of the manubrium and body of the sternum. These parts meet at an angle resulting in the prominence. This ridge also indicates the level of the second costal cartilages, which in turn are attached to the second ribs. By palpating the ridge it is possible to locate the second ribs adjacent to it, and then count the ribs below this one.
(3) **The xiphoid process** of the sternum is the inferior part of the bone. Composed of cartilage in the young it becomes ossified and often fuses with the body above. (Xiphoid = a sword or sword tip).
The costal notches are niches on the lateral margins of the sternum that articulate with the medial ends of the costal cartilages. The manubrium has one complete

notch, and one half-notch on either side. The other half of the lower notch is on the adjacent upper margin of the body. The body has four complete notches and an upper and a lower half-notch on either side. The xiphoid has a half-notch to complete the notch for the seventh rib bilaterally.

2. The ribs: F. 9-1, 2,4
L. costa — a rib, pl. costae, adj. costal.

There are twelve pairs of ribs, twelve left, and twelve right. Each rib is attached to a costal cartilage at its anterior end. The ribs are long, curved, flat bones that form the posterior wall, the lateral walls, and part of the anterior wall of the thorax. From above down the ribs are named as the first, second, third, etc., and as right or left. Each rib has a dorsal or vertebral extremity or end, a body, and a ventral or sternal extremity or end.
(1) **The dorsal extremity** has a head, a neck, and a tubercle.
The head of a rib is its slightly expanded posterior end. It has one or two small articular surfaces (facets) according to whether it articulates with one or two vertebral bodies.
The neck of a rib is its slightly constricted part connecting the head to the body.
The tubercle of a rib is a small prominence on the dorsal surface at the junction of the neck and body. It has a small articular surface (facet) that forms a joint with a similar surface on a transverse process of a vertebra.
(2) **The body** of a rib is the long, flat , curved part that extends from the neck around the chest wall. At its lower margin there is a groove — the costal groove — for an intercostal artery, vein and nerve.
The angle of a rib is a prominence in front of the neck where the rib becomes bent anteriorly.
(3) **The ventral or sternal extremity** is the anterior end that articulates with its costal cartilage.

The ribs become progressively longer from the first to the seventh, then progressively shorter so that the eleventh and twelfth end in the lateral or posterior abdominal wall. These are only palpable towards the back.

The first rib has two grooves running transversely on its upper anterior surface for the subclavian artery and vein.

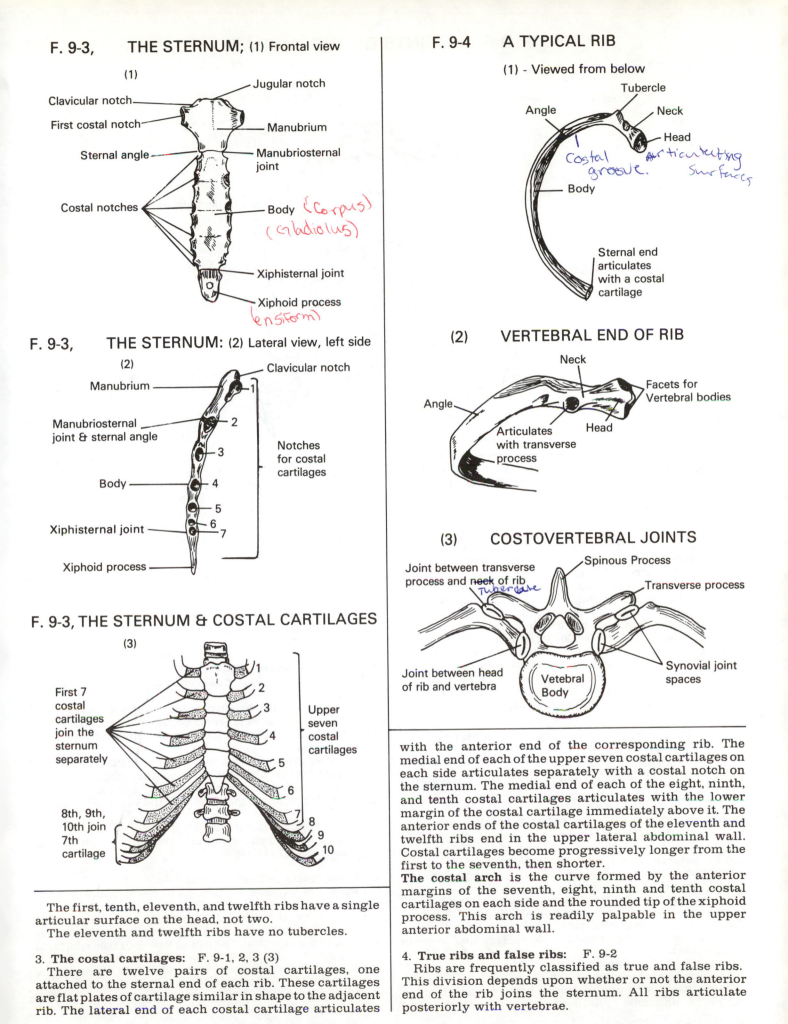

F. 9-3, THE STERNUM; (1) Frontal view

(1)

- Jugular notch
- Clavicular notch
- First costal notch
- Sternal angle
- Costal notches
- Manubrium
- Manubriosternal joint
- Body (Corpus) (Gladiolus)
- Xiphisternal joint
- Xiphoid process (ensiform)

F. 9-3, THE STERNUM: (2) Lateral view, left side

(2)

- Clavicular notch
- Manubrium
- Manubriosternal joint & sternal angle
- Body
- Xiphisternal joint
- Xiphoid process
- Notches for costal cartilages
- 1
- 2
- 3
- 4
- 5
- 6
- 7

F. 9-3, THE STERNUM & COSTAL CARTILAGES

(3)

- First 7 costal cartilages join the sternum separately
- 8th, 9th, 10th join 7th cartilage
- Upper seven costal cartilages
- 1
- 2
- 3
- 4
- 5
- 6
- 7
- 8
- 9
- 10

The first, tenth, eleventh, and twelfth ribs have a single articular surface on the head, not two.

The eleventh and twelfth ribs have no tubercles.

3. **The costal cartilages:** F. 9-1, 2, 3 (3)

There are twelve pairs of costal cartilages, one attached to the sternal end of each rib. These cartilages are flat plates of cartilage similar in shape to the adjacent rib. The lateral end of each costal cartilage articulates

F. 9-4   A TYPICAL RIB

(1) - Viewed from below

- Tubercle
- Angle
- Neck
- Head
- Costal groove.
- Articulating surfaces
- Body
- Sternal end articulates with a costal cartilage

(2)   VERTEBRAL END OF RIB

- Neck
- Angle
- Facets for Vertebral bodies
- Articulates with transverse process
- Head

(3)   COSTOVERTEBRAL JOINTS

- Joint between transverse process and neck of rib
- Tubercle
- Spinous Process
- Transverse process
- Joint between head of rib and vertebra
- Vetebral Body
- Synovial joint spaces

with the anterior end of the corresponding rib. The medial end of each of the upper seven costal cartilages on each side articulates separately with a costal notch on the sternum. The medial end of each of the eight, ninth, and tenth costal cartilages articulates with the lower margin of the costal cartilage immediately above it. The anterior ends of the costal cartilages of the eleventh and twelfth ribs end in the upper lateral abdominal wall. Costal cartilages become progressively longer from the first to the seventh, then shorter.

The costal arch is the curve formed by the anterior margins of the seventh, eight, ninth and tenth costal cartilages on each side and the rounded tip of the xiphoid process. This arch is readily palpable in the upper anterior abdominal wall.

4. **True ribs and false ribs:**   F. 9-2

Ribs are frequently classified as true and false ribs. This division depends upon whether or not the anterior end of the rib joins the sternum. All ribs articulate posteriorly with vertebrae.

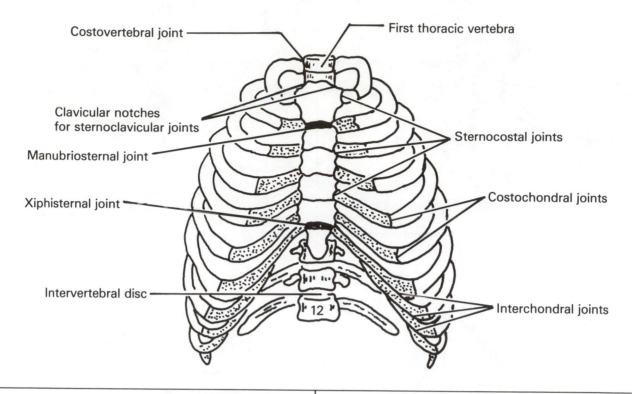

True ribs (NA); the upper seven pairs articulate by means of their costal cartilages with the sternum at a costal notch. Because of this sternal articulation and the fact that they all join the vertebrae behind they are named **vertebrosternal ribs** (vertebra and sternum).

The false ribs (NA); the lower five pairs do not articulate with the sternum. The upper 3 pairs of false ribs, the eighth, ninth, and tenth ribs have costal cartilages that are joined to the costal cartilage immediately above, the eight to the seventh, the ninth to the eighth, and the tenth to the ninth costal cartilage. Their connection with the sternum is indirect, by the seventh costal cartilage. These have been named **vertebrochondral** ribs; (vertebra + chondros, a cartilage). The lower two pairs, the eleventh and twelfth have no connection with the sternum. The anterior ends of their costal cartilages end in the abdominal muscles laterally. They can be palpated through the lateral wall of the abdomen. These ribs have been named **vertebral** ribs, or floating ribs.

S. 100   STRUCTURES FORMING THE JOINTS OF THE THORAX

Refer to S. 97 for a list of the joints and the structures forming each joint.

S. 101   DETAILED STUDY OF THE JOINTS OF THE THORAX

1. **The intervertebral and interarticular joints** of the vertebrae have been described in S. 91 along with the other joints of the vertebral column.

The student should review the following terms:

Fovea = a small pit or depression at a joint,

Facet = a small smooth articular surface,

Articular surface = a smooth bone surface that helps to form a joint.

2. **Costovertebral joints:** F. 9-4 (3)

There are two types of joints between the thoracic vertebrae and ribs: joints of the heads of ribs and vertebral bodies; costotransverse joints between the tubercles of ribs and transverse processes of vertebrae.

(1) **Joints of heads of ribs** and vertebral bodies.

The articular surface of the head of a rib articulates with a pit (fovea) on the lateral surface of the body of a vertebra. The first, tenth, eleventh, and twelfth vertebral bodies have a single pit on the lateral surface of the body on each side, that articulates with the head of a single rib.

The other vertebral bodies have two pits on each side, one close to the upper border, the other at the lower margin. Here the head of a rib articulates with the inferior pit of one vertebra, and the superior pit of the vertebra below, as well as with the adjacent intervertebral disc.

Synovial joints, permitting slight movement.

(2) **Costotransverse joints:** F. 9-4 (3)

Location — between ribs and transverse processes of vertebrae.

Formed by pit or fovea on anterior surface of transverse process of thoracic vertebra, and a small facet (articular surface) on the tubercle of a rib, except for 11th and 12th ribs.

Synovial gliding joints, permitting slight movement.

The movements of the ribs increase the anteroposterior and lateral diameters of chest at inspiration, while the descent of the diaphragm increases the vertical diameter.

3. **Costochondral joints:** F. 9-5

Located at the anterior, or sternal, end of a rib and the adjacent costal cartilage.

4. **Sternocostal or Costosternal joints:** F. 9-3, 5

Located at the lateral margins of the sternum on each side.

Formed by the sternal (anterior) end of a costal cartilage articulating with a costal notch on the lateral margin of the sternum. The upper seven or true ribs form this type of joint. The second articulates with half-notches on the manubrium and body of the sternum, the seventh with half-notches on the lower body and xiphoid.

Sliding synovial joints, except first, which is cartilaginous.

5. Interchondral joints: F. 9-5

Located between adjacent anterior margins of the seventh, eight, ninth, and tenth costal cartilages and the rib cartilage above each one.

6. Sternoclavicular joints: F. 9-3, 5

Two, one right, one left.

These have been described along with the superior limb, located between the sternal end of the clavicle and clavicular notch of the sternum. They are synovial gliding joints.

7. Manubriosternal joint: one only.   F. 9-5

Located between the inferior margin of the manubrium and the upper margin of the body of the sternum, at the sternal angle.

A cartilaginous joint, frequently ossified in later life.

8. Xiphisternal or Xiphosternal Joint: one only.   F. 9-5

Located between the inferior end of the body of the sternum and the xiphoid process of the sternum.

A cartilaginous joint, which becomes ossified in adolescence in the second decade of life.

S. 102   CONGENITAL ANOMALIES OF THE BONES OF THE THORAX

Cervical ribs: an extra rib may be present on one or both sides above the first rib. It joins the seventh cervical vertebra. It may be short or long. It may press upon the large nerve trunks that pass from the brachial plexus in the neck to the upper limb.

Lumbar ribs: short extra ribs may be present below the twelfth ribs, they join the first lumbar vertebra.

Absence of ribs: one or more ribs may be absent on one or both sides. This frequently accompanies anomalies of adjacent vertebrae.

Forked rib: the anterior or sternal end of a rib may divide into two parts.

Fused ribs: two adjacent ribs may be joined for a part of their length.

Pectus excavatum: the sternum may be depressed so that it lies closer to the thoracic vertebrae than usual. The anteroposterior diameter of the thorax is decreased, and the heart may be pressed upon with some disability. The reverse condition, that of a pigeon chest, with prominent sternum may be present.

S. 103   LANDMARKS — PROMINENCES OF THE BONES OF THE CHEST

The jugular notch can be felt as a depression on the upper end of the sternum and is often visible. It lies opposite the joint between the second and third thoracic vertebrae. Previously named suprasternal notch.

The sternal angle is often visible and can be palpated as a ridge passing transversely across the upper sternum at the junction of the manubrium and body of the sternum. It lies opposite the costal cartilages of the second ribs. It lies opposite the joint between the fourth and fifth thoracic vertebrae. (OT) angle of Louis or Ludwig.

The xiphisternal joint can be felt at the lower end of the body of the sternum. It lies opposite the joint between the ninth and tenth thoracic vertebrae.

The costal arch is formed by the anterior margins of the 8th, 9th, and 10th costal cartilages and the tip of the xiphoid. It is readily palpable. The gall bladder often lies posterior to the right upper costal margin, i.e. the right side of the arch.

S. 104   THORACIC PECULIARITIES AND RADIOGRAPHY

A mounted skeleton should be viewed in the supine and prone positions to demonstrate the observations outlined here.

The sternum in the prone position is completely hidden by the thoracic vertebrae. By turning the subject into either oblique position the vertebrae are displaced from the underlying sternum. With the subject in the lateral position the sternum is visible but is partly covered by the anterior rib margins.

The sternoclavicular joints and medial ends of the clavicles are also obscured by the vertebrae unless an oblique position is assumed.

The ribs: if viewed from the front or back the lateral parts appear to be foreshortened as they are directed anteroposteriorly. By rotating the subject into an oblique position these parts are visible face on. In the lateral position the ribs on opposite sides of the body overlap. In the living subject the lower ribs are overlapped by the dense liver.

The costal cartilages although visible in the skeleton are translucent and therefore not visible in radiographs unless they are calcified.

If the facts outlined above are remembered radiography of the thorax may be improved. It must be emphasized, however, that radiography of the chest with its organs is completely different from radiography of the ribs sternum and thoracic vertebrae.

S. 105   TERMS USED IN STUDY OF THE THORAX

angle of sternum	intercostal space
angle of a rib	jugular notch
body of sternum	manubrium sterni
cervical rib	neck of rib
chondrosternal joint	rib or costa
corpus sterni	shaft or rib
costa, pl. costae	sternum, pl. sterna
costal, adj.	sternal, adj.
costal cartilage	sternal angle
costotransverse joint	sternochondral joint
costosternal joint	sternoclavicular joint
costovertebral joint	sternocostal joint
costal arch	sternomanubrial joint
ensiform process	suprasternal notch
facet	thorax, adj. thoracic
false rib	true rib
fovea, foveae, a pit	tubercle of rib
gladiolus of sternum	xiphoid process
head of rib	xiphisternal joint

NOTES

NOTES

F. 10-1   ASPECTS (VIEWS) OF THE SKULL: SUTURES:

(1) Frontal view;

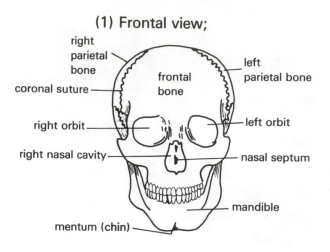

right parietal bone
coronal suture
right orbit
right nasal cavity
frontal bone
left parietal bone
left orbit
nasal septum
mandible
mentum (chin)

(2) Lateral view;

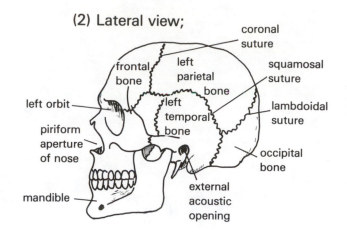

coronal suture
frontal bone
left parietal bone
squamosal suture
left orbit
left temporal bone
lambdoidal suture
piriform aperture of nose
occipital bone
mandible
external acoustic opening

(3) Vertex;

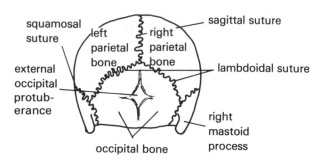

frontal bone
coronal suture
left parietal bone
right parietal bone
sagittal suture
lambdoidal suture
occipital bone

(4) Posterior view;

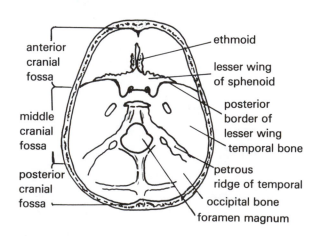

squamosal suture
left parietal bone
right parietal bone
sagittal suture
external occipital protuberance
lambdoidal suture
right mastoid process
occipital bone

(5) Basal view;

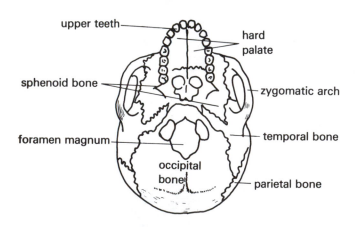

upper teeth
hard palate
sphenoid bone
zygomatic arch
foramen magnum
temporal bone
occipital bone
parietal bone

(6) Basal view;
— from inside of skull —

anterior cranial fossa
ethmoid
lesser wing of sphenoid
middle cranial fossa
posterior border of lesser wing
temporal bone
posterior cranial fossa
petrous ridge of temporal
occipital bone
foramen magnum

10. THE SKULL

BONES OF THE SKULL AND THEIR PARTS

I. CRANIUM: 8 BONES

frontal (1)   temporal (2)
parietal (2)   sphenoid (1)
occipital (1)   ethmoid (1)

II. VISCERAL CRANIUM: 14 BONES

maxillae (2)   palatine (2)
zygomatic (2)   inferior conchae
nasal (2)   or turbinates (2)
lacrimal (2)   vomer (1)
                     mandible (1)

III. OTHER BONES:

hyoid bone (1)   (malleus (2)
auditory ossicles;   (incus (2)
                     (stapes (2)

IV. TEETH — (32)

REGIONS OF THE SKULL (ASPECTS) 6

frontal, anterior or facial (1)
lateral, right and left (2)
posterior or occipital (1)
vertical, superior or vertex (1)
basal or inferior (1)

SUTURES OF THE SKULL:

coronal suture (1)
sagittal suture (1)
lambdoidal suture (1)
squamosal sutures (2)
less obvious sutures — several

PARANASAL SINUSES: 4 pairs

frontal sinuses (2)
ethmoidal sinuses (2)
maxillary sinuses (2)
sphenoidal sinuses (2)

CRANIAL FOSSAE: #3

anterior cranial fossa (1)
middle cranial fossa (1)
posterior cranial fossa (1)

OTHER GENERAL TERMS:

bregma (1)
lambda (1)
orbits (2) — cavities
nasal opening, piriform aperture
nasal septum (1)
nasal cavities (2)
mouth or oral cavity
mentum, chin (1)
zygomatic arch (2)
external acoustic opening
external acoustic meatus
mastoid process
external occipital protuberance
squama or squamous part
processes of cranial bones

I. CEREBRAL CRANIUM: 8 bones

FRONTAL BONE: (1)

1. **Squamous part; — squama**
   frontal eminence
   supraorbital margins (2)
   supraorbital notches (2)
   glabella (1)
   frontal sinuses (2)

2. **Orbital parts**
   ethmoidal notch
   orbital plates

3. **Nasal part**
   nasal spine

PARIETAL BONES (2)
parietal eminences

OCCIPITAL BONES (1)
foramen magnum
1. **Squamous part**
   external occipital protuberance
   internal occipital protuberance

2. **Lateral parts; (2)**
   occipital condyles (2)
   hypoglossal canals (2)

3. **Basilar part** (basal) (1)

TEMPORAL BONES: (2)
mastoid process
styloid process
1. **Squamous part;**
   zygomatic process
   mandibular fossa

2. **Petrous part;**
   internal ear
   internal acoustic opening
   internal acoustic meatus
   carotid canal

3. **Tympanic part;**
   external acoustic opening
   external acoustic meatus

SPHENOID BONE: (1)

1. **Body;**
   optic canal (foramen)
   sella turcica
   dorsum sella
   pituitary fossa (hypophyseal)
   posterior clinoid processes
   middle clinoid processes
   carotid grooves (2)
   sphenoidal sinuses (2)

2. **Lesser wings of sphenoid** (2)
   anterior clinoid processes

3. **Greater wings** of sphenoid (2)
   foramen rotundum (2)
   foramen ovale (2)

4. **Pterygoid processes** (2)
   medial; lateral plates

Superior orbital fissure (2)

ETHMOID BONE: (1)

1. **cribriform** or horizontal plate
   crista galli

2. **Perpendicular plate** (1)

3. **Ethmoidal labyrinths** (2)

II. VISCERAL CRANIUM: 14 bones

MAXILLAE: (2) each has;—
(1) **Body;**
    maxillary sinus
    infraorbital margin
    infraorbital foramen
    anterior nasal spine
    maxillary tuberosity
    nasal opening, aperture

(2) **Processes;**
    frontal process
    zygomatic process
    alveolar process
    palatine process

ZYGOMATIC BONES: (2);—
(1) Body
(2) Processes;   maxillary
                frontal
                temporal

NASAL BONES: (2)

LACRIMAL BONES: (2)
    lacrimal groove

PALATINE BONES: (2)
(1) **Horizontal plate**
(2) **Perpendicular plate**

INFERIOR CONCHAE (2)
    inferior turbinate bones

VOMER: (1)

MANDIBLE: (1)

(1) **Body;**
    angles
    symphysis
    mental protuberance
    mental foramina (2)
    alveolar part

2. **Ramus** (2)
    coronoid processes
    condylar process (neck & head or condyle)
    mandibular notches
    mandibular foramina
    mandibular canals

III. OTHER BONES:
    Hyoid (1) body & cornua
    auditory ossicles; 3 pairs
    malleus, incus, stapes

IV. THE TEETH: 20 deciduous
                      32 permanent
    Parts of a tooth;—
    crown, neck, root
    Structure of a tooth;—
    pulp cavity, dentine, enamel, cement, apical
    foramen

Classification of teeth; permanent —

Molar	12	canine	4
premolar	8	incisor	8

S. 107   THE DIVISIONS OF THE SKULL

The skeleton of the head has been named the skull, or cranium, or calvaria. (G. kranion, L. calvaria). It includes twenty-two bones, some paired and some single. For descriptive purposes it has been divided into two parts, the cerebral cranium and the visceral cranium or facial skeleton. These parts are not separate from each other, some bones of the cerebral cranium being joined to some of the facial bones.

The cerebral cranium is the part of the skull that encloses and protects the brain. It is sometimes referred to as the brain case. It has eight bones:

one frontal bone	two temporal bones
two parietal bones	one sphenoid bone
one occipital bone	one ethmoid bone

The visceral cranium or facial skeleton includes the fourteen bones of the face. It forms cavities (orbits) that protect the eyes, nasal passages, and mouth. (L. viscus = organ), adj. visceral — referring to an organ. There are:

two maxillae	two palatine bones
two zygomatic bones	two inferior conchae
two nasal bones	one vomer
two lacrimal bones	one mandible

The student should obtain a skull with the vertex or crown sawn through so that it may be removed and replaced as desired. The sutures should be outlined with India ink or other suitable marker. Using this skull and the diagrams included in the chapter the locations of the various bones and joints that are visible on the surface should be identified and memorized.

Some of the cranial bones are not visible or labelled on the surface; these have been left for the detailed study to follow.

S. 108   THE JOINTS OF THE SKULL

1. These lie between adjacent bones and are of a special type with interlocking or overlapping edges named **sutures**. In addition a synovial joint is present between the mandible and cerebral cranium on each side. The main sutures are:

The coronal suture, between the frontal and two parietal bones; corona = a crown.

The sagittal suture, between the two parietal bones; sagitta = an arrow.

The lambdoidal suture, between the occipital and parietal bones; (G. letter lambda).

The squamosal sutures, right and left, between the parietal and temporal bones on each side; squamous = a scale. A thin flat plate.

Other minor sutures, named from the bones they separate are not included in this study.

S. 109   ASPECTS, SURFACES, OR REGIONS OF THE SKULL

When viewed from the outside there are six surfaces or aspects presented: 10-1

1. The frontal, anterior, or facial aspect is the part that is visible when viewed from the front — frontal view. Some of the facial and the frontal bones are visible.

2. **The lateral aspect,** surface or region, is the left or right surface visible when the skull is viewed from the side.

3. **The posterior** or occipital aspect, region or surface, is the posterior surface of the skull which is visible from the rear.

4. **The vertical aspect,** surface or region is the top or crown of the skull, visible from above (also named the vertex).

5. **The basal aspect,** or inferior surface, is the part visible from below, a basal view.

The student should become familiar with the bones and sutures visible on each of these projections, since illustrations of the skull are often shown from these surfaces.

In the detailed study of the bones of the skull certain structures are referred to again and again. In order to avoid repetition some of them are defined in this section.

The orbits are the two cone shaped cavities for the eyeballs.

The lateral canthus, or external canthus, is the point at the outer margin of the eye where the upper and lower eyelids meet.

The nasal cavities are the two chambers of the nose; a right and a left.

The nasal septum is the partition between the two nasal cavities.

The piriform opening is the pear shaped hole between the two maxillary bones into the nasal cavities. (The anterior nasal opening.)

The paranasal sinuses are cavities within some of the cranial bones. There are four pairs. They are located close to the nasal cavities so are named paranasal (para = beside). Each of them opens into one of the nasal cavities and each contains air.

The mouth is the expanded cavity at the upper end of the digestive tract. It opens behind into the oral pharynx (throat). (L. os = an opening, a mouth).

The mentum is the chin, the prominence on the anterior part of the mandible at the midline.

The external acoustic meatus is the passage leading into the ear and ear drum.

The external acoustic opening is the opening leading into the ear and ear drum.

The zygomatic arch is a slender bridge of bone that extends on each side from in front of the external acoustic opening to the zygomatic bone below, and lateral to, the orbit.

The mastoid process is a bony prominence that extends down from the lateral surface of the skull behind the ear.

The external occipital protuberance is a small but definite prominence on the occipital bone in the midline, where this bone begins to curve forwards to help form the base of the skull.

The squamous part of a bone is the flat part. The term is applied to the flat parts of the frontal, occipital, and temporal bones. (L. squama = a scale, i.e. flat).

The processes of the facial bones are named from the bone with which that process forms a joint, e.g. the frontal process of the maxilla, articulates with the frontal bone.

The fontanelles are the unossified area at the junctions where three bones of the skull meet in the infant. The two important ones lie one at each end of the sagittal suture. These areas are soft to the touch.

The bregma is that point on the vertex of the skull where the sagittal suture meets the coronal suture. Three cranial bones, the frontal and two parietals, are in contact here.

The lambda is that point at the posterior end of the sagittal suture where the occipital and two parietal bones are in contact.

The cranial fossae are three depressions on the inner surface of the floor of the skull: the anterior, middle, and posterior cranial fossae. Each extends from one lateral wall of the cranium to the other lateral wall. Specific lobes of the cerebrum occupy these fossae.

The basal foramina are openings in the floor or base of the skull through which nerves and veins leave the cranial cavity or arteries enter it.

F. 10-2 THE PARANASAL SINUSES:

(1) — Frontal view —

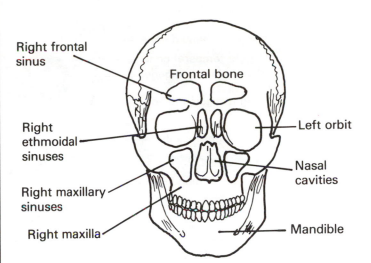

Right frontal sinus
Frontal bone
Right ethmoidal sinuses
Left orbit
Right maxillary sinuses
Nasal cavities
Right maxilla
Mandible

(2)- Sagittal section — from left side

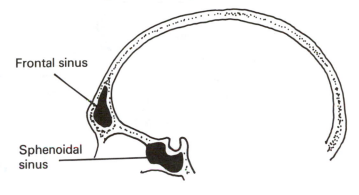

Frontal sinus
Sphenoidal sinus

(3) Paranasal sinuses — relation to skin surface

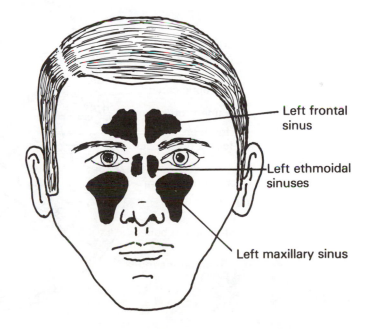

Left frontal sinus
Left ethmoidal sinuses
Left maxillary sinus

F. 10-3   THE SKULL
— frontal view —

** = a bone

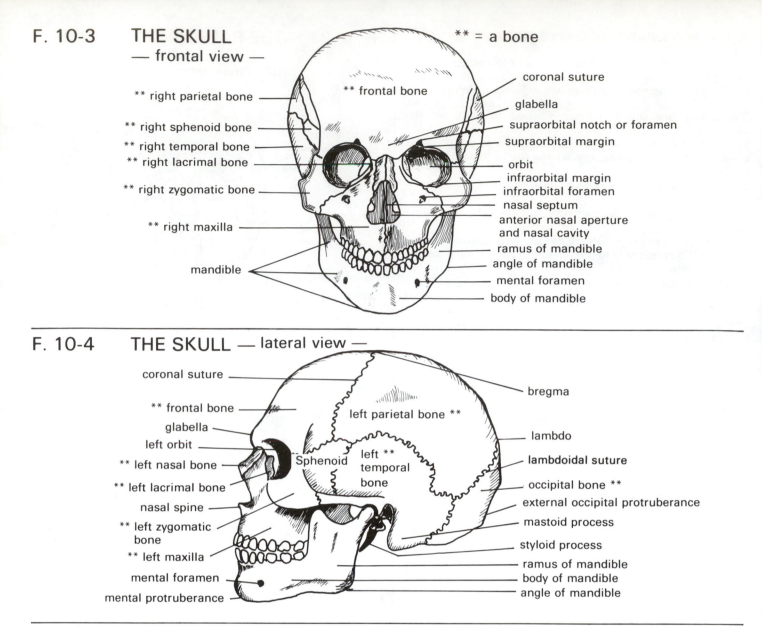

- ** right parietal bone
- ** right sphenoid bone
- ** right temporal bone
- ** right lacrimal bone
- ** right zygomatic bone
- ** right maxilla
- mandible

- ** frontal bone
- coronal suture
- glabella
- supraorbital notch or foramen
- supraorbital margin
- orbit
- infraorbital margin
- infraorbital foramen
- nasal septum
- anterior nasal aperture and nasal cavity
- ramus of mandible
- angle of mandible
- mental foramen
- body of mandible

F. 10-4   THE SKULL — lateral view —

- coronal suture
- ** frontal bone
- glabella
- left orbit
- ** left nasal bone
- ** left lacrimal bone
- nasal spine
- ** left zygomatic bone
- ** left maxilla
- mental foramen
- mental protruberance

- left parietal bone **
- Sphenoid
- left ** temporal bone
- bregma
- lambdo
- lambdoidal suture
- occipital bone **
- external occipital protruberance
- mastoid process
- styloid process
- ramus of mandible
- body of mandible
- angle of mandible

F. 10-5   THE SKULL — base from inside of the skull —

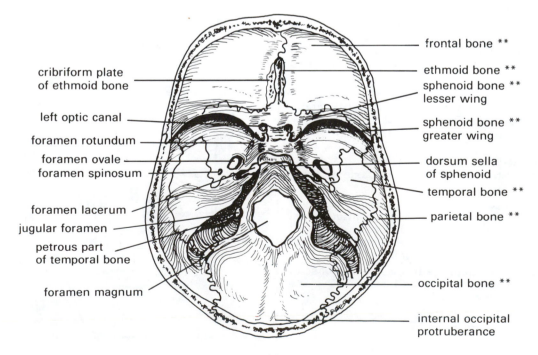

- cribriform plate of ethmoid bone
- left optic canal
- foramen rotundum
- foramen ovale
- foramen spinosum
- foramen lacerum
- jugular foramen
- petrous part of temporal bone
- foramen magnum

- frontal bone **
- ethmoid bone **
- sphenoid bone ** lesser wing
- sphenoid bone ** greater wing
- dorsum sella of sphenoid
- temporal bone **
- parietal bone **
- occipital bone **
- internal occipital protruberance

DETAILED STUDY OF BONES OF THE SKULL

The skull or cranium includes the twenty-two bones that form the skeleton of the head. The terms skull and cranium are used interchangeably. The skull, for descriptive purposes, is divided into two parts — the cerebral cranium and visceral cranium. These parts are not separated but some of the bones of one part articulate with bones of the other section. The bones of the vault are classified as flat bones.

The cerebral cranium is that part of the skull that encloses and protects the brain. It is sometimes referred to as the brain case or calvaria. It has eight bones: one frontal, two parietal, one occipital, two temporal, a sphenoid, and an ethmoid bone.

The visceral cranium or facial skeleton includes the fourteen bones of the face. It forms cavities that protect the eyes, nasal passages, and mouth. (L. viscus = an organ) adj. visceral = relating to an organ. There are two maxillae, two zygomatic, two nasal, two lacrimal, two palatine bones, two inferior conchae, a single vomer, and a mandible.

S. 112 **THE CEREBRAL CRANIUM: 8 CRANIAL BONES**

1. **THE FRONTAL BONE**   F. 10-6 (1), (2)

This single bone forms the forehead and part of the vertex of the skull. In the fetus there are two frontal bones that meet at the midline. A suture named the **metopic** suture lies between the two parts. It disappears as development progresses. The frontal bone has three parts: a squamosal part, an orbital part, and a nasal part.
(1) **The squamous part** forms the forehead and part of the vertex. Its special features include the frontal eminences, supraorbital margins, supraorbital notches, glabella, and frontal sinuses.
The frontal eminences are two rounded prominences on the squamous part of the frontal bone, one on each side of the midline, visible on the forehead.
The supraorbital margins, right and left, are ridges of bone that form the superior borders of the circular opening into each orbit. They are covered by the eyebrows.
The supraorbital notch (or foramen) is a small notch or sometimes a foramen located towards the medial end of each supraorbital margin. A nerve to the face is transmitted through this.
The glabella is a smooth flat area between the medial ends of the supraorbital margins, and above the upper ends of the nasal bones.
The frontal sinuses are paired cavities in the squamous part of the frontal bone above the supraorbital margins. If present they can be small, large, or unequal in size and may extend into orbital plates of the frontal bone above each orbit.
(2) **The orbital part** of the frontal bone consists of two flat plates of bone, called the orbital plates. They form the roofs of the orbits and the floor of the anterior cranial fossa. They extend posteriorly from the supraorbital margins of the squamous part and join the sphenoid bone. The orbital part has an ethmoidal notch.
The ethmoidal notch is a definite gap between the medial borders of the two orbital plates. The cribriform plate of the ethmoid bone fits into this space. On the inferior surfaces of each margin of the notch are several partial cavities. These together with similar cells of the ethmoid bone help to form some of the cells of the ethmoidal sinuses.

F. 10-6   THE FRONTAL BONE
(1) anterior view

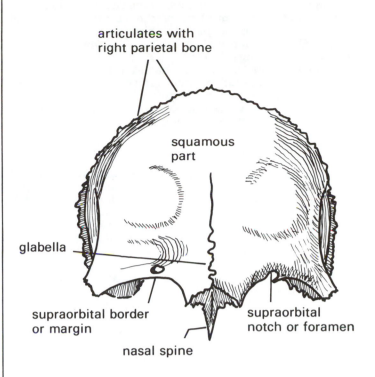

articulates with right parietal bone

squamous part

glabella

supraorbital border or margin

nasal spine

supraorbital notch or foramen

(2) viewed from inside and below

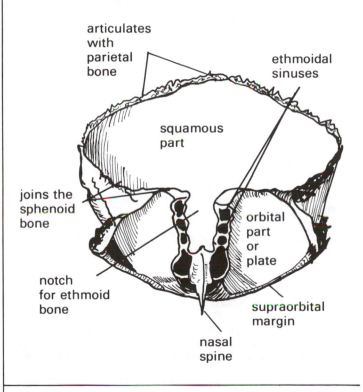

articulates with parietal bone

ethmoidal sinuses

squamous part

joins the sphenoid bone

orbital part or plate

notch for ethmoid bone

supraorbital margin

nasal spine

(3) **The nasal part** (third part) of the frontal bone extends down in the midline between the supraorbital margins. This extension articulates with the nasal and lacrimal bones and frontal processes of the maxillae.
The nasal spine is a pointed process that forms a small part of the nasal septum behind the nasofrontal articulations.

2. THE PARIETAL BONES   F. 10-7

The two parietal bones, left and right, form part of the lateral walls and vertex or roof of the skull. These paired bones meet at the midline of the vertex to form the sagittal suture. Each parietal bone is rectangular in shape. On its inner surface are grooves for branches of the middle meningeal artery.

The parietal eminence, or parietal tuberosity, is a rounded prominence on the lateral surface of each parietal bone. The distance between the two eminences is the greatest transverse diameter of the skull, important in childbirth. (L. paries = a wall).

3. THE OCCIPITAL BONE   F. 10-8 (1) (2)

This single bone forms part of the posterior wall, and the posterior part of the base of the cranium. The upper margin on each side joins a parietal bone to form the lambdoidal suture. (L. occiput = back part)

The foramen magnum is a large opening in the inferior part of the occipital bone through which the medulla oblongata of the brain joins the spinal cord. It is formed entirely by the occipital bone.

The occipital bone consists of four parts, the squamous part, two lateral parts, and the basilar part.

(1) **The squamous part** (squama = flat plate L.) is the flat curved part of the occipital bone. It forms the inferior part of the posterior wall of the cranium and curves forward to help form the base.

The external occipital protuberance is a definite rounded bony prominence. It is located at the midline of the external surface of the squamous part where it begins to curve forwards. It is easily palpable.

The internal occipital protuberance is a similar prominence on the internal surface of the squamous part opposite the external protuberance.

(2) **The two lateral parts** of the occipital bone lie one on either side of the foramen magnum between the squamous part posteriorly and the basal part in front. The occipital condyles and the hypoglossal canals are located on these lateral parts.

The two occipital condyles, right and left, are oval shaped prominences located on the inferior surfaces of each lateral part. They have smooth surfaces that articulate with the superior articular processes of the atlas.

The two hypoglossal canals are short passages from each lateral margin of the foramen magnum, close to its front end, that open below in front of the condyles. The hypoglossal or twelfth cranial nerves leave the cranium by these canals.

(3) **The basilar (basal) part** of the occipital bone is that part anterior to the foramen magnum. It unites with the sphenoid bone anteriorly and helps to form the base of the skull.

4. THE TEMPORAL BONES   F. 10-9 (1) (2)

The two temporal bones, a right and a left, form a part of each lateral wall and part of the base of the skull. (L. tempus = time). The name is perhaps applied to this bone as the hair becomes grey over these parts of the scalp first signifying the passage of time. Each temporal bone consists of a squamous part, a petrous part, and a tympanic part.

The mastoid process is a large bony prominence that extends down from the temporal bone posterior to the ear. It contains small cavities, or cells, that open into the middle ear by which infection of the middle ear may sometimes spread to the mastoid. This process is palpable on each side.

The styloid process is a long slender sliver-like process that extends down from each temporal bone. It is not

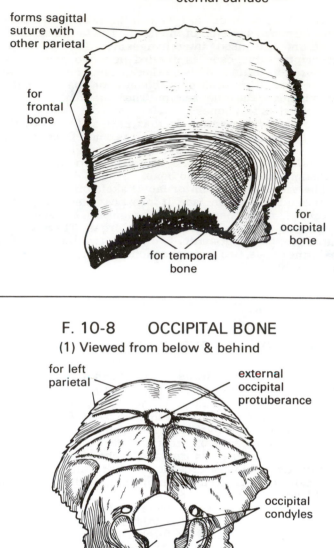

F. 10-7   THE LEFT PARIETAL BONE
— eternal surface —

forms sagittal suture with other parietal

for frontal bone

for occipital bone

for temporal bone

F. 10-8   OCCIPITAL BONE
(1) Viewed from below & behind

for left parietal

external occipital protuberance

occipital condyles

foramen magnum

articulates with sphenoid bone

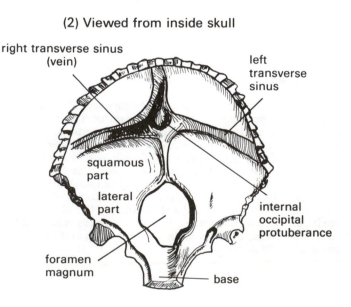

(2) Viewed from inside skull

right transverse sinus (vein)

left transverse sinus

squamous part

lateral part

internal occipital protuberance

foramen magnum

base

F. 10-9　LEFT TEMPORAL BONE

(1) lateral view

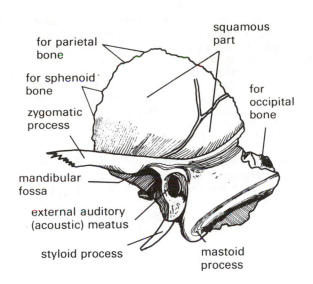

for parietal bone

squamous part

for sphenoid bone

zygomatic process

for occipital bone

mandibular fossa

external auditory (acoustic) meatus

styloid process

mastoid process

(2) left temporal bone from inside skull

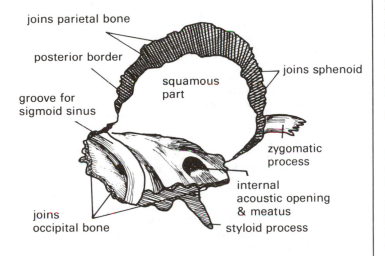

joins parietal bone

posterior border

squamous part

joins sphenoid

groove for sigmoid sinus

zygomatic process

internal acoustic opening & meatus

joins occipital bone

styloid process

palpable but is often visible in lateral radiographs of the skull lying behind the ramus of the mandible and parallel to this bone.

(1) **The squamous part** of the temporal bone is the flat part on the lateral side of the skull above the ear. This part has a zygomatic process and a mandibular fossa.

The zygomatic process is a slender bony process that extends horizontally forward from the opening of the ear. It arises from the squamous part of the temporal bone. It meets a similar process that extends back from the zygomatic bone to form the zygomatic arch. It is named zygomatic because it joins the zygomatic bone.

The mandibular fossa is a depression on the temporal bone in front of the opening into the ear and below the origin of the zygomatic process. The condyle (head) of the mandible fits into it to form the temporomandibular joint. (Joint of the lower jaw).

(2) **The petrous part** of the temporal bone extends medially in the floor of the cranium from the squama to the basal part of the occipital bone.

It is shaped somewhat like a pyramid and is sometimes called the petrous pyramid. Its medial end, or apex, is pointed. Its posterior margin articulates with the squamous and the basal parts of the occipital bone. Its anterior margin joins the greater wing of the sphenoid. When viewed from the inside of the skull it presents a sharp ridge reaching from the squamous part to its apex. This ridge marks the separation between the middle and posterior cranial fossae on either side. The petrous part has within it the parts of the ear concerned with hearing and equilibrium. It is composed of hard dense bone (G. petra = a rock, i.e. a hard object). This petrous part has an internal acoustic opening, an internal acoustic meatus, and a carotid canal.

The internal acoustic opening, or porus, is an opening on the dorsal surface of the petrous part close to its medial end. Its leads into the internal acoustic canal.

The internal acoustic meatus is a canal leading from the internal acoustic opening to the inner ear. The vestibulocochlear, or acoustic nerve, enters this passage.

The carotid canal is a short passage that begins as a circular opening on the inferior surface of the petrous pyramid close to its medial end. It passes through the petrous part to enter the cranium at the foramen lacerum in front of the apex of the petrous part. The internal carotid artery passes through this canal to reach the cranial cavity.

(3) **The tympanic part** of the temporal bone forms the anterior and inferior walls of the external acoustic meatus as well as a part of the posterior wall. It also helps to form the mandibular fossa. Its junction with the mastoid process is marked by a vertical line in front of the mastoid process. (G. tympanon = a drum).

The mastoid process is sometimes described as a fourth part of the temporal bone. Its anterior part is derived from the squama, its posterior part from the petrous part.

5. **THE SPHENOID BONE**　F. 10-10 (1) (2)

This single bone helps to form a small part of the lateral wall on each side, as well as part of the base of the skull. It is difficult to visualize as the front of it is hidden by the facial bones. It is wedged between the orbital plates of the frontal bone anteriorly and the squamous and petrous parts of the temporal bones posteriorly. The basal part of the occipital joins it behind. The sphenoid bone has been compared to a bat with two pairs of extended wings. It must be examined from the inside and outside of the skull, along with diagrams F. 10 - (1) and (2). (G. sphen = a wedge + oid = like, i.e. like a wedge).
The sphenoid bone has a body, a pair of greater wings, a pair of lesser wings, and two pterygoid processes.

(1) **The body** of the sphenoid bone is the central part and lies in the midline of the base of the skull, in front of the basal part of the occipital bone. Its anterior part lies horizontally and joins the ethmoid bone. The body of this bone has several features that require further description:

The chiasmatic groove is a horizontal depression that crosses the body of the bone in front of the sella turcica. The optic chiasma (crossing of optic nerves) occupies this groove.

The optic canals (OT. optic foramina) are two short passages extending anteriorly and laterally one from each side of the body of the sphenoid bone. A lesser wing lies lateral to each of them. The optic canals open into the posterior ends of the orbits and carry the optic nerves and ophthalmic vessels from the cranium to the orbits. (G. opitkos = eye).

The sella turcica is a saddlelike bony depression on the upper surface of the body of the sphenoid. Its deep central

F. 10-10   THE SPHENOID BONE

(1) viewed from behind & above

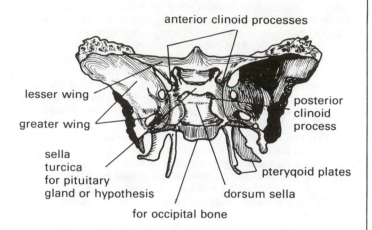

anterior clinoid processes

lesser wing

greater wing

sella
turcica
for pituitary
gland or hypothesis

for occipital bone

dorsum sella

posterior
clinoid
process

pterygoid plates

(2) sphenoid bone to outline its foramina

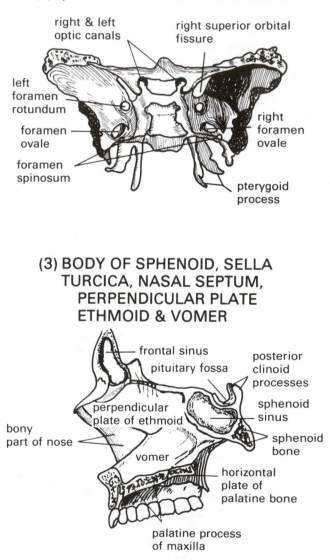

right & left
optic canals

right superior orbital
fissure

left
foramen
rotundum

foramen
ovale

foramen
spinosum

right
foramen
ovale

pterygoid
process

(3) BODY OF SPHENOID, SELLA TURCICA, NASAL SEPTUM, PERPENDICULAR PLATE ETHMOID & VOMER

frontal sinus

pituitary fossa

posterior
clinoid
processes

perpendicular
plate of ethmoid

bony
part of nose

vomer

sphenoid
sinus

sphenoid
bone

horizontal
plate of
palatine bone

palatine process
of maxilla

cavity contains the pituitary gland. (L. sella = saddle + Turcica = Turkish; a Turkish saddle).
The dorsum sella is the posterior wall of the sella turcica. It curves upwards similar to the back of a saddle.

(L. dorsum = back + sella = saddle; i.e. back of a saddle).
The pituitary fossa (NA. hypophyseal fossa) is the central cavity of the sella turcica. It contains the pituitary gland (hypophysis) (Hypo = below + physis, to grow; something growing down).
The posterior clinoid processes are two small rounded prominences that are located one at each lateral end of the upper margin of the dorsum sella.
The middle clinoid processes lie one on each side of the upper anterior surface of the sella turcica, medial to and posterior to the opening of the optic canal. These may be difficult to locate except on fresh specimens.
The carotid groove is an often poorly defined groove that passes vertically along each lateral surface of the body of the sphenoid, and accommodates the internal carotid artery.
The sphenoidal air sinuses are two cavities within the body of the sphenoid bone one on either side of the midline, and separated by a bony septum. They lie posterior to the nasal cavities and open into these. In a lateral radiograph of the skull they will be located under the sella turcica, one superimposed upon the other.

(2) **The lessser wings** of the sphenoid bone, i.e. the minor wings, right and left, extend laterally from the horizontal anterior part of the body of this bone. Each is triangular in shape and each joins an orbital plate of the frontal bone anteriorly. The posterior margin of each lesser wing is its free border. This free margin forms the anterior limit of the middle cranial fossa. The inferior surface of each lesser wing forms the upper margin of the superior orbital fissure at the posterior end of each orbit.
The anterior clinoid processes are two bony processes that extend posteriorly from the medial end of the posterior free margin of each lesser wing. They lie lateral to the openings of the optic canals. Students should note that there are three pairs of clinoid processes.

(3) **The greater wings** (major wings), right and left, are two winglike bony plates that originate one from each lateral surface of the body of the sphenoid. Each one forms a small part of the floor and lateral wall of the skull. Each one is visible on the lateral aspect of the cranium anterior to the squamous part of the temporal bone. The frontal and zygomatic bones lie anterior to it. If viewed from the inside of the skull the squamous part of the temporal bone is lateral to it while the petrous pyramid lies posterior to it. The foramen rotundum and foramen ovale lie in this greater wing.
The foramen rotundum (round opening) is a small opening in the medial basal part of the greater wing just lateral to the superior orbital fissure. It transmits the maxillary nerve (a branch of the trigeminal or fifth cranial nerve) out of the cranium.
The foramen ovale (oval opening) is a larger opening in the basal part of the greater wing lateral to and behind the foramen rotundum. It transmits the mandibular branch of the trigeminal nerve out of the cranium.
The superior orbital fissure is a comma shaped opening on the medial margin of each greater wing. The lesser wing forms a roof, and the body forms a medial margin. It opens in front into the posterior end of the orbit lateral to the optic canal. Three cranial nerves pass by this opening to the muscles of the eye.

(4) **The pterygoid processes** right and left extend down vertically from the inferior surface of the body of the sphenoid bone on each side. Each consists of a medial and a lateral plate of bone that are joined together anteriorly. A well marked groove separates the two plates behind. They are about one inch in vertical length and lie one on each lateral wall of a nasal cavity posteriorly. Each articulates with the posterior margin of the adjacent maxilla. To locate them the skull must be examined on its posterior aspect. (G. pteryx = a wing + oid = like).

F. 10-11   THE ETHMOID BONE

(1) Diagram of ethmoid to show parts

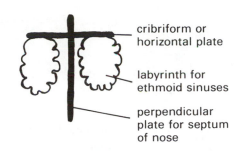

cribriform or horizontal plate

labyrinth for ethmoid sinuses

perpendicular plate for septum of nose

(2) Ethmoid bone from behind

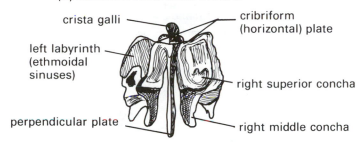

crista galli

left labyrinth (ethmoidal sinuses)

perpendicular plate

cribriform (horizontal) plate

right superior concha

right middle concha

(3) Ethmoidal bone — from above

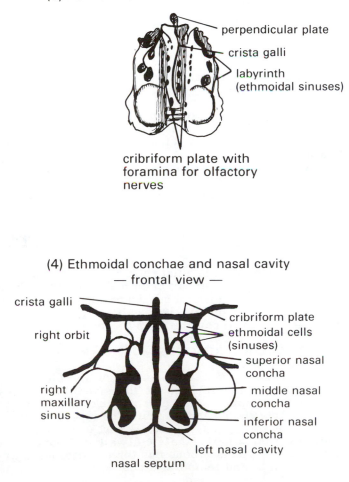

perpendicular plate

crista galli

labyrinth (ethmoidal sinuses)

cribriform plate with foramina for olfactory nerves

(4) Ethmoidal conchae and nasal cavity
— frontal view —

crista galli

right orbit

right maxillary sinus

cribriform plate

ethmoidal cells (sinuses)

superior nasal concha

middle nasal concha

inferior nasal concha

left nasal cavity

nasal septum

6. THE ETHMOID BONE   F. 10-11 (1, 2, 3, 4)

The single ethmoid bone lies in the anterior part of the base of the skull between the orbits. Like the sphenoid bone it is only partly visible because of adjacent bones. It has a cribriform plate, a perpendicular plate, and two labyrinths. (G. ethmos = a sieve + oid = like; i.e. sievelike).

(1) **The cribriform** plate of the ethmoid bone must be studied using a skull with skull cap removed, as it is visible only from inside the skull. It is wedged between the two orbital plates of the frontal bone in the ethmoidal notch on the floor of the anterior cranial fossa. It is a small area and contains several minute openings by which the branches of the olfactory nerves (nerves of smell) enter the cranium. These openings can be seen if the skull is held so that light enters the nasal cavities from the facial aspect. These small openings serve to identify the cribriform plate positively. (L. cribrum = a sieve + form = like; i.e. like a sieve).
The crista galli is a very thin plate of bone that passes vertically up from the upper surface of the cribriform plate. The falx cerebri is attached to it. (L. crista = a crest + Gallus = a rooster; i.e. a rooster's comb).

(2) **The perpendicular plate** of the ethmoid is a flat sheet of bone that extends **down** in the midline from the cribriform plate and forms a part of the nasal septum. It can be visualized by looking into the nasal cavities from the facial aspect. F. 10-11.

(3) **The ethmoidal labyrinths**, right and left, are suspended from the inferior surface of the cribriform plate. The lateral wall of each labyrinth helps to form the medial wall of the orbit.
The ethmoidal sinuses, right and left are multiple small cavities, or cells in the labyrinths. There are three groups — anterior, middle, and posterior bilaterally. A part of the wall enclosing these cells is formed by one or other adjacent bone. These sinuses are apparent in radiographs of the skull between the medial surface of the orbit and lateral wall of the nasal cavity on each side. The sinuses open into the nasal cavities.
The superior and middle nasal conchae are bony shelves that extend into the nasal cavities from the lateral walls. They are parts of the ethmoid bone. See also inferior concha, under — the facial bones. (L. concha = a shelf, a curved shelf).

S. 113   THE VISCERAL CRANIUM

The visceral cranium or facial skeleton has fourteen bones, six paired, and two single bones, See Section 110.

1. THE MAXILLAE   F. 10-12 (1, 2, 3)

Maxilla = jaw bone; pl. maxillae, adj. maxillary.
There are two maxillae, a right and a left. The two maxillary bones unite at the midline in early life and only remnants of the suture line can be identified. Since the appearance in the adult skull suggests a single bone the student must remember that there are two of them joined together. Both have similar parts, identified as right or left. Together the maxillae make up a large part of the facial skeleton. They reach from the infraorbital margins to the upper teeth and roof of the mouth. They form parts of:
    the floor of each orbit
    the medial part of each infraorbital margin
    the lateral wall of each nasal cavity
    the floor of each nasal cavity
    the roof of the mouth
Each maxilla has a body and four processes.

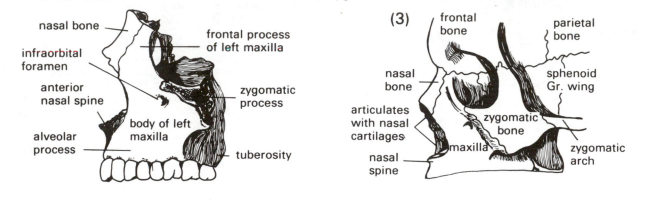

F. 10-12 (2) Lateral views left maxilla, zygomatic bone; nasal and lacrimal bone

(1) **The body** of each maxilla is a boxlike structure forming part of each cheek, and lying on the lateral side of the nose. Its special features include the maxillary sinus, the infraorbital margin, the infraorbital foramen, the anterior nasal spine, and the maxillary tuberosity.

The maxillary sinus is a large cavity within the body of the maxilla. It contains air and has an opening into the corresponding nasal cavity. It is visible as a dark shadow on films.

The infraorbital margin is a ridge of bone that forms the lower border of the orbital opening. Its medial half is formed by the maxilla, its lateral part by the zygomatic bone.

The infraorbital foramen is a small opening on the anterior surface of the maxilla just below the infraorbital margin.

The anterior nasal spine is a single small pointed projection in the midline at the base of the nasal septum. It can be palpated with the thumb in one nasal cavity and the index finger in the other one.

The maxillary tuberosity is a small rounded bony prominence posterior to the third upper molar tooth. (Remove mandible for proper viewing).

(2) **Processes of the maxilla:** there are four processes, frontal, zygomatic, alveolar, and palatine. Like those of the frontal bone, these are named from the bone with which each articulates.

The frontal process extends up along the lateral side of the nose posterior to the nasal bone, and along the medial margin of the orbit. It joins the frontal, nasal and lacrimal bones.

The zygomatic process extends up and laterally from the body of the maxilla to meet the maxillary process of the zygomatic bone. It helps to produce the prominence of the cheek here.

The alveolar process extends down from the inferior part of the body of the maxilla to form sockets for eight upper teeth. With its mate it forms the horseshoe shaped alveolar arch.

The palatine process is a flat shelf of bone that extends medially from the lower part of the body of the maxilla to meet its mate and form all but the posterior part of the hard palate. The palate is the bony partition that separates the nasal cavities from the mouth.

2. THE ZYGOMATIC BONES   F. 10-12 (1, 2, 3)

There are two zygomatic bones, a right and a left. Each of them is quadrilateral in shape (four sided). Each forms the prominent upper lateral part of each cheek, lateral to the orbit. It is visible and palpable. The superior margin is concave and forms the lateral part of each orbital margin as well as part of the lateral wall and floor of each orbit. Each zygomatic bone has three processes, maxillary, frontal, and temporal. Each is named from the bone to which it is joined.

The **maxillary process** at its lower margin joins the maxilla.

The **frontal process** joins the frontal bone at the upper lateral margin of the orbit.

The **temporal process** joins the zygomatic process of the temporal bone to form the long slender zygomatic arch.

3. THE NASAL BONES   F. 10-12 (1, 2, 3)

There are two nasal bones, a right and a left. The two of them form the bony part of the bridge of the nose. Each is a small flat rectangular shaped bone that joins the nasal part of the frontal bone above, and the nasal cartilages below. The two nasal bones meet at the midline and extend laterally to join the frontal process of the maxilla. The septum of the nose lies behind the joint between the two nasal bones. The distal part of the nose is not bony, but is formed of cartilages which join the bony parts along the margins of the piriform opening into the nasal cavities.

4. THE LACRIMAL BONES   F. 10-12 (1, 2, 3)

The paired right and left lacrimal bones are two small flat bones located in the medial walls of the orbits. Each lies posterior to the frontal process of the maxilla with which it articulates. There is a groove where the two join called the lacrimal groove. This groove is occupied by the lacrimal sac above and below by the nasolacrimal duct which opens into a nasal cavity. Tears are carried by this channel into the nasal cavity. (L. lacrima — a tear).

5. THE PALATINE BONES   F. 10-13, 14

There are two palatine bones, a right and a left. They are named palatine as they help to form the hard palate, the bony part of the roof of the mouth. In order to examine them the skull should be held so that the posterior openings of the nasal cavities can be seen from the rear. Each palatine bone is an "L" shaped structure, and includes horizontal and vertical plates of bone.

The **vertical part** forms the dorsal part of the lateral wall of the nasal cavity.

The **horizontal plate** extends medially from the inferior part of the vertical plate to meet the horizontal plate of the other palatine bone. Together they join the palatine processes of the maxillae to form the posterior part of the hard palate. The suture line may be visible when the palate is examined from below.

6. THE INFERIOR CONCHAE   F. 10-11, 15

Inferior concha or inferior turbinate bone; (L. concha = a shell) s. concha, pl. conchae.

There are two inferior conchae, a right and a left, one within each nasal cavity. Each is a thin curved plate of bone that extends medially like a shelf into the nasal cavity from the lateral wall. Its medial margin is free in the nasal cavity. There are also superior and middle conchae in each nostril. These are not separate bones but form parts of the ethmoid. The three shelves so formed in the lateral part of each nasal cavity divide this cavity into 3 partial compartments or meatuses. The conchae can be visualized from the facial aspect of the skull. They can also be identified in radiographs of the skull in posteroanterior projections.

7. THE VOMER   F. 10-11

L. vomer = a ploughshare. The vomer is a single flat four sided bone plate that resembles a ploughshare. It lies vertically within the nose and forms the lower posterior part of the nasal septum. Its lower margin rests upon the upper surface of the hard palate. It lies below and partly behind the perpendicular plate of the ethmoid bone that

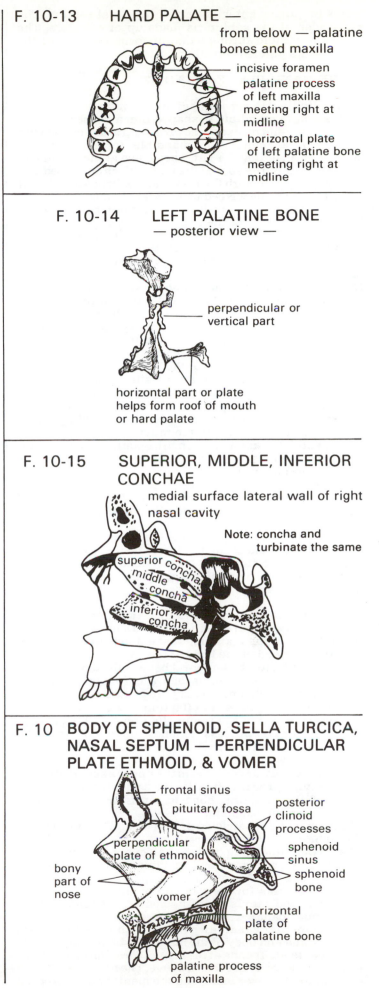

F. 10-13   HARD PALATE —
from below — palatine bones and maxilla

- incisive foramen
- palatine process of left maxilla meeting right at midline
- horizontal plate of left palatine bone meeting right at midline

F. 10-14   LEFT PALATINE BONE
— posterior view —

- perpendicular or vertical part
- horizontal part or plate helps form roof of mouth or hard palate

F. 10-15   SUPERIOR, MIDDLE, INFERIOR CONCHAE

medial surface lateral wall of right nasal cavity

Note: concha and turbinate the same

superior concha
middle concha
inferior concha

F. 10   BODY OF SPHENOID, SELLA TURCICA, NASAL SEPTUM — PERPENDICULAR PLATE ETHMOID, & VOMER

- frontal sinus
- pituitary fossa
- posterior clinoid processes
- sphenoid sinus
- sphenoid bone
- perpendicular plate of ethmoid
- bony part of nose
- vomer
- horizontal plate of palatine bone
- palatine process of maxilla

forms the upper anterior part of the nasal septum. The remainder of the septum is made up of cartilage. The vomer is visible through the piriform aperture of the nose, i.e. its anterior opening.

8. THE MANDIBLE F. 10-16 (1) (2)

L. mandibula = lower jaw; adj. mandibular.

This single mandible is shaped like a horseshoe with its posterior end on each side turned up to articulate with the temporal bone. The mandible forms from two segments, a right and a left. These fuse at the midline in the first year of life so that the mandible is described as a single bone. Although the maxillae go through a similar fusion they are described as separate bones.

The mandible has a body, and two rami, right and left, that meet the body at the angles.
The angle of the mandible on each side is the prominent rounded posterior part where the horizontal section turns upward to become the vertical ramus.

(1) **The Body** of the mandible is the horizontal part extending from one angle, around to the midline and back to the angle on the other side, forming a horseshoe curve. The body has several features: the symphysis, mental protuberance, mental foramen, and the alveolar part.
The symphysis is a faint ridge running vertically in the midline anteriorly where the two segments of the mandible became fused to form a single bone.
The mental protuberance is the flat prominence at the inferior part of the symphysis anteriorly — the chin. (mentum = chin, adj. mental). This must be distinguished from L. mens, adj. mental, the mind.
The mental foramen is a small opening on the outer surface of the body of the mandible on each side below the second premolar tooth. The mental branch of the mandibular nerve emerges through this opening to reach the face. The student should note here that the opening of the parotid duct within the mouth is opposite the upper second molar tooth.
The alveolar part or the body is the upper part or border with sixteen sockets for the lower teeth. Note that this part of the mandible has been named "alveolar part" while the similar parts of the maxillae for the upper teeth are named "alveolar processes".

(2) **The ramus:** pl. rami; (L. ramus = a branch).
There are two rami here, a right and a left. The ramus is the flat part of the mandible that extends upwards from the posterior end of the body at the angle on each side. It forms a joint with the temporal bone — the temporomandibular joint. Each ramus has a coronoid process, a mandibular notch, and a condylar process.
The coronoid process is a flat thin upward extension of the anterior part of the ramus. Its upper end is pointed.
The condylar process extends upwards from the posterior part of the ramus to the adjacent joint. It consists of two parts, a head (condyle) and a neck.
The head, caput, or condyle, is the upper expanded end of the condylar process. It articulates with the mandibular fossa of the temporal bone to form the temporomandibular joint on each side.
The neck is the constricted part that connects the head to the remainder of the ramus.
The mandibular notch is a half-moon shaped notch between the coronoid and condylar processes of the ramus on each side.
The mandibular foramen is an opening on the medial surface of the ramus about its center. It opens into the mandibular canal.
The mandibular canal is a passage within the lower part of the ramus and body of the mandible, below the roots of the lower teeth. It ends at the mental foramen on the outer surface of the body described above. The nerves and blood vessels to the lower teeth, and the mental nerve proceed

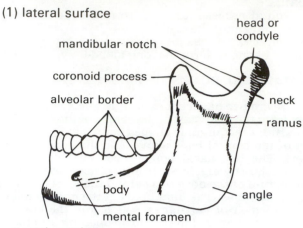

F. 10-16   LEFT HALF OF MANDIBLE
(1) lateral surface

head or condyle
mandibular notch
coronoid process
neck
alveolar border
ramus
body
angle
mental foramen
mental protuberance

(2) Medial surface left half of mandible —

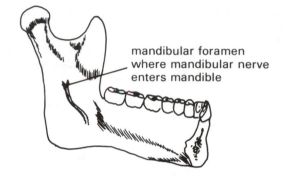

mandibular foramen where mandibular nerve enters mandible

F. 10-17   HYOID BONE
— anterior view —

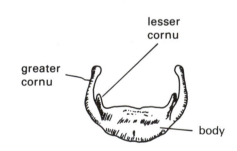

lesser cornu
greater cornu
body

THE AUDITORY OSSICLES
— GROSSLY ENLARGED —

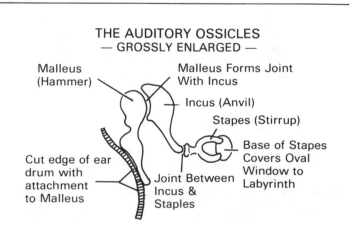

Malleus (Hammer)
Malleus Forms Joint With Incus
Incus (Anvil)
Stapes (Stirrup)
Base of Stapes Covers Oval Window to Labyrinth
Cut edge of ear drum with attachment to Malleus
Joint Between Incus & Stapes

to their terminations through this canal. The mandible is palpable throughout its length, and its protuberance is often quite prominent.

OTHER BONES RELATED TO THE SKULL

S. 114   THE HYOID BONE   F. 10-17

The hyoid bone is a single "U" shaped bone that is located in the anterior part of the neck. It lies horizontally about half way between the body of the mandible and the thyroid cartilage (Adam's apple). It has a body, and two greater and two lesser cornua (horns).
The body is the curved anterior part.
The greater cornu extends back from the posterior end of the body on each side as a bent process. There are two, a right, a left.
The lesser cornu is much smaller and extends up from the greater cornu on each side. (L. cornu = a horn, pl. cornua). This bone is visible in lateral radiographs of the cervical area, below the mandible.

S. 115   THE AUDITORY OSSICLES

L. os = bone + icle = small - a small bone.
The auditory ossicles are three small paired bones located in the middle ear within the petrous parts of the temporal bones. These bones have been named:
the malleus = a hammer
the incus = an anvil
the stapes = a stirrup
They have been named from their resemblance to a hammer, an anvil, and a stirrup. The malleus is attached at one end to the inner surface of the ear drum (tympanic membrane). The stapes has its foot rest in contact with a membrane covering an opening in the internal ear. The incus is placed between the malleus and stapes in such a manner that together these bones form a system of levers. Sound vibrations are conveyed by them from the ear drum to the inner ear and magnified in the process.

S. 116   THE ORBIT   F. 10-3

In Section 109 the orbits were defined as the cavities in the facial skeleton for the eyeballs. There are two, a right and a left. They lie below the frontal bone and above the maxillae and the zygomatic bones. Each cavity is cone shaped, its base lying in front and its apex at the pointed posterior end. The **supraorbital margin** of the base is formed by the frontal bone, the **infraorbital margin,** by the maxilla and zygomatic bones. The zygomatic bone also forms the lateral margin, while the medial margin is formed by the frontal and maxillary bones.
The roof of the orbit is formed by the frontal and sphenoid bones, and the floor by the maxilla, zygomatic and palatine bones. The medial wall is formed by the

maxilla, lacrimal, ethmoid, and body of the sphenoid, and the lateral wall by the zygomatic and greater wing of the sphenoid. At the posterior end the opening of the optic canal is located. The comma shaped superior orbital fissure lies lateral to the optic opening. The inferior orbital fissure lies at the junction of the lateral wall and floor.

S. 117   THE NASAL CAVITIES   F. 10-3, 11 (4), 15

There are two nasal cavities, right and left. They are the chambers of the nose, extending from the nostrils or openings (nares), on the face to the nasal pharynx, or throat, behind. Some authorities describe this structure as a single nasal cavity with a septum dividing it into two parts. The lateral walls were described with the ethmoid bone as having three shelves: the superior, middle, and inferior conchae. These shelves partly divide each nasal cavity into three passages: the superior, middle, and inferior meatuses. Above the superior concha is a narrow space that has the nerve endings of the olfactory nerve of smell in its roof. The nasal cavities are very narrow so that air entering or leaving them is in contact with the walls, where it is warmed and moistened.
The nasal septum is a partition between the two nasal cavities extending vertically from front to back. It is formed by the perpendicular plate of the ethmoid and the vomer. A gap anteriorly is filled in by cartilage. There is a crest above where the two nasal bones meet at the midline, and another crest below where the two palatine processes of the maxillae and the two palatine bones meet at the midline. (crest — from crista = a ridge). F. 10-11
The four paranasal sinuses on each side open into the corresponding nasal cavity.

S. 118   THE PARANASAL SINUSES   F. 10-2(1, 2, 3)

The paranasal sinuses or accessory nasal sinuses are cavities within some of the cranial bones. The diploe in these bones is replaced by a cavity between the inner and outer tables of bone. They are adjacent to the nasal cavities, hence paranasal. Each of them communicates with (opens into) a nasal cavity. The lining membrane of each nasal cavity is continued into each of the sinuses. As a result of the communication with the nasal cavity each sinus contains air. There are four pairs, described as right and left, frontal, ethmoidal, maxillary and sphenoidal sinuses.
1. **The frontal sinuses,** right and left are paired cavities within the squamosal part of the frontal bone. One lies on each side of the midline in the forehead. They may be large, or small, or unequal in size. Each opens into the corresponding nasal cavity. In some subjects they are absent.

---

(1) — Frontal view —

right parietal bone
left parietal bone
coronal suture
Frontal bone
right orbit
left orbit
right nasal cavity
nasal septum
mandible
mentum (chin)

THE PARANASAL SINUSES:

(1) — Frontal view —

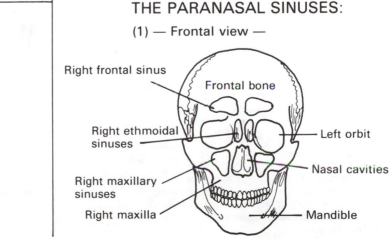

Right frontal sinus
Frontal bone
Right ethmoidal sinuses
Left orbit
Right maxillary sinuses
Nasal cavities
Right maxilla
Mandible

101

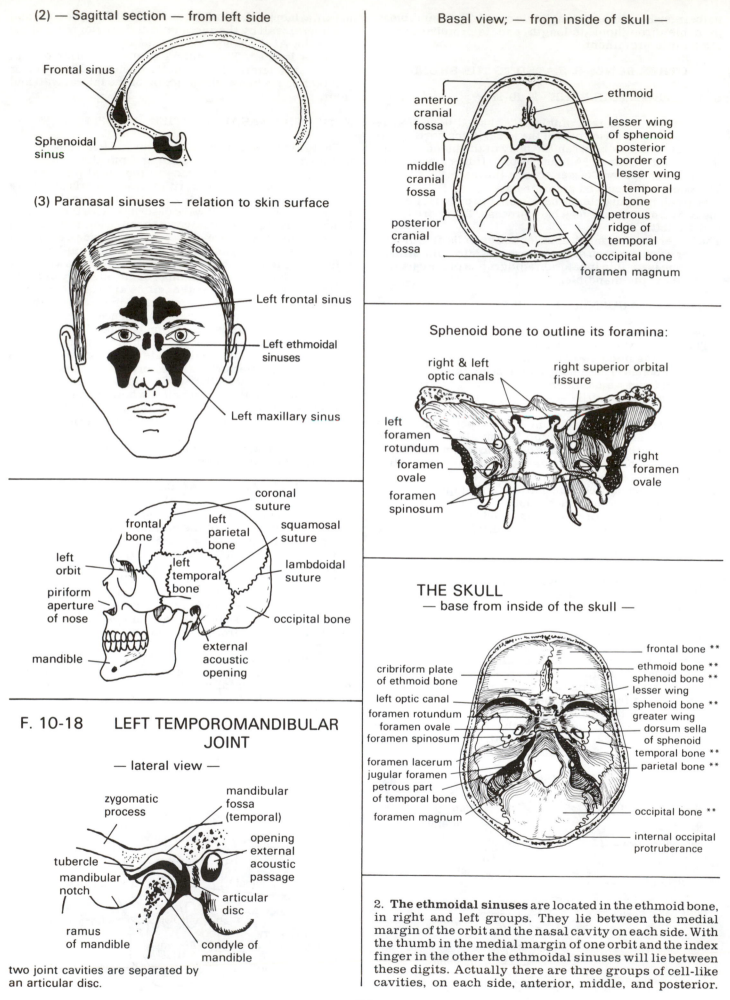

(2) — Sagittal section — from left side

Frontal sinus

Sphenoidal sinus

(3) Paranasal sinuses — relation to skin surface

Left frontal sinus

Left ethmoidal sinuses

Left maxillary sinus

coronal suture

frontal bone

left parietal bone

squamosal suture

left temporal bone

lambdoidal suture

left orbit

piriform aperture of nose

occipital bone

mandible

external acoustic opening

F. 10-18   LEFT TEMPOROMANDIBULAR JOINT

— lateral view —

zygomatic process

mandibular fossa (temporal)

tubercle

mandibular notch

opening external acoustic passage

articular disc

ramus of mandible

condyle of mandible

two joint cavities are separated by an articular disc.

Basal view; — from inside of skull —

anterior cranial fossa

ethmoid

lesser wing of sphenoid

posterior border of lesser wing

middle cranial fossa

temporal bone

petrous ridge of temporal

posterior cranial fossa

occipital bone

foramen magnum

Sphenoid bone to outline its foramina:

right & left optic canals

right superior orbital fissure

left foramen rotundum

foramen ovale

foramen spinosum

right foramen ovale

THE SKULL
— base from inside of the skull —

cribriform plate of ethmoid bone

frontal bone **

ethmoid bone **

sphenoid bone **

lesser wing

left optic canal

foramen rotundum

foramen ovale

foramen spinosum

sphenoid bone **

greater wing

dorsum sella of sphenoid

temporal bone **

parietal bone **

foramen lacerum

jugular foramen

petrous part of temporal bone

foramen magnum

occipital bone **

internal occipital protruberance

2. **The ethmoidal sinuses** are located in the ethmoid bone, in right and left groups. They lie between the medial margin of the orbit and the nasal cavity on each side. With the thumb in the medial margin of one orbit and the index finger in the other the ethmoidal sinuses will lie between these digits. Actually there are three groups of cell-like cavities, on each side, anterior, middle, and posterior.

Parts of the walls of some of these cells are formed by adjacent bones, such as the frontal.

3. **The maxillary sinuses** are two large cavities, one in each of the left and right maxillary bones behind the cheeks. Each has an opening into a nasal cavity. They are also named the antra (G. antron = a cave).

4. **The sphenoidal sinuses**, a right and a left, are cavities within the body of the sphenoid bone. They are separated by a bony septum and each one opens into a nasal cavity.

THE JOINTS OF THE SKULL

The joints of the skull include the sutures and the temporomandibular joints. The sutures were listed and defined in S. 108. A more detailed study has been undertaken in the following paragraphs.

S. 119 THE SUTURES OF THE SKULL F. 10-1

The sutures of the skull are the special interlocking joints between the various bones of the cerebral and visceral skeleton. They are classified as fibrous joints and are immovable. The more prominent sutures are defined below. Additional minor sutures named from the bones bordering them are not listed in this study.

1. **The coronal suture** crosses the vertex of the skull transversely (from side to side) from one lateral surface of the skull to the other. The frontal bone is in front of it and the anterior borders of the two parietal bones form its posterior margin. (L. corona = a crown)

2. **The sagittal suture** extends from front to back along the median line of the skull, between the adjacent superior margins of the two parietal bones. It passes from the posterior margin of the frontal bone to the pointed upper end of the occipital bone. (L. sagitta = arrow)

3. **The lambdoidal suture** is shaped like an inverted letter "V" on the posterior aspect of the skull. The posterior margins of the parietal bones lie in front of it, the occipital bone forms its posterior margin. (G. letter lambda - λ ; English letter "L").

4. **The squamosal suture** is visible on the lateral surfaces of the skull. It forms a curved line, convex above, between the inferior margin of the parietal bone above and the upper margin of the temporal bone below. There are two squamosal sutures, a right and a left. (L. squama = a scale — a thin plate).

S. 120 THE TEMPOROMANDIBULAR JOINTS
F. 10-18

There are two temporomandibular joints, a right and a left. They are synovial or diarthrodial joints, both hinge and gliding type. The head or condyle of each ramus of the mandible articulates with the mandibular fossa of the corresponding temporal bone. This synovial joint differs from some others in that an articular disc composed of cartilage is interposed between the adjacent bones. The synovial cavity is thus separated into two parts, each with a synovial lining. When the mouth is opened the articular disc moves forward to lie below the tubercle in front of the mandibular fossa. The head of the mandible also moves forward. As the mouth is closed the disc and head move backwards together. The condyle of the mandible can be palpated if the tip of a finger is placed in front of the external opening of the ear, and the mouth is opened and closed.

Note: Some temporary sutures are visible on radiographs of infants' skulls.
The metopic suture extends vertically on the squamous part of the frontal bone in the midline between the two halves of this bone. It fills in with bone and eventually disappears, as development progresses.

Two mendosal sutures pass obliquely upwards, one in each half of the squamous part of the occipital bone. These fill in with osseous tissue and disappear during development.

The top of the skull should now be removed and the cranial fossae and basal foramina can be demonstrated effectively.

S. 121 CRANIAL FOSSAE — A DETAILED STUDY
F. 10-1(6); F. 10-5

The cranial fossae were briefly described in S. 110. The base of the skull when examined from the inside presents three levels that resemble a house with split levels. Proceeding from front to back each lies at a lower level than the one in front of it. They are named the anterior, middle, and posterior cranial fossae. Each extends from the lateral skull wall on one side to the other lateral wall. Certain parts of the brain occupy each of these fossae. Certain foramina are located in the floor of each fossa, allowing the entrance of arteries to the cranial cavity, and the exit of some cranial nerves and veins.

1. **The anterior cranial fossa** lies above the orbits. Its floor is formed by the orbital plates of the frontal bone, the cribriform plate of the ethmoid, and the anterior parts of the body and the two lesser wings of the sphenoid. The orbital plates bulge into the floor on either side. The posterior border is formed by the free edges of the lesser wings. The frontal lobes of the cerebrum occupy this fossa.

2. **The middle cranial fossa** lies at a lower level posterior to the anterior one. It extends from the free margins of the lesser wings of the sphenoid posteriorly to the ridges of the petrous parts of the temporal bones. Its floor is formed by the greater wings and sella turcica of the sphenoid, and the squamous and petrous parts of the temporal bone. On either side of the body of the sphenoid it forms a cup shaped cavity that is occupied by the temporal lobes of the brain. These two cup-shaped cavities and the sella form a single middle cranial fossa, not two.

3. **The posterior cranial fossa** is a large cavity located behind the middle fossa and at the lowest level of the three, with the foramen magnum at its base. Its floor is formed by the squamous, lateral, and basal parts of the occipital bone, and the posterior surfaces of the petrous pyramids. The hindbrain occupies this large cavity.

S. 122 THE BASAL FORAMINA F. 10-5

There are several openings in the base of the skull, most of them paired, right and left. As indicated above they provide openings through which the cranial nerves and blood vessels pass to or from the cranial cavity. Some of them have been described with the individual bones of the skull.

1. **The optic canals** (OT. foramina) right and left, form short passages on the lateral margins of the body of the sphenoid under cover of the lesser wings. They are located at the posterior end of each orbit on its medial wall. Each transmits an optic nerve and ophthalmic artery to the orbit. Note that the term "canal" replaces the older term foramen.

2. **The superior orbital fissures**, right and left are comma shaped slits in the lateral wall of each orbit close to its posterior end. They are formed by the margins of the greater and lesser wings of the sphenoid bone. Several of the cranial nerves pass through them to reach the orbit. They are often visible in A.P. or P.A. views of the skull.

3. **The inferior orbital fissures,** a right and left are narrow slits between the lateral wall and the floor of the orbit. Each lies between the orbital part of the maxilla and greater wing of the sphenoid.

4. **The foramen rotundum** is a small circular opening at the medial margin of the greater wing of the sphenoid close to the medial end of the superior orbital fissue. It opens on the anterior surface of the sphenoid posterior to the maxilla, and transmits a maxillary nerve, a branch of the trigeminal, to the face. There are two of these foramina, right and left. (rotundum = round).

5. **The foramen** ovale is an oval opening in the base of the greater wing of the sphenoid bone posterior and lateral to the foramen rotundum. It is larger than the round opening. It transmits the mandibular division of the trigeminal nerve to the space in front of the ramus of the mandible. There are two of these foramina a right and a left.

6. **The foramen spinosum** is a very small circular opening in the greater wing of the sphenoid bone posterior and lateral to the large foramen ovale. On the external surface it lies close to the styloid process, a small sharp process. It transmits the middle meningeal artery to the inner surface of the skull. There are two, a right and a left.

7. **The foramen lacerum** is a ragged opening at the apex of the petrous pyramid between it and the posterolateral margin of the body of the sphenoid bone. Its medial margin is continuous with the carotid groove on the sphenoid. The internal carotid artery enters the cranium by this opening after it has passed through the carotid canal. This artery enters the base through the carotid opening below. There are two of these foramina, a right and a left.

8. **The carotid opening** is a hole on the inferior surface of the petrous pyramid about 1 cm posterolateral to the apex. It opens into the carotid canal, which in turn opens into the foramen lacerum as described above. There are two carotid openings, right and left.

9. **The carotid canal** passes from the carotid opening on the inferior surface of the petrous pyramid through this bone to the foramen lacerum. It carries the internal carotid artery. There are two, a right and a left.

10. **The jugular foramen** is a large irregular opening located between the lateral part of the occipital bone and the petrous pyramid. It is below the internal acoustic opening. The internal jugular vein leaves the cranium through this opening. There are two, right and left.

11. **The foramen magnum** is a single very large opening in the basal part of the occipital bone. The medulla oblongata of the hindbrain joins the spinal cord at this opening.

12. **The internal acoustic opening** (NA. porus) is an opening on the posterior surface of the petrous pyramid above the jugular foramen. It is about 2 cm or ¾ of an inch lateral to the petrous apex. It forms an opening into the internal acoustic or auditory meatus. The vestibulocochlear or acoustic nerve passes through the meatus to reach the internal ear. There are two such openings, right and left.

13. **The internal acoustic meatus** is a canal that passes from the internal acoustic opening into the internal ear through the petrous part of the temporal bone, carrying the vestibulocochlear nerve.

In this study the orbital fissures, carotid canal, and internal acoustic meatus have been included because of their intimate relationship to the foramina.

All of these foramina are visible in the base of the skull on the outer surface and on the inner surface when the skull cap has been removed. The student should position the skull for each foramen so that each is plainly seen. For instance the optic canal shows up most satisfactorily when the skull is turned to the oblique position.

S. 123 **DEVELOPMENT OF THE CRANIAL BONES**

In the section on development of bones intramembranous ossification was defined. The bones of the skull, with the exception of those of the base and the facial bones, are formed first as membrane followed by intramembranous type of ossification. These bones of the vault are ossified before birth. Some ossify from more than one center; the frontal bone, for instance, has a right and a left ossification center. The parts formed by each of these bones are separate at birth but fuse early and the metopic suture line disappears. Between adjacent bones the membrane persists and is visible on radiographs as a dark line. These parts become the sutures. Because of these membranes between bones the cranial bones can overlap (moulding) during the birth process so that the head becomes smaller as it passes through the birth canal.

The bones at the base of the skull and the facial bones form first as cartilages, and intracartilaginous ossification takes place in them.

S. 124 **THE FONTANELLES** F. 10-20

The fontanelles are gaps in bone formation at the angles where several bones meet. The membranes here persist and do not become converted into bone until after birth. Soft spots are palpable in these areas since membrane rather than bone is present. (Fontanel or fontanelle or fonticulus = a fountain). F. 10-20

The Anterior fontanelle or frontal fontanelle is located in the midline of the vertex of the skull where the posterior surface of the frontal bone meets the anterior and upper margins of both parietal bones. This is palpable as a diamond-shaped soft area. It fills in with bone at about eighteen months of age and is then named **the bregma.** F. 10-20

The posterior fontanelle or occipital fontanelle lies in the midline on the upper posterior surface of the skull where the pointed end of the occipital bone meets the postprior margins of the two parietal bones. It is not as large as the anterior one and ossifies at about six months of age. It becomes the **lambda.** F. 10-20

There are four other fontanelles at the inferior angles of the parietal bones where they meet the temporal bones on the same side.

Palpation of the fontanelles may be done on sick children. They become sunken when for some reason the child is not retaining fluid as in diarrhoea or vomiting. If they are bulging outwards they suggest increased pressure within the cranium.

S. 125 **CONGENITAL ANOMALIES OF SKULL**

Kephale = head; (G), hence cephale and cephalic
Acephalus = without a head
Dicephalus, bicephalus = two heads, dicephalic
Macrocephalus = a large head, macrocephalic
Hydrocephalus = fluid in the head
Microcephalus = a small head: microcephalic
Occasionally a monster of one of these types occurs and may be demonstrated on films taken during pregnancy.

F. 10-20 SKULL OF AN INFANT — VERTEX —
— for fontanelles —

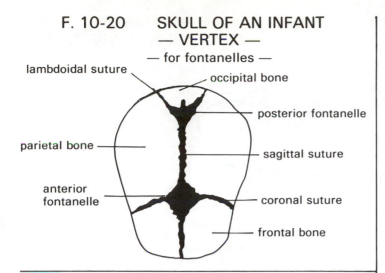

F. 10-21 LANDMARKS OF THE SKULL

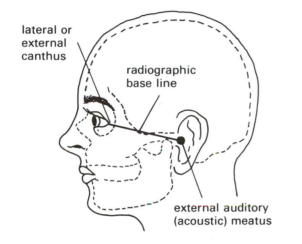

F. 10-22 INTRACRANIAL VENOUS SINUSES
— lateral view —

— from above —

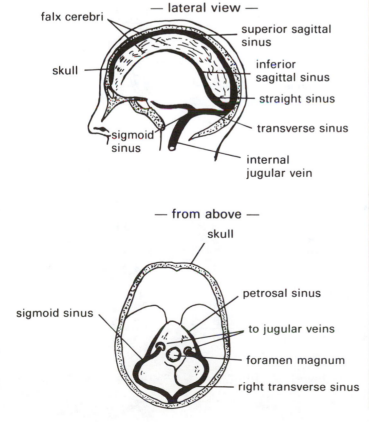

Cleft palate: during development the two halves of the soft or hard palate may fail to unite at the midline in the roof of the mouth. Fluid or food taken into the mouth will enter the nasal cavities and come out through the nostrils.

The anterior part of the alveolar process of the maxilla containing the incisor teeth forms from a separate epiphysis. This may fail to unite with the remainder of the hard palate, or may be a part of the cleft palate.

The upper lip may also fail to unite on one side or other and result in a hare lip deformity. This may accompany a cleft palate. Note: hare lip = a fissured lip like a rabbit.

S. 126 LANDMARKS AND PROMINENCES OF THE SKULL

The external occipital protuberance, mastoid processes, openings of the acoustic meatuses, the temporomandibular joints, orbits, zygomatic bones and arches, and the mandible are either palpable or visible.

The lateral canthus of the eye is the point lateral to the eye where the upper and lower eyelids meet, a landmark in radiography.

The radiographic base line extends from the lateral canthus to the opening (porus) of the external acoustic meatus.

The interpupillary or interorbital line is a line drawn from one pupil to the other.

S. 127 ANATOMICAL PECULIARITIES OF THE SKULL APPLIED TO RADIOGRAPHY

Intracranial grooves are present on the internal surface of the lateral wall of the skull extending up and laterally from the foramen spinosum. These are occupied by branches of the middle meningeal artery which supply blood to the dura mater and the internal surface of cranial bones. The middle meningeal artery is a branch of the external carotid artery. As the skull is thinned out along the paths of the branches the grooves show up as dark lines on films.

Much larger grooves are present along the paths of the venous sinuses of the skull. These are well marked, passing transversely across the occipital bone and down posteriorly to the petrous ridge of the temporal bone. In radiographs they should be well outlined as dark images.

The student should obtain a skull with skull cap removed and verify the observations listed here.

As the skull is a hollow globe, it is impossible to obtain radiographs of most parts without superimposition of other parts. There are a few structures however that may be filmed without overlapping provided that the head is placed in positions that will allow this. The mandible, mastoid cells, zygomatic arches and the teeth are included in this group. The petrous parts of the temporal bones, the external occipital protuberance and the ridges on either side of it are particularly troublesome because of their density and position.

With the skull cap removed the student can study the effect on the frontal and maxillary bones of having the nose and forehead touching the table top, then the nose and chin on the table top.

In skull radiography positions have been devised to minimize the degree of overlapping and to prevent the thick parts from obscuring the structures being filmed. The student may with the help of the skull visualize why some of these positions have been adopted.

A true lateral view of the skull:
place the cranium in the lateral position by having the interpupillary or interorbital line perpendicular to the table top,
the median line of the face parallel to the table top,
the sagittal suture of the skull viewed from the head end parallel to the table top.

A true anteroposterior or posteroanterior view will be obtained if:

the sagittal suture (median line) on the top of the head is perpendicular to the table top;

the radiographic base line (orbitomeatal line) is also perpendicular to the table top.

Using the naked skull these observations can be verified and followed.

S. 128 THE TEETH F. 10-19 (1) (2) (3)

A.S. toth = a tooth; L. dens = tooth, adj. dental. The average subject acquires two sets of teeth in a life time. Twenty baby or deciduous or milk teeth are replaced by a further set of thirty-two permanent teeth, sixteen in the maxillae, and sixteen in the mandible.

S. 129 THE PARTS OF A TOOTH F. 10-19, 2. 3

Each tooth has **a crown**, a neck and one or more roots. **The crown** is the exposed part, visible in the mouth.

Cusps are small rounded prominences on the surfaces of the premolar and molar teeth. They are separated from each other by shallow grooves. The number of cusps varies from one to five on each crown. The cusps grind food.

The neck is the constricted part where the crown and root meet.

The root, the smaller tapered end, is located in the alveolar process of the maxilla or the alveolar part of the mandible. The number of roots varies with the type of tooth. Most teeth have one root but some have two and some three roots. The pointed end of each root is named the apex, and has an opening or foramen by which a nerve and artery enter and a vein leaves the cavity within the tooth.

The lingual surface of a tooth is that aspect facing the tongue. (L. lingua = a tongue).

The labial or buccal surface is its outer aspect facing the lip or cheek. (L. labium = lip + bucca = cheek).

S. 130 THE STRUCTURE OF A TOOTH F. 10-19, 2, 3

Each tooth has a pulp cavity, a dentine layer and a covering of enamel or cement.

(1) **The pulp cavity**, (NA, dental cavity) is the central cavity within a tooth extending from the crown down through the root. It ends at a small opening at the apex (tip) of the root. The cavity contains connective tissue, a nerve and blood vessels. It could be compared to the medullary cavity in long bones with bone marrow and vessels.

(2) **The dentine** is modified bone that surrounds the pulp cavity in both the crown and root. It makes up most of the solid part of each tooth.

(3) **The enamel** is a thin covering that forms an outer coat for **the crown**. Enamel is the hardest substance in the human body.

(4) **The cement** (cementum) is a layer of modified bone that covers the dentine of the root.

(5) **The apical foramen** is a minute opening in the tip of the root by which a nerve and an artery enter, and a vein leaves the pulp cavity. In the mandible these structures reach the tooth by passing through the mandibular canal to reach the small foramina.

(6) **The peridental membrane** is a layer of fibrous tissue that surrounds the root of each tooth and attaches it to the wall of the socket in which the tooth lies.

F. 10-19   RIGHT PERMANENT
(1) UPPER & LOWER TEETH
— outer surfaces —

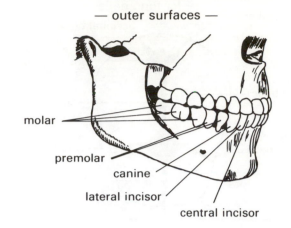

molar
premolar
canine
lateral incisor
central incisor

(2) PARTS & STRUCTURE OF A TOOTH

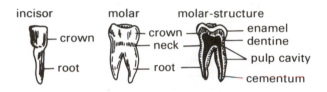

incisor — crown, root

molar — crown, neck, root

molar-structure — enamel, dentine, pulp cavity, cementum

(3) INCISOR TOOTH-STRUCTURE
— lateral view —

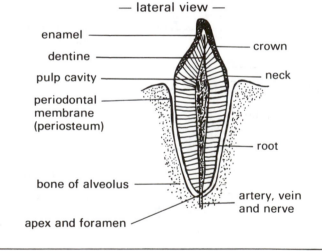

enamel
dentine
pulp cavity
periodontal membrane (periosteum)
bone of alveolus
apex and foramen
crown
neck
root
artery, vein and nerve

S. 131 CLASSIFICATION OF TEETH F. 10-19 (1)

	Right	Left	Total
DECIDUOUS:	M P C I	I C P M	
Upper —	2 0 1 2	2 1 0 2	10
Lower —	2 0 1 2	2 1 0 2	10
	4 0 2 4	4 2 0 4	20
PERMANENT:	M P C I	I C P M	
Upper —	3 2 1 2	2 1 2 3	16
Lower —	3 2 1 2	2 1 2 3	16
	6 4 2 4	4 2 4 6	32

M = molar, P = premolar, C = canine, I = incisor

S. 132 THE DECIDUOUS TEETH

Each child develops a set of diciduous or temporary or baby teeth. No premolars and only two molars develop on each side, upper and lower. These deciduous teeth develop in the sockets of the maxillae and mandible, and 5 (five) to each side upper and lower. Each of them erupts or breaks through the bone and gum. The approximate time of eruption is fairly constant for each tooth, and the sequence of the eruption is fairly definite. There is however some variation in the time of eruption for both sets. While these deciduous teeth will fall out eventually they should be cared for in order to have them remain in place, preserving the space until the permanent tooth is ready to erupt.

S. 133 THE PERMANENT TEETH   F. 10-19

Sixteen of the thirty-two permanent teeth are imbedded in the alveolar processes of the maxilla, and sixteen in the alveolar part of the mandible. There are eight right and eight left in each instance. The permanent teeth gradually replace the deciduous teeth by pushing them out. The sequence of eruption is fairly constant. The first molars, the so-called six year molars, appear behind the second deciduous, or baby molars. These are permanent teeth not deciduous and should not be neglected.

The pattern is similar on both sides and in both upper and lower jaws with two incisors, one canine, two premolars, and three molars in each group totalling eight teeth or thirty-two in all.

The incisor teeth are in the front of the mouth. There are eight incisors, four upper and four lower, with two on either side of the midline. The incisor teeth adjacent to the midline are named **central incisors**, while the incisors lateral to the central are named **lateral incisors**. The incisors have a sharp cutting edge in order to bite through food. Each has a single root.

The canine, cuspid or dog teeth lie one on the lateral side of each lateral incisor tooth. There are four canine teeth, a single right and left in both upper and lower jaws. Each canine has a single root. The edge of the crown has a blunt, rounded, pointed cusp. In carniverous animals the canine teeth are longer than the incisors to help grasp and secure their prey. (L. canius = dog).

The premolar teeth, or bicuspid teeth, lie posterior to the canines, and anterior to the molars. There are eight premolars, two right and two left in both upper and lower jaws. The anterior one is named the **first premolar**, the hind one the **second premolar**. Each premolar tooth has two cusps, hence the name bicuspid. Each has usually one root. (pre = in front of + molar).

The molar teeth lie posterior to the premolars. (L. mola = a mill, i.e. a grinder) There are twelve of them, three right and three left in both upper and lower jaws. Commencing with the molar next to the second premolar they are named as first, second, and third molars, right or left, upper or lower. The first molar is often the first permanent tooth to erupt. It appears behind the last deciduous molar tooth of the infant. It is frequently mistaken for a deciduous tooth and allowed to decay without attention.

The third molar or wisdom tooth is the last permanent tooth to erupt and may appear at 20 to 25 years of age. Sometimes it is directed towards the front of the mouth and impinges on the second molar. It may not erupt. The number of cusps on the molars varies from three to five. The anterior molars may have the most. The upper molars usually have three roots, the lower molars have two. The roots may spread out or may be fused together.

The molar and premolar teeth because of their cusps chew or grind food. The tongue pushes food between the upper and lower jaws when the mouth is partly open. Closing the mouth grinds or masticates food.

Occlusion refers to the position of the upper and lower teeth in relation to each other when the mouth is closed. The upper incisors normally project in front of the lower incisors when the mouth is closed. The cusps of the upper premolar and molar teeth fit into the grooves between the cusps of the lower opposing teeth, or the gaps between them. In malocclusion the cusps and grooves do not fit into each other, or in some instances opposing teeth may not touch each other. This deformity may occur following a fracture of the mandible or maxilla and must be corrected if possible.

S. 134 THE FUNCTIONS OF TEETH

To summarize;
incisors bite off mouthfuls of food;
premolars and molars grind, or masticate food.
The object is to break food into small pieces so that the digestive juices may come into intimate contact with food particles when it reaches the digestive organs.

S. 135 TERMS USED TO DESCRIBE THE SKULL

alveolar process	frontal bone
angle of mandible	glabella
antrum, antra	greater wing of sphenoid
anterior clinoid process	hydrocephalus
anterior cranial fossa	hyoid bone
anterior nasal spine	incus
accessory nasal sinuses	incisor teeth
acoustic canal	infraorbital margin
anterior fontanel	infraorbital foramen
(fontanelle)	internal acousitc
auditory canal	opening (porus)
auditory ossicles	internal acoustic meatus
base of skull	internal ear
basilar part of	interorbital line
occipital bone	interpupillary line
basilar foramina	inferior concha or
bregma	turbinate
canine tooth	labyrinth of ethmoid bone
capitulum of mandible	lacrimal bone
carotid canal	lambda
foramen	lambdoidal
groove	lambdoidal suture
canal optic	lateral canthus of eye
cerebral cranium	lesser wing of the
cheek bone	sphenoid bone
clinoid process	malar bone
condylar process of	malleus
mandible	mandible
coronal suture	mandibular
cranial fossae	mandibular canal
cranium	foramen
crista galli	notch
cusp	mastoid cells
deciduous teeth	mastoid process
dorsum sella	maxilla, maxillae
ethmoid bone	maxillary
ethmoidal labyrinth	maxillary sinus
ethmoidal notch	maxillary tuberosity
ethmoidal sinuses	median line
external acousitc	mentum - chin
meatus	mental - re: chin
external ear	mental foramen
external occipital	mental protubernace
protuberance	middle clinoid processes
facial bones	middle cranial fossa
fontanel or	middle ear
fontanelle	midsagittal plane
foramen magnum	midsagittal section
ovale	molar tooth
rotundum	nasal bones
spinosum	nasal septum

nasal cavities
nasolacrimal groove
occipital bone
occipital condyles
optic canal (foramen)
orbit
orbital
orbitomeatal line
orbital part of
   frontal bone
palatine bones
paranasal sinuses
parietal bones
parietal eminence
perpendicular plate
   of ethmoid
petrous part of
   temporal bone
piriform opening
pituitary fossa
pituitary gland
posterior clinoid
   processes
premolar teeth
ramus of mandible

sagittal suture
squamous part of
   frontal bone
   occipital bone
   temporal bone
squama
squamosal suture
stapes
submental
supraorbital margin,
   border
suture
symphysis of mandible
temporomandibular joint
posterior cranial fossa
temporal bone
tooth - dens
tympanic part of
   temporal bone
vertex of skull
visceral cranium
vomer
zygomatic bone
zygomatic arch
zygomatic process

S. 136 ANALYSIS OF SOME SKELETAL ANATOMICAL TERMS

*   Scaphoid = boatlike
    Lunate; luna = moon
    Triquetrum = triangle
    Pisiform = pea shaped
    Trapezium = 4 sided, as a table
    Trapezoid = like the trapezium
    Capitate = head shaped
    Hamate = a hook
    Radius = a spoke
    Clavicle = a key
    Phalanx = a line of soldiers, rectangle
    Fibula = a clasp or buckle
    Tibia = a flute
    Cuboid = cubelike
    Cuneiform = wedge shaped
*   Navicular = boat shaped
    Vertebra = a joint, a segment
    Manubrium = a handle
    Xiphoid = sword like, point of sword
    Parietal; paries = a wall
    Temporal; tempus = time
    Occipital; from caput = head
    Sphenoid = wedgelike
    Ethmoid = sievelike
*   Scaphoid (NA) bone of wrist;
*   Navicular (NA) bone of foot

S. 137 SOME BONES WITH MORE THAN ONE NAME

Scaphoid (NA), carpal navicular (OT)
Lunate (NA), semilunar (OT)
Triquetrum (NA), triangular (OT)
Trapezium (NA), greater multangular (OT)
Trapezoid (NA), lesser multangular (OT)
Capitate (NA), os magnum (OT)
Hip, os coxae, innominate
Calcaneus or os calcis
Talus (NA), astragalus (OT)
Navicular (NA), scaphoid of foot (OT)
First cervical = atlas
Second cervical = axis
Seventh cervical = vertebra prominens
Zygomatic (NA), malar (OT)
Inferior concha (NA), inferior turbinate (OT)
Maxillary sinus (NA), antrum, pl. antra. (OT)

Thumb or pollex
Great toe or hallux

S. 138 PLURAL FORMS AND ADJECTIVES

Scapula	scapulae	scapular
Clavicle	clavicles	clavicular
Brachium	brachia	brachial
Humerus	humeri	humeral
Antebrachium	antebrachia	antebrachial
Radius	radii	radial
Ulna	ulnae	ulnar
Carpus	carpi	carpal
Metacarpal b	metacarpal bs	metacarpal
Phalanx	phalanges	phalangeal
Pelvis	pelves	pelvic
Sacrum	sacra	sacral
Coccyx	coccygeal bone	coccygeal
Ilium	ilia	iliac
Ischium	ischia	ischial
Pubis	pubes	pubic
Femur	femora	femoral
Tibia	tibiae	tibial
Fibula	fibulae	fibular
Tarsus	tarsi	tarsal
Metatarsal b	metatarsal bs	metatarsal
Phalanx	phalanges	phalangeal
Vertebra	vertebrae	vertebral
Costa	costae	costal
Sternum	sterna	sternal
Mandible	mandibles	mandibular

S. 139 UPPER AND LOWER LIMBS COMPARED

SHOULDER GIRDLE
   clavicle
   scapula

PELVIC GIRDLE
   right hip bone
   left hip bone
PELVIS:
   sacrum & coccyx
   hip bones; Rt. & Lt.

ARM: BRACHIUM
   humerus

THIGH:
   femur
   patella

FOREARM:
ANTEBRACHIUM:
   radius
   ulna

LEG:
   tibia
   fibula
TARSUS:
TARSAL BONES

WRIST: CARPAL BONES:
   scaphoid bone
   lunate bone
   triquetral bone
   pisiform bone
   trapezium
   trapezoid bone
   capitate bone
   hamate bone

   calcaneus, os calcis
   talus, astragalus
   medial cuneiform, 1st
   intermediate
     cuneiform, 2nd
   lateral cuneiform, 3rd
   cuboid bone

INSTEP:
METATARSAL B

PALM: METACARPAL B
   first metacarpal
   second metacarpal
   third metacarpal
   fourth metacarpal
   fifth metacarpal

   first metatarsal
   second metatarsal
   third metatarsal
   fourth metatarsal
   fifth metatarsal
DIGITS TOES
PHALANGES

DIGITS & PHALANGES
   thumb, pollex or
     first digit
   index finger, or
     second digit
   middle finger, or
     third digit
   ring finger, or
     fourth digit

   great toe, hallus,
     or first digit
   second toe or
     second digit
   third toe or
     third digit
   fourth toe
     or fourth digit

little finger, or	fifth or little toe or	shoulder (humeral)	hip
fifth digit	fifth digit	elbow	knee
thumb - proximal &	great toe - proximal	proximal radioulnar	proximal tibiofibular
distal phalanges	and distal phalanges	distal radioulnar	distal tibiofibular
fingers - proximal,	other toes - proximal	wrist (radiocarpal)	ankle
middle, distal	middle, distal	intercarpal	intertarsal
phalanges	phalanges	carpometacarpal	tarsometatarsal
SESAMOID BONES	**SESAMOID BONES**	intermetacarpal	intermetatarsal
		metacarpophalangeal	metatarsophalangeal
JOINTS:	**JOINTS:**	interphalangeal;-	interphalangeal; -
sternoclavicular	sacroiliac	proximal	proximal
acromioclavicular	symphysis pubis	distal	distal

S. 140   MEDICAL TERMS — altered by a prefix — (for reference only)

articulation	—periarticular	—around, about a joint; periarticular hemorrhage
articulation	—interarticular	—between joints, interarticular hemorrhage
os	—periosteum	—covering of bone (peri= around + os=bone)
os	—interosseous	—between (among) bones — interosseius membrane
median	—paramedian	—beside the median line of the body
spine	—supraspinous	—above a spine or spinous process
spine	—infraspinous	—below a spine or spinous process
spinatus	—supraspinatus	—above a spine or spinous process; supraspinatous muscle
spinatus	—infraspinatus	—below a spine or spinous process; infraspinatous muscle
scapula	—suprascapular	—above the scapula; suprascapular notch
scapula	—infrascapular	—below or under the scapula
scapula	—periscapular	—around the scapula
clavicle	—supraclavicular	—above the clavicle; supraclavicular lymph nodes
clavicle	—infraclavicular	—below the clavicle; infraclavicular lymph nodes
clavicle	—subclavian	—below the clavicle; subclavian artery
acromion	—subacromial	—below the acromion; subacromial bursa
glenoid	—subglenoid	—below the glenoid cavity of the humerus
coracoid	—subcoracoid	—below the coracoid process
caput	—subcapital	—below a head; subcapital fracture
tubercle	—intertubercular	—between tubercles; intertubercular groove of humerus
deltoid	—subdeltoid	—below the deltoid
condyle	—epicondyle	—on or upon a condyle; epicondyle of humerus
condyle	—intercondylar	—between condyles; intercondylar fossa of femur
condyle	—supracondylar	—above a condyle; supracondylar fracture of humerus
trochlea	—supratrochlear	—supratrochlear; above a trochlea (humerus or talus)
carpal	—metacarpal	—beyond the carpus; metacarpal bones
carpus	—intercarpal	—between the carpal bones; intercarpal joints
phalanges	—interphalangeal	—between the phalanges; proximal interphalangeal joint
crest	—intercristal	—between two crests; intercristal line
trochanter	—intertrochanteric	—between the two trochanters; intertrochanteric crest
pubis	—suprapubic	—above the pubic bones
pubis	—subpubic	—below the pubic bones; subpubic arch
patella	—suprapatellar	—above the patella suprapateller bursa
patella	—prepatellar	—in front of the patella; prepatellar bursa
patella	—infrapatellar	—below the patella; infrapatellar bursa
tarsus	—metatarsal	—beyond the tarsal bones; metatarsal bones
tarsus	—intertarsal	—between the tarsal bones; intertarsal joints
vertebra	—intervertebral	—between the vertebrae; intervertebral disc
vertebra	—paravertebral	—beside a vertebra; paravertebral ligaments
sacrum	—presacral	—in front of the sacrum; presacral air insufflation
sternum	—suprasternal	—above the sternum; suprasternal notch
sternum	—retrosternal	—behind the sternum; retrosternal goitre
sternum	—substernal	—below the sternum
costa	—intercostal	—between ribs; intercostal muscle, artery etc.
costa	—subcostal	—below a rib, subcostal plane
thorax	—transthoracic	—across the thorax
thorax	—hemithorax	—half the thorax; right hemithorax
nose	—paranasal	—beside the nose; paranasal sinuses
maxilla	—submaxillary	—below the maxilla
mandibular	—submandibular	—below the mandible; submandibular gland (salivary gland)
mentum	—submental	—under the chin; submental lymph glands:
orbit	—supraorbital	—above the orbit; subraorbital margin (border)
orbit	—infraorbital	—below the orbit; infraorbital foramen
dens	—peridental	—around the tooth; peridental membrane
parietal	—biparietal	—two parietal bones; biparietal diameter of skull
occipital	—suboccipital	—below the occipital bone; suboccipital nerve
sella turcia	—suprasellar	—above the sella turcia; suprasellar tumour
temporal	—infratemporal	—below the temporal bone; infratemporal fossa

NOTE: FOR REFERENCE AND PRACTICE IN USING PREFIXES————DO NOT MEMORIZE.

NOTES

11. THE MUSCULAR SYSTEM

GENERAL TERMS DEFINED

Muscular tissue was described in Section 24; this should be reviewed. Muscles may be classified according to location in the body, microscopic appearance, or nervous control. In the latter two there is overlapping, resulting in some confusion. The classification according to location is the most reasonable one and is employed in this study.

CLASSIFICATION OF MUSCLES BY LOCATION

1. Skeletal muscles — usually attached to bones.

2. Visceral muscles — in the walls of some organs.

3. Cardiac muscles — in the heart, the myocardium.

CHARACTERISTICS

1. Skeletal, striated, voluntary

2. Visceral, nonstriated, involuntary

3. Cardiac, striated, involuntary

Visceral and cardiac muscles are discussed under the study of the heart, and of the many organs in their proper place.

Skeletal muscles, the red or lean meat of the body, make up almost half of the body weight. They are usually attached at their ends to bones and usually cross a joint. Their contractions cause movement at that joint. Skeletal muscles are usually controlled by the cerebrum, the conscious part of the brain. Nerve fibers pass from the cerebrum through the spinal cord and out to the muscles. By impulses relayed from the cerebrum these muscles are made to contract at will; hence the name **voluntary**, (from voluntas = will). In addition, when examined through a microscope, alternating light and dark bands can be seen passing across the muscle fibers transversely; hence the name **striated**, (Stria = a band or line).

All muscles are composed of cells, which are also called muscle fibers. In skeletal muscles the fibers are cylindrical. Each fiber may be very short or may measure up to 4-5 cm or about 1.5 inches in length. Each fiber has a cell membrane surrounding it, with several nuclei lying on the inner surface of the cell membrane; hence the term multinuclear. Each fiber has many fine threadlike structures called fibrils or myofibrils running lengthwise within the cytoplasm. These fibrils are capable of contracting to become shortened and thickened. When muscle fibers shorten they exert a pull, causing movement.

The muscle fibers are joined end to end and side by side. They form muscle bundles similar to the sticks in an armful of firewood with the sticks parallel. Connective tissue lies between adjacent fibers and also surrounds each bundle binding the fibers together. The entire muscle is also surrounded by a layer of connective tissue (a muscle sheath). The sheath is also termed fascia.

A slice of raw beefsteak is made by cutting one or more muscles at right angles to the length of the muscle fibers. The flat cut surface represents the cut ends of the muscle fibers. If the fibers are pulled apart, fine glistening strands of connective tissue will be seen between the fibers. A thick layer of connective tissue may be outlined surrounding the muscle, and separating it from other muscles contained in the cross section. F.11-1(3)

Skeletal muscles vary in shape, size and length. They

TWO SKELETAL MUSCLE FIBERS

(1) Longitudinal view

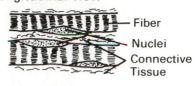

— Fiber
— Nuclei
— Connective Tissue

(2) Cross section — muscle bundle

— Fibers
— Nuclei

(3) Cross section of arm

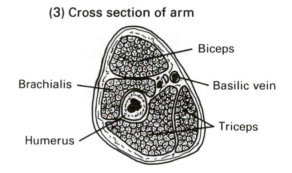

Biceps
Brachialis
Basilic vein
Triceps
Humerus

F. 11-2 **MUSCLES AS LEVERS — ELBOW**

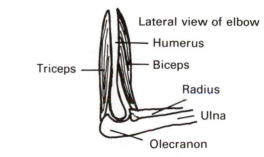

Lateral view of elbow
— Humerus
Triceps
— Biceps
Radius
Ulna
Olecranon

F. 11-3 **TENDONS — BACK OF HAND**

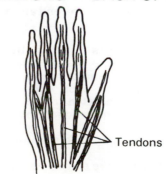

Tendons

F. 11-4 BURSAE AT THE KNEE

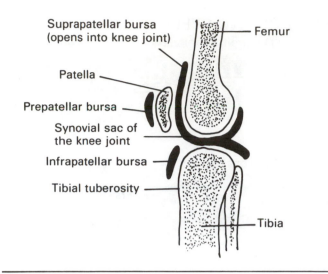

Suprapatellar bursa
(opens into knee joint)
Femur
Patella
Prepatellar bursa
Synovial sac of
the knee joint
Infrapatellar bursa
Tibial tuberosity
Tibia

may be very short, long, flat, strap-shaped, spindle-shaped, broad sheets, or bulky masses, triangular, or quadrilateral, etc. Longer muscles require many fibers connected end to end to give the required length.

Skeletal muscles are usually attached at each end to a bone, but some are attached to a cartilage, a ligament, another muscle or to the skin. The attachment to bone is by connective tissue, usually a tendon or aponeurosis, but sometimes directly to bone. This attachment may be very small or to a large surface, sometimes to an entire bone surface.

The origin of a muscle is its more fixed, less movable attachment, usually its proximal end.

The insertion of a muscle is its more movable end, usually its distal end. The bone or other point of insertion is pulled upon during contraction and is the part that moves.

A tendon sometimes termed a sinew, cord, or leader, is a cordlike fibrous connective structure that extends from the end of a muscle to a bony attachment. Tendons, since they contain fibrous tissue, are neither stretchable nor contractile. If the dorsal surface of the hand is examined with the fingers flexed at the proximal interphalangeal joints, the tendons can be seen as cords extending to the proximal ends of the digits on the dorsum of the hand. F.11-3, 12, 13

An aponeurosis is a sheet of fibrous connective tissue that is often attached at one end to a muscle, often a flat muscle, and by the other end to a bone, cartilage, ligament or other muscle. See diagram of the muscles of the anterior abdominal wall. F.11-5

Tendons and aponeuroses are actually attached to the connective tissue sheaths of muscles rather than to muscle fibers. Some muscles have a tendon at each end, others have a tendon at one end and an aponeurosis at the other end.

A tendon sheath is a tunnel-like channel that surrounds a tendon. The sheath has a lining of synovial membrane and the tendon is covered by a similar layer. These are joined at each end to form a closed sac. Synovial fluid is secreted into the sac and this helps to reduce friction as the tendon moves.

Tenosynovitis is an inflamation of a tendon sheath.

A bursa is a saclike structure lying between a muscle or tendon and an adjacent bony prominence over which the muscle or tendon moves. Bursae may also be located between the skin and an underlying bony prominence, as for example between the olecranon and tip of the elbow. The synovial sac is lined by a synovial membrane and secretes synovial fluid into the sac to decrease friction between the two surfaces. Bursae are frequently located near a joint. Occasionally a bursa opens into a joint cavity. The suprapatellar bursa, for instance, in front of and above the knee joint communicates with the knee joint, while the prepatellar and infrapatellar bursae are entirely separate from it. F.11-4

Bursitis is an inflammation of a bursa and it is frequently very painful.

Calcification, the deposition of calcium, may occur in a bursa or tendon sheath, and may be demonstrated radiographically.

S. 142 SOME CHARACTERISTICS OF SKELETAL MUSCLES

Muscles are described as irritable, conductive, extensible, elastic, and contractile.

Irritability is the property of being able to respond to stimuli.

Conductivity is the ability to conduct impulses from nerves, from electrical stimuli, etc.

Extensibility is the facility of stretching. This occurs by a lengthening of the fibrils of each muscle fiber.

Elasticity is the ability to return to the original length following stretching.

Contractility is the ability to become shorter, and is due to a shortening with thickening of each fibril of each muscle fiber.

S. 143 FUNCTIONS OF MUSCLES

1. Skeletal muscles contract and cause movement.
2. They maintain position (posture) in the upright and other positions of the body.
3. They give support to joints by maintaining a partial state of contraction.

Muscle tone: muscles do not completely relax when at rest, but remain partly contracted. This partial contraction helps to support joints, to maintain posture and hastens muscular contraction and movement when movement is required, thus avoiding jerky movements.

Contraction: the main function of all muscles is to contract and cause movement of the body or a part of it. When contractions occur the muscle fibers become shorter and thicker. The muscle becomes firm and visible. To demonstrate muscular contraction the part must be moved against some resistance. For example — place the wrist and the hand under a table top and attempt to bend the elbow. The muscles of the front of the arm will become firm, visible and palpable as a hard mass.

In Section 56 the various types of movements at joints were listed and defined as:

abduction	flexion	pronation
adduction	extension	internal
circumduction	inversion	rotation
rotation	eversion	external
	supination	rotation

These movements are brought about by contraction of muscles. So there are adductor muscles, abductor muscles, etc.

It is seldom that one muscle is responsible for any movement. Usually a group of muscles contract together. For example the several flexors of the wrist and hand function together to flex the wrist and digits.

Muscles that cross a joint to insert into a bone and cause movement are located proximal to that joint. Thus the flexors of the forearm lie in front of the humerus, above the elbow joint.

Muscles are sometimes described as prime movers, antagonists, synergists, and fixation muscles.

1. **Prime movers** are muscles that initiate and carry out some movement, such as flexion of the forearm. Frequently several muscles, rather than a single muscle, contract together to perform a movement.

2. **Antagonists** are muscles that perform some movement opposite to that caused by the prime movers. Most muscles are arranged in opposing groups, e.g. on opposite surfaces of a joint.

If one group causes flexion the opposing group will cause extension. Similarly adductors oppose abductors, supinators oppose pronators, internal rotators oppose external rotators, etc. When one group contracts the opposing group will relax gradually so that the movement is smooth and graceful. These antagonists also act to prevent excessive or superfluous movement, by remaining partly contracted.

3. **Synergists** are muscles that act with the prime movers to accomplish some movement but prevent unwanted movement. In flexion of the fingers, the extensors of the wrist contract to prevent flexion of the wrist as well. Syn = together + ergon = work; i.e. working together.

4. **Fixation muscles** are those that hold the adjacent bones in a fixed position so that the prime movers may accomplish some certain movement. In flexion of the forearm at the elbow the scapula is held rigid by the shoulder muscles.

Posture — the maintenance of the upright position of the body consists of a balanced contraction of some muscle groups and the partial relaxation of opposing groups. The head must be kept upright, the spine rigid, the femora extended at the hips, the legs extended at the knees, and the feet dorsi-flexed.

Muscular movements are often complex and must be acquired by practising over and over in order to become proficient. So the infant may learn to put food in the mouth, to crawl, to walk, to run and jump. Most of these movements become automatic with practice and then do not require conscious effort.

S. 144 MUSCLE SPASM

Spasm is a contraction of muscles that may persist for a long period of time, without relaxation. Spasm of the muscles of a limb, or a part of the trunk often accompanies injury to the part. This injury may be a simple sprain, a dislocation or a fracture. The spasm or muscular contraction holds the limb in a fixed position, usually partial flexion. This may be nature's method of preventing further injury. Every technician has seen elbows, wrists, hips, and knees held rigid and partly flexed after injury. The technician must not attempt to force the limb into the usual positions for routine views because apart from causing pain this might cause further injury. The technic must be adapted to the case in hand.

Sometimes spasm with partial flexion is present when there is inflammation about a joint. With no history of injury the technician may attempt to position the limb, if this is not too painful. Pressure if exerted gradually will often overcome the spasm.

S. 145 PARALYSIS OF MUSCLES

Paralysis of muscles follows injury to, or destruction of, the nerves supplying that muscle. Poliomyelitis (infantile paralysis) may attack cells in the spinal cord that supply motor nerves to skeletal muscles. The nerve cells die and do not transmit impulses to the muscle supplied by the affected nerve. The muscle becomes flabby and will not contract. The opposing muscles, if not affected then have no opposition and will pull the limb into the position normally assumed by their contraction. In another type of paralysis the paralysed muscles become contracted and rigid, deforming the limb by their contraction.

Paralysis which is due to a nerve injury or destruction must be distinguished from the inability to move a part following an injury in which a muscle or tendon is cut in two. If the muscle or tendon is repaired function will be restored.

S. 146 DISTRIBUTION OF MUSCLES

The student should compare the appearance of a mounted skeleton with that of a well-developed living subject. It becomes apparent that there must be a great deal of soft tissue filling in the spaces and rounding out the body and limbs. Some of this is due to subcutaneous fat lying under the skin. Most is due to the muscles of the part since they account for almost half the body weight. There are many muscles distributed throughout the body. In the head alone there are more than forty muscles. These are responsible for the movements of the eyes, eyelids, ears, nose, mouth and jaws.

Frequently there are two or more layers of muscles; the superficial muscles lying close to the skin, and the deep muscles. On the sole of the foot, for example, there are four layers of muscles, each layer consisting of several muscles.

As there are identical bones in the right and left halves of the body, except for the sternum vertebrae, etc., so also is there a similar distribution of skeletal muscles. Each side of the head, neck, trunk, upper and lower limb has similar muscles — bilaterally similar. There are a few exceptions such as the tongue and diaphragm. Physicians, and particularly orthopedic surgeons, must know the location and action of the various muscles. With this knowledge the surgeon knows what movement to undertake to correct a deformity caused by a fractured bone, as the attached muscles will pull upon the fractured bone fragments in a definite direction.

S. 147 MUSCLE GROUPING BY LOCATION

Muscles may be grouped according to the part of the body in which they are located, and their position in that part as anterior, posterior, medial, lateral, superficial or deep, etc. Sometimes they are grouped according to their action as flexors, extensors, abductors, etc. The list included here is for general information only and no student should be required to learn these details.

1. **Muscles of the head:**
(1) scalp muscles
(2) muscles of the eyelids
(3) muscles of the eyes
(4) muscles of the nose
(5) muscles of the mouth
(6) muscles of mastication (chewing)

2. **Muscles of the neck:**
(1) superficial cervical muscles
(2) lateral cervical muscles
(3) suprahyoid, and infrahyoid muscles (anterior neck)
(4) anterior vertebral muscles
(5) lateral vertebral muscles

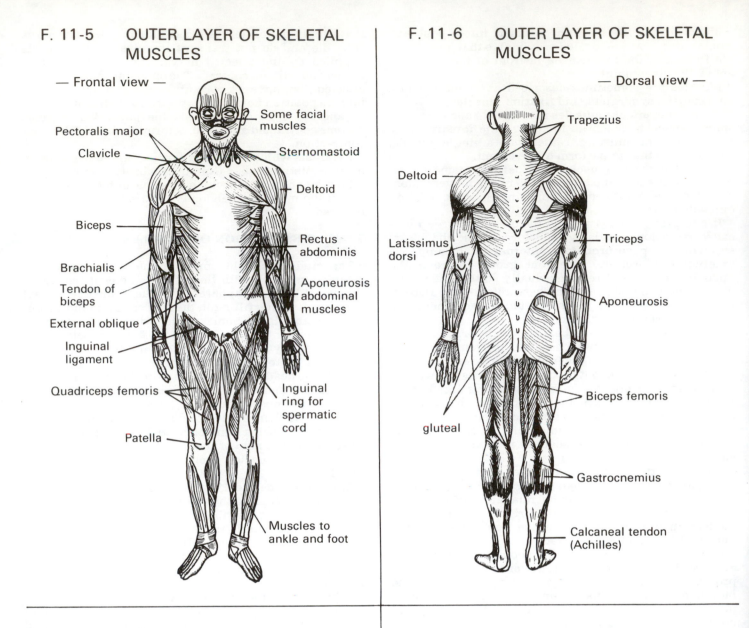

F. 11-5 OUTER LAYER OF SKELETAL MUSCLES

— Frontal view —

- Pectoralis major
- Clavicle
- Some facial muscles
- Sternomastoid
- Deltoid
- Biceps
- Rectus abdominis
- Brachialis
- Tendon of biceps
- Aponeurosis abdominal muscles
- External oblique
- Inguinal ligament
- Quadriceps femoris
- Inguinal ring for spermatic cord
- Patella
- Muscles to ankle and foot

F. 11-6 OUTER LAYER OF SKELETAL MUSCLES

— Dorsal view —

- Trapezius
- Deltoid
- Latissimus dorsi
- Triceps
- Aponeurosis
- gluteal
- Biceps femoris
- Gastrocnemius
- Calcaneal tendon (Achilles)

---

3. Muscles of the trunk:
(1) deep muscles of the back
(2) suboccipital muscles
(3) muscles of the thorax
(4) muscles of the abdomen
   (a) anterolateral
   (b) posterior
(5) muscles of the pelvis
(6) muscles of the perineum

4. Muscles of the upper limb or member (NA):
(1) connecting the limb to the vertebral column
(2) connection of the upper limb to the chest wall
(3) muscles of the shoulder
(4) muscles of the arm — brachial muscles
(5) muscles of the forearm — antebrachial
(6) muscles of the hand

5. Muscles of the lower limb or member:
(1) muscles of the iliac region
(2) muscles of the thigh;
   (a) anterior femoral muscles
   (b) medial femoral
   (c) gluteal muscles — buttocks
   (d) posterior femoral
(3) muscles of the leg, crural muscles;
   (crus = leg)

   (a) anterior crural muscles
   (b) lateral crural
   (c) posterior crural (superficial) deep
(4) muscles of the foot;
   (a) dorsal muscles of the foot
   (b) plantar muscles of the foot
      first, second, third, fourth, layers.

The head: these are responsible for movements of the eyelids, eyes, nose, mouth, jaws, ears.

The neck: the rounded cylindrical appearance of the neck is due to the muscles located here. Those in front control flexion of the head and neck, as well as movements of the tongue and trachea during swallowing and respiration. The lateral groups control lateral flexion of the head and cervical vertebrae. The posterior group, the suboccipital, controls backward bending, or hyperextension of the head and neck.

The thorax: several of the muscles of the anterior and posterior chest wall pass laterally to the shoulder girdle or upper arm. These provide for movements of the shoulder and arm. Other muscles, such as the intercostals fill in the gaps between adjacent ribs. These elevate the ribs during inspiration (breathing in). The diaphragm will be considered separately.

The abdomen: the abdominal cavity, unlike the thoracic cavity, has no bony wall except posteriorly, where the lumbar vertebrae and their transverse processes afford some protection to contained organs. The remainder of the wall at the front and laterally is formed by muscles. The upper abdomen does receive some protection from the lower ribs and costal cartilages. The posterior abdominal wall on each side of the vertebrae is formed by the psoas major and quadratus lumborum muscles. The fibers of the latter extend between the twelfth rib and the iliac crest. (Quadratus = four sided) The psoas major muscle is described separately below.

The lateral and anterolateral parts of the wall are formed by three layers of flat, sheet-like muscles and their aponeuroses of connective tissue. The fibers of each layer are directed at an angle to the fibers of each of the other layers to give added support. They are named external oblique, internal oblique, and transversus, bilaterally. Close to the midline a muscle passes vertically on each side from the xiphoid process and costal cartilages to the symphysis and pubic bones — **the rectus abdominis muscle.**

The pelvis: the muscles of the pelvis form a sling across the floor of the pelvic cavity, with openings for the anal, urinary, and female passages.

The perineum: that space between the two ischial tuberosities, the pubic arch and coccyx is filled in by the perineal muscles.

The upper limb: the muscles extending from the chest wall to the shoulder girdle have been noted.

The **shoulder muscles** pass over the shoulder to the humerus and are responsible for the movements of the upper humerus.

The muscles of the arm, the brachial muscles, lie on the anterior and posterior surfaces of the humerus. These cross the elbow joint to insert into the radius or ulna, and they cause movements of the forearm. On the front of the arm are the flexors of the forearm, the biceps and the brachialis. On the posterior surface there is the triceps brachii, the extensor of the forearm.

The **muscles of the forearm,** the antebrachial muscles control movements of the wrist and hand. The flexor muscles of the wrist and hand lie on the front of the forearm and extend down from the medial epicondyle of the humerus. These end in long tendons that insert into the carpal bones or phalanges. There are superficial and deep flexors of the digits.

The extensor muscles of the wrist and hand pass down from the lateral epicondyle of the humerus along the dorsal surface of the forearm. They end in tendons that insert into the carpal bones or phalanges. The tendons on the back of the hand can be visualized when the fingers are flexed. By grasping the forearm with the fingers of one hand, and flexing and extending the digits the flexors and extensors can be felt as firm masses of tissue. This movement is often employed to outline muscles.

The lower limb: as with the upper limb muscles connect the lower limb to the trunk. These pass from the pelvic bones to the femur across the hip joint to provide for flexion, extension, abduction, and adduction of the femur, as well as internal and external rotation. The adductors lie on the medial side of the femur from the pubic bone to the medial surface of the femur.

The abductors, the gluteal muscles, lie on the lateral surface and form the mass of the buttock. The flexors are in front, and the extensors are behind the hip joint.

The thigh: the muscles that make up the thigh control movements at the knee. In flexion at the knee the movement is posteriorly, not anteriorly as at the elbow. The **flexor muscles** for the leg lie on the posterior surface of the femur, not the anterior. They cross the posterior surface of the knee to insert into the tibia or fibula. The **extensor** muscles of the leg (anterior femoral muscles) lie anterior to the femur. The principal one, the quadriceps femoris extends down in front of the femur. It ends in a tendon, the quadriceps tendon, that crosses the anterior surface of the knee to insert into the tibial tuberosity. The patella is contained within this quadriceps tendon as a sesamoid bone. This muscle straightens the leg at the knee.

The leg: the muscles of the leg overlying the tibia and fibula control movements of the ankle and foot. The foot is placed at right angles to the leg. This arrangement is very different from that of the hand. This has resulted in much confusion as to what is flexion and extension at the ankle. Raising the forefoot away from the floor is called **dorsiflexion.** Bending the foot downwards in the opposite direction is termed **plantar flexion.**

The **anterior crural** muscles, on the front of the tibia and fibula, end in tendons that cross the anterior surface of the ankle to insert into the upper surface of the bones of the foot. These extend (dorsiflex) the tarsal bones and digits. (Crus = leg, pl. crura, adj. crural, referring to the leg).

The **posterior crural** muscles form a superficial and a deep layer posterior to the tibia and fibula. They constitute the calf of the leg. The outer layer consists of three muscles, the outer one extending down across the knee joint from the femoral condyles and ending in a tendon that inserts into the tuberosity of the calcaneus. This tendon, the Achilles tendon, is also formed from the other main muscles of the calf. They are responsible for plantar flexion at the ankle. The deep muscles extend down on the back of the leg, enter the foot below the medial malleolus and insert into the plantar surfaces of the toes. The plantar muscles flex the toes.

The foot: short muscles lie on the dorsal and plantar surfaces of the foot. The plantar muscles form four layers. These muscles are used to cause plantar flexion of the phalanges of the toes.

The much-condensed description of the muscle groups included above is incomplete and attempts to explain only the large muscle groups. Thus the mass of muscle anterior to the femur causes extention of the leg at the knee. The muscles on the front of the humerus cause flexion of the forearm at the elbow, etc. The movements of adduction, abduction, inversion, eversion, etc. have been disregarded.

HOW MUSCLES ARE NAMED

Each individual muscle has a name. This name has been assigned because of some characteristic of the muscle, such as its location, shape, direction, action or parts.

1. By location — pectoralis major; the larger muscle of the chest or breast; pectus = the breast.
2. By shape — quadratus = four sided, rectus = straight; deltoid = triangular.
3. By direction of fibers — transversus = cross ways; oblique = at an angle.
4. By action = part affected — flexor digitorum = bender of fingers.
5. Number of parts — biceps = two heads; triceps = three heads; quadriceps = four heads.

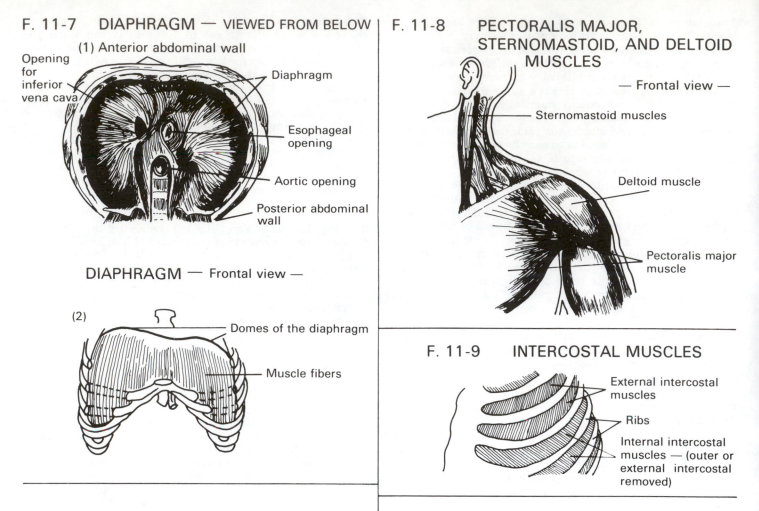

F. 11-7 DIAPHRAGM — VIEWED FROM BELOW

(1) Anterior abdominal wall

Opening for inferior vena cava

Diaphragm

Esophageal opening

Aortic opening

Posterior abdominal wall

DIAPHRAGM — Frontal view —

(2)

Domes of the diaphragm

Muscle fibers

F. 11-8 PECTORALIS MAJOR, STERNOMASTOID, AND DELTOID MUSCLES

— Frontal view —

Sternomastoid muscles

Deltoid muscle

Pectoralis major muscle

F. 11-9 INTERCOSTAL MUSCLES

External intercostal muscles

Ribs

Internal intercostal muscles — (outer or external intercostal removed)

S. 148 SOME MUSCLES IMPORTANT IN RADIOGRAPHY

There are four muscles that are of paramount importance in radiography: the diaphram, the pectoralis major, the psoas major, and the intercostals.

1. The diaphram; F.11-7

G. diaphragma = a partition; adj. diaphragmatic
G. phren = diaphragm; adj. phrenic.

The diaphragm is a dome-shaped muscular partition that separates the thorax and abdomen. It has actually two domes, a right and a left. The right dome lies at a slightly higher level than the left. It may reach the level of the fifth or even fourth rib anteriorly. The left dome is flattened to accommodate the inferior part of the heart. This muscle is attached along its base to the inner surface of the thoracic cage, the lower sternum, lower ribs, costal cartilages and upper lumbar vertebrae, (like a bell tent pegged to the ground). The muscle fibers pass cranially (up) from the base to end in a **central tendon** that forms the roof of the domes. This tendon is a sheet of fibrous tissue.

There are three large openings in the diaphragm:
1. The aortic hiatus, through which the aorta enters the abdomen from the thorax;
2. The esophageal hiatus, through which the esophagus enters the abdomen;
3. The opening (foramen) of the inferior vena cava, through which this vein enters the thorax.

The convex upper surface of the diaphragm forms the floor of the chest cavity. The concave inferior surface forms the roof of the abdomen.

The base of the diaphragm along its attachment to the thoracic wall lies at a level definitely lower than the top of the domes.

The costophrenic sinus, (NA costodiaphragmatic recess).

Above the attachment of the diaphragm to the chest wall, where the diaphragm begins to separate from the wall, there is a narrow space called the costophrenic sinus or recess. In front it is at a higher level than laterally, and it is still lower posteriorly. The lung does not extend down into this space. Free fluid that has accumulated in the chest cavity will collect at the lowest level possible. It will fill the costophrenic sinus posteriorly first, then the lateral part, and finally if there is sufficient fluid the anterior part.

The domes of the diaphragm move down with each inspiration (breath taken), as the muscle fibers contract. The domes must move down since the base is fixed to the thoracic wall, and the pull is exerted on the domes. This downward movement increases the vertical diameter of the chest. At expiration (breathing out) the muscle fibers relax and the domes ascend to a higher level, decreasing the vertical diameter of the chest cavity.

Phrenic arteries, superior and inferior, and right and left are branches of the aorta supplying the diaphragm with blood. The right and left phrenic nerves send fibers to it. Respiratory movements are automatic, and occur during sleep and unconsciousness. The cerebrum, however, can control the rate and depth when so directed. The rhythmic contraction of the diaphragm occurs about eighteen times per minute.

The diaphragm and radiography: the positions that the diaphragm may assume are particularly important in radiography.
1. The tops of the domes lie well above the costal margins and may reach the fourth ribs.
2. Following inspiration they lie at a lower level than when following expiration.
3. With the subject lying down the diaphragm lies at a higher level than when upright.

F. F. 11-10 DELTOID, BICEPS, AND BRACHIALIS MUSCLES
— FRONTAL VIEW —

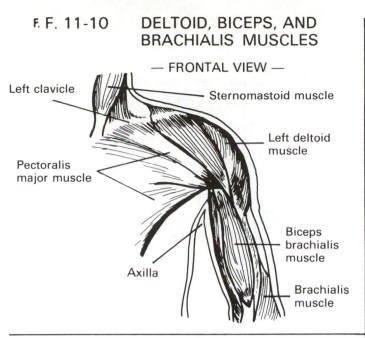

Left clavicle
Sternomastoid muscle
Left deltoid muscle
Pectoralis major muscle
Biceps brachialis muscle
Axilla
Brachialis muscle

F. 11-11 PSOAS MAJOR MUSCLE
— Frontal view —

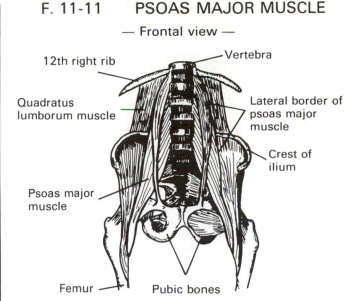

12th right rib
Vertebra
Quadratus lumborum muscle
Lateral border of psoas major muscle
Crest of ilium
Psoas major muscle
Femur
Pubic bones

---

4. Air escaping from a hole in the wall of the stomach or intestine will rise to the highest possible level in the abdomen. It will collect under the diaphragm if the subject is sitting upright or standing.

The diaphragm must be included in both chest and abdomen radiography.

2. The pectoralis major muscle   F. 11-8 & 10

L. pectus = chest or breast; adj. pectoral.

There are two pectoralis major muscles, a right and a left. Each is a thick fan-shaped muscle that covers the upper anterior chest wall. It is named "major" as there is a further smaller pectoral muscle beneath it — the pectoralis minor. It originates from attachments to the sternum and anterior chest wall. The fibers extend laterally and converge to form a tendon that after folding over upon itself crosses the axilla. It inserts into the intertubercular groove of the humerus. Contraction adducts the arm.

Its importance in radiography is due to the fact that in subjects doing heavy manual work it is large and thick and will absorb x-ray radiation. The chest underlying the muscle will appear grey. Lung detail will then be obscured. Beware of the wrestler, etc.

When a breast is removed because of cancer the pectoralis major muscle is often removed as well. With no breast or muscle to absorb radiation the film may be dark.

3. The psoas major muscle: F.11-11

This muscle lies lateral to the lumbar vertebrae in the posterior wall of the abdomen. It originates by attachments to the lumbar vertebrae and their transverse processes. The fibers form a thick muscle mass in the posterior abdomen. They descend in the abdominal wall and in front of the ilium. They pass around the pelvic inlet then behind the inguinal ligament to the thigh. Passing anterior to the hip joint, they insert into the lesser trochanter of the femur. This muscle helps to flex the thigh. It is named "major" since there is a smaller muscle, the psoas minor, which when present lies anterior to the psoas major.

The psoas major muscle is sufficiently thick to absorb some of the x-rays that enter it in radiography of the abdomen. In anteroposterior views the lateral border of the psoas major muscle will be visible as an oblique line passing down lateral to the lumbar vertebrae.

A good radiograph of the abdomen will demonstrate this shadow. Its absence may be due to faulty technic, to a tumor or mass in the kidney, to a large blood clot, or to an abscess of the kidney. The visualization of this muscle border is therefore important.

There are right and left psoas major muscles.

4. The intercostal muscles: F.11-9

Inter = between + costa = a rib; between ribs. The intercostal muscles fill in the spaces between adjacent ribs, and their costal cartilages — the intercostal spaces. There are two layers, the external intercostals or outer layer, and the internal intercostals or inner layer. The fibers of the external layer pass down and forwards, those of the inner layer down and posteriorly between adjacent ribs. These muscles elevate the ribs during inspiration, thus increasing the size of the chest cavity. Some of the muscles in the neck aid in this movement.

S. 149   OTHER EXAMPLES OF SKELETAL MUSCLES

The muscles described below have been added so that the student may obtain some idea of the names, shapes, and extent of muscles. The origins and insertions of these muscles have been listed since some Syllabi may require this knowledge.

1. The sternomastoid muscle: F.11-7

This is a strap-shaped muscle that passes obliquely down the neck from the mastoid process to the upper sternum and medial part of the clavicle. It divides the neck into anterior and posterior triangles. There are two of these muscles, a right and a left. If one of them contracts the head is turned to the opposite side, and the mastoid process on that side is tipped towards the shoulder. Wry neck is a contraction of one of the sternomastoid muscles that results in a characteristic deformity. The head is turned to the opposite side and the chin elevated. This sometimes is caused by injury to the neck during childbirth. This muscle is also known as the sternocleidomastoid muscle.

2. The pectoralis minor muscle:

When a name such as pectoralis major muscle is used it implies that there are two pectoral muscles, one of which will be named pectoralis minor = the smaller.

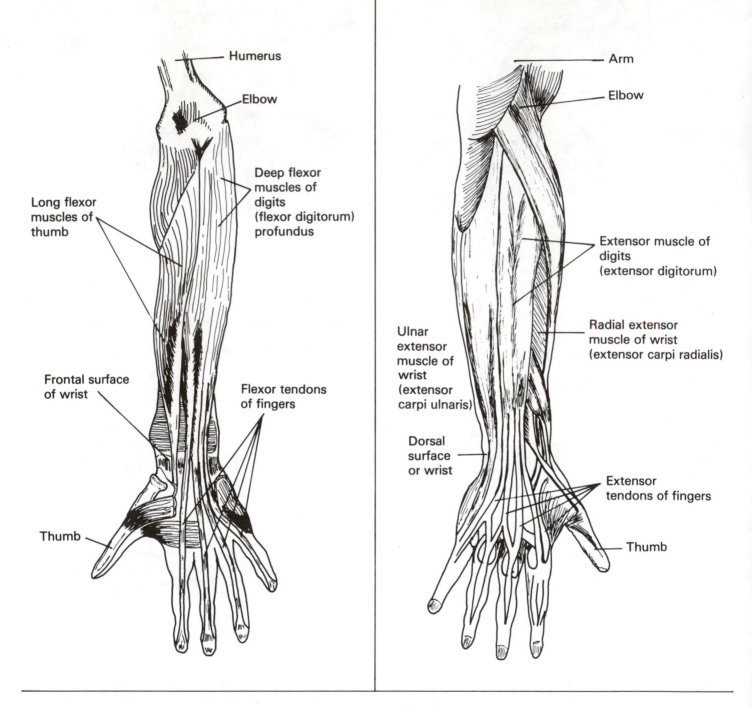

F. 11-12 DEEP FLEXOR MUSCLES OF FOREARM

—Flexor of digits, etc.
Frontal view right hand

Humerus

Elbow

Long flexor
muscles of
thumb

Deep flexor
muscles of
digits
(flexor digitorum)
profundus

Frontal surface
of wrist

Flexor tendons
of fingers

Thumb

F. 11-13 EXTENSOR MUSCLES OF FOREARM

—Extensors of digit, etc.
Dorsal view right hand

Arm

Elbow

Extensor muscle of
digits
(extensor digitorum)

Ulnar
extensor
muscle of
wrist
(extensor
carpi ulnaris)

Radial extensor
muscle of wrist
(extensor carpi radialis)

Dorsal
surface
or wrist

Extensor
tendons of fingers

Thumb

The pectoralis minor lies behind the pectoralis major muscle and extends from the anterior chest wall to the coracoid process of the scapula. There are two pectoralis minor muscles, a left and a right.

3. The deltoid muscle: F.11-6, 7, 8, 10

G. delta = Greek Letter Δ , D; triangular.
The deltoid muscle forms the rounded curve of the shoulder. It connects the humerus to the shoulder. It extends from the lateral end of the clavicle and the scapula to the deltoid tubercle of the humerus. This tubercle is located on the lateral surface of this bone at its midpoint. The muscle covers the shoulder joint. Its contraction causes abduction of the humerus. There are

right and left deltoids.

4. The biceps brachii: F.11-5, 10

Bi = two+ceps = head + brachium = arm. The biceps muscle is located in front of the humerus. It is spindle-shaped, and becomes visible when the forearm is flexed at the elbow. Two tendons at its upper end originate from the scapula, hence the name biceps — two heads. Another tendon at its lower end crosses the elbow joint and inserts into the radial tuberosity. It flexes the forearm. The student may note that this muscle mass is in the arm, not in the forearm. The bicipital or intertubercular groove on the front of the humerus contains one of the biceps tendons.

5. The brachialis muscle: F.11-10

This muscle of the arm (brachium) lies in front of the lower humerus, but behind the biceps muscle, which hides most of it. From its origin on the anterior surface of the humerus it passes down, crosses the elbow joint and inserts into the coronoid process of the ulna. Like the biceps it flexes the forearm.

6. The triceps brachii: F.11-6

Tri = three + ceps = head + brachium, arm. The triceps as the name implies has three heads at its upper end, where it originates from the scapula and humerus. It extends down the back of the humerus, crosses the elbow joint posteriorly and inserts into the olecranon of the ulna. Contraction extends the forearm, just the opposite to the action of the biceps and the brachialis.

7. The flexor digitorum profundus: F.11-12

This muscle has been included as an example of a flexor of the digits. Its name means the deep flexor of the digits; (profundus = deep). There is also a superficial flexor of the fingers. The muscle has its origin from the anterior, lateral, and posterior surfaces of the ulna. It passes down upon the anteromedial surface of the forearm, and divides into four tendons, one for each finger. These cross the wrist and palm of the hand and end by inserting into the terminal phalanges of the fingers. Contraction causes flexion of the fingers.

8. The extensor digitorum: F.11-13

This muscle lies on the lateral margin of the forearm whereas the flexors lie on the medial and anterior margins. Originating at the lateral epicondyle of the humerus it extends down the forearm to the back of the wrist. It ends here by dividing into four tendons, one for each finger. The tendons insert into the posterior surfaces of the phalanges. Contraction causes extension of the fingers, the opposite action to that of the flexors.

While these flexor and extensor muscles have been included as examples many other muscles are also present on the anterior and the posterior surfaces of the forearm. These control movements of the wrist, thumb and fingers, including pronation and supination, and abduction and adduction.

9. The trapezius muscle: F.11-6

There are two trapezius muscles, a right and a left. Each is triangular in shape, the two together form a four-sided figure, diamond-shaped, on the posterior surface of the neck and thorax. (See diagram). The broad base of each muscle is attached medially to the base of the skull and the spinous processes of the cervical and thoracic vertebrae. The fibers converge at the shoulder and insert into the acromion, the lateral part of the clavicle and the spine of the scapula. Its action is upon the shoulder and arm. (Trapezium = four sides).

10. The latissimus dorsi muscle: F.11-6

Latus = broad; latissimus = broadest. There are two of these muscles. Each is triangular in shape and is located on the back of the lower thorax and abdomen. The base of the muscle is attached to the spinous processes of the lower thoracic and lumbar vertebrae, the posterior part of the iliac crest and the sacrum. The fibers pass up and laterally and cross the axilla to insert into the upper humerus. The action is to rotate the arm.

The trapezius and latissimus dorsi muscles of both sides form an outer muscular layer that covers almost the entire back of the trunk. Other muscles form deeper layers here as well.

11. The sacrospinalis muscle: F.11-11

The right and left sacrospinalis muscles extend vertically from the occipital bone to the sacrum on either side of the spinous processes of the vertebrae. They are attached to these bones. Contraction of one of them causes lateral flexion of the spine. Contraction of both of them causes dorsiflexion of the spine. If one becomes paralysed, the other, having no opposition, will flex the spine towards the noraml side. This results in scoliosis, and is sometimes present in poliomyelitis.

12. The psoas minor muscle: F.11-11

The psoas minor muscle, when present, lies in front of the psoas major muscle of the same side. It is long and slender and is located in the posterior abdominal wall and upper pelvis upon the corresponding psoas major muscle.

13. The gluteus maximus muscle: F.11-6

Gluteus = buttock. This muscle forms most of the tissue mass of the buttock on either side. It originates from the posterior ilium, the lower sacrum and coccyx. Its fibers pass laterally to insert into the lateral surface of the upper femur. Also present beneath it are gluteus medius and minimus muscles. Hence the name maximus = the largest of three.

14. The quadriceps femoris muscle: F.11-15

Quadri = four + ceps = head + femoris. The quadriceps femoris muscle of the thigh is actually four muscles located anterior to the femur. It forms a large muscle mass here that is palpable and often visible. One part, the rectus femoris, arises from the ilium, while the other three arise from the body of the femur anteriorly. All four of them unite to form a common tendon, the **quadriceps tendon**, above the knee joint. This tendon inserts into the tibial tuberosity. The tendon encloses within it the patella, a sesamoid bone. Below the patella the tendon is named the **patellar ligament**. This muscle extends the leg at the knee.

Three bursae are usually present about this tendon:

The suprapatellar bursa lies between the upper part of the quadriceps tendon and the lower part of the femur, above the patella. It opens into the knee joint.

The prepatellar bursa lies in front of the petalla. It does not open into the knee joint.

The infrapatellar bursa lies between the lower part of the tendon and the upper tibia.

15. The biceps femoris muscle: F.11-6

Bi = two + ceps = head + femoris. This muscle is located posterior to the femur in the thigh. It arises from the ischial tuberosity, and the posterior surface of the femur. It crosses the knee joint posteriorly and inserts into the upper end of the fibula. It flexes the leg at the knee, and opposes the quadriceps femoris. There are right and left muscles.

16. The gastrocnemius muscle: F.11-6

Gastro = belly (stomach) + kneme = leg. This muscle forms most of the mass of the calf muscles on the back of the leg. It has two heads with their origins in the medial and lateral condyles of the femur. It crosses the posterior surface of the knee joint, and passes down in the calf of the leg. With the soleus muscle that lies beneath, it forms the Achilles tendon (NA. calcaneal tendon). This tendon inserts into the calcaneal tuberosity on the back of the os calcis. It causes plantar flexion of the foot, and helps flex the leg at the knee.

S. 150 EXAMPLES OF LIGAMENTS AND TENDONS

1. **The inguinal ligament** lies obliquely in the crease between the anterior surface of the upper thigh and the lower abdomen on each side. It is the thickened lower border of the outer anterior abdominal muscle. It consists of fibrous connective tissue and extends from the anterior superior spine of the ilium to the tubercle of the pubic bone. The femoral artery and vein pass posterior to its midpoint to enter the thigh. There is also a ringlike opening present above the medial end of the inguinal ligament through which the spermatic cord containing the deferent duct passes on its way to the scrotum. These two areas form openings out of the abdomen through which other structures may herniate (rupture). F.11-5.

2. **The patellar ligament** is that part of the quadriceps femoris tendon that extends from the apex of the patella to the tibial tuberosity. See F.11-15.

3. **The ligamentum teres** (round ligament) or ligamentum capitis femoris (ligament of femoral head) lies within the hip joint. It extends from a pit (fovea) on the head of the femur to the lower margin of the rim of the acetabulum. F.11-16

4. **The calcaneal tendon** (tendo Achilles or Achilles tendon) extends down from the calf muscles to the tuberosity of the calcaneus, posterior to the ankle and heel. It is formed by the gastrocnemius and soleus muscles. Its margins are palpable at the heel. F.11-6

Note that **tendons** are cordlike bands of fibrous connective tissue that connect muscle to bone, and are nonstretchable.

Ligaments are cords, bands, or sheets of fibrous connective tissue often extending from one bone to another across a joint. They help to hold the bone ends together at a joint.

WEAK ABDOMINAL AREAS

Under this heading are included umbilical, inguinal, and femoral deficiencies. Sometimes the abdominal wall at these locations is not well developed or has become weakened. If there is an increase in intraabdominal pressure such as occurs in straining to lift heavy objects, some abdominal contents, bowel or other structure, may be pushed out as a bulge at the weak area. This bulging indicates a hernia — a rupture.

1. **Umbilical area** — the abdominal wall at the navel may be incompletely formed or weak and may result in an **umbilical hernia**. In newborn infants there sometimes appears to be a hernia at this location but it usually disappears without surgery.

2. **Inguinal area** – inguen = groin, indicated by the oblique crease where the lower abdomen and upper thigh meet anteriorly. A canal and ring above the medial part of the inguinal ligament normally provides a passage for the deferent duct in the spermatic cord to leave the abdomen to enter the scrotum. The support here may be lacking and an **inguinal hernia** may result.

3. **Femoral area** — the femoral artery and vein pass out of the abdomen and enter the thigh behind the inguinal ligament at about its midpoint. Sometimes there is a weak area beside these vessels which may result in a **femoral hernia**.

S. 151 TERMS RELATING TO THE MUSCULAR SYSTEM

skeletal muscle	muscles of the head
visceral muscle	muscles of the neck
cardiac muscle	muscles of the trunk
voluntary muscle	muscles of the upper limb
involuntary muscle	muscles of the lower limb
striated muscle	diaphragm
nonstriated muscle	diaphragmatic
smooth muscle	phren, phrenic
muscle cell	aortic haitus
muscle fiber	esophageal hiatus
muscle fibril	foramen of the inferior
multinucleated	vena cava
muscle bundle	costophrenic sinus (recess)
muscle sheath	pectoralis major muscle
origin of a muscle	pectoralis minor muscle
insertion of a	psoas major muscle
muscle	psoas minor muscle
aponeurosis	sacrospinalis muscle
tendon	deltoid muscle
tendon sheath	biceps brachii muscle
bursa, bursae	triceps brachii muscle
bursitis	quadriceps femoris muscle
irritability	inguinal ligament
conductivity	patellar ligament
extensibility	Achilles tendon
contractility	flexor digitorum profundus
muscle tone	extensor digitorum
muscle spasm	biceps femoris muscle
muscle paralysis	gastrocnemius muscle
calcaneal tendon	soleus muscle
brachialis muscle	tendinitis

F. 11-14   THE SACROSPINALIS MUSCLES

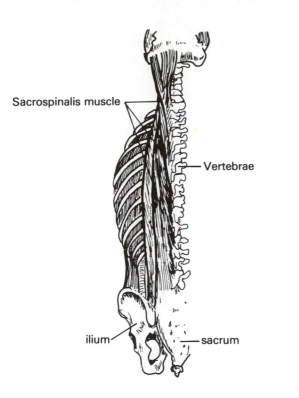

Sacrospinalis muscle

Vertebrae

ilium

sacrum

F. 11-15   QUADRICEPS FEMORIS MUSCLE
—FRONTAL VIEW—

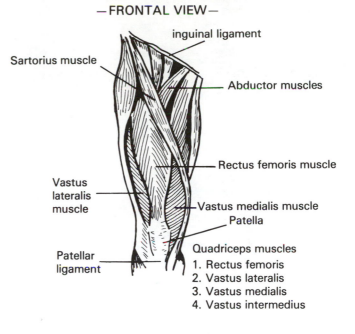

inguinal ligament

Sartorius muscle

Abductor muscles

Rectus femoris muscle

Vastus lateralis muscle

Vastus medialis muscle
Patella

Patellar ligament

Quadriceps muscles
1. Rectus femoris
2. Vastus lateralis
3. Vastus medialis
4. Vastus intermedius

NOTES

12. THE BLOOD

Blood is composed of a liquid, the plasma, and of formed elements, the blood cells. The volume of the plasma and cells is about equal - 50% of each.

S. 152 CONSTITUENTS OF THE BLOOD

1. THE BLOOD PLASMA F.12-1

(1) Water
(2) Inorganic salts;
  sodium   phosphates
  potassium   bicarbonates
  calcium   iodine
  magnesium   iron
  chlorides   copper
(3) Blood Proteins;
  albumin
  fibrinogen
  globulins — antibodies
(4) Digested food products;
  amino acids from proteins,
  glucose from carbohydrates
  sugars and starches.
  fatty acids, glycerine from lipids (fats)
(5) Hormones from endocrine glands;
  from: pituitary gland
        thyroid gland
        parathyroi;s
        pancreas
        suprarenal glands
        gonads:
          ovaries
          testes
(6) Vitamins; A.B.C.D. etc. and subgroups
(7) Antibodies
(8) Gases in solution
(9) Waste Products

2. BLOOD CELLS F.12-2

(1) Red blood cells or erythrocytes
(2) White blood cells, leucocytes or leukocytes
  Granulocytes:
  a. neutrophils (polymorphonuclear)
  b. eosinophils
  c. basophils
  Agranulocytes or nongranulocytes:
  a. lymphocytes, large and small
  b. monocytes
(3) Blood platelets or thrombocytes

DERIVATIVES OF TERMS

Cytos - kytos - a cell
Erythros - red
Leukos, leuco = white
Corpus - body
Corpuscle - corpusculus
Corpusculus - corpus + culus (little) a little body
Erythrocyte - erythros + cytos - a red cell
leukocyte - leukos + cytos - a white cell (also leucocyte)

S. 153 THE BLOOD PLASMA F.12-1

The plasma, the liquid part of the blood, is a colorless fluid. Ninety percent of it is water. In this are dissolved many different substances which are described below.

1. **Water** — the solvent; from ingested fluids and foods - 90%
2. **Inorganic salts**, etc., in solution. These include sodium, potassium, calcium, and magnesium as chlorides,

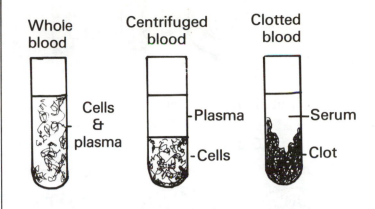

F. 12-1 THE BLOOD

Whole blood — Cells & plasma

Centrifuged blood — Plasma — Cells

Clotted blood — Serum — Clot

phosphates, bicarbonates, etc. These are electrolytes, and form positive and negative ions. Also present are iodine, iron, and copper.

3. **Blood proteins** — Three proteins circulate in solution in plasma: albumin, fibrinogen, and globulin in its several forms. These are large molecules that cannot pass through membranes so help to retain water in the blood. Gammaglobulin is formed in the lymphocytes, and is concerned with the production of antibodies, immunity to disease, and the destruction of substances foreign to the body such as transplanted organs. The others are formed in the liver. Fibrinogen assists in blood clotting.

4. **Digested food products** — These are amino acids from digested proteins, glucose from digested carbohydrates (including starches and sugars), and the fatty acids and glycerine from lipids (simple fats).

5. **Products of endocrine glands — hormones** — These include secretions formed by the pituitary, thyroid, parathyroid, and suprarenal glands, the pancreas and the gonads (ovaries and testes). See chapter on the endocrine glands.

6. **Vitamins** — The many vitamins absorbed with food such as A. B. C. D. etc., with their subgroups.

7. **Antibodies** — These combat disease, and get rid of foreign proteins.

8. **Gases in solution** — Nitrogen, oxygen and hydrogen circulate in the plasma.

9. **Waste products** — Those waste products formed from cellular activity, such as muscular action. They are transported in solution.

S. 154 THE FUNCTIONS OF THE PLASMA

The blood vessels will be studied in the chapter following, but mention must be made here of the capillaries. These are the smallest blood vessels. Their walls are very thin and consist of a single layer of cobblestone flat cells. Water and its dissolved components can pass into these capillaries or out of them through the thin walls.

Blood plasma with all its solutes (dissolved substances) is transported by blood vessels to all parts of the body. The digested food products, vitamins, and inorganic salts, plus water are taken from the digestive tract into the plasma by capillaries. The products are transported throughout the body to reach body tissues and cells. Hormones in solution are similarly transported. Oxygen is absorbed by capillaries in the lungs for distribution, (see red blood cells). Waste material in solution in plasma is carried to the kidneys

F. 12-2   BLOOD CELLS:

1. Red blood cells or erythrocytes;

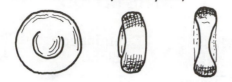

2. White blood cells or leucocytes

(1) Granulocytes;

Neutrophil   Eosinophil   Basophil

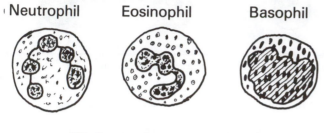

(2) Agranulocytes = non

Lymphocytes   Monocyte
Small   Large

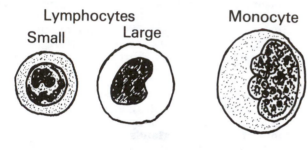

3. Platelets or thrombocytes

and other excretory organs to be eliminated. Carbon dioxide is conveyed to the lungs for excretion. The process includes first the taking up by the plasma of the solutes, then their transportation, and finally their passage into body cells, lungs, skin, kidneys, liver, etc. (Fluoridation of the water supply to a city consists of the dissolving of it in water, then pumping it throughout the water system. Blood plasma like the city water plays a similar role in the body)

Note: If a test tube filled with blood, to which a chemical has been added to prevent clotting, is allowed to stand (or is centrifuged) the plasma will rise to the top. This fluid will be colorless. The blood cells will drop to the bottom of the tube. F.12-1

S. 155   THE BLOOD CELLS   F.12-2

Blood cells are produced by the blood forming organs: the bone marrow, liver, spleen, lymph nodes, and thymus gland. They are set free upon maturity to enter blood capillaries and circulate in the blood. There are three groups: red blood cells, white blood cells, and platelets.

1. RED BLOOD CELLS or CORPUSCLES — ERYTHROCYTES

Corpuscle - corpus = body + culum = little; F.12-2. Erythrocyte - erythros = red + kytos (cytos) = cell. Red blood cells, R.B.C., are disc shaped cells, concave on both surfaces when viewed on edge. They are formed in the bone marrow. During the process of maturing they lose their nuclei, and are discharged into blood capillaries. They are minute, measuring only about seven microns in diameter. There are about 4,500,000 to 6,000,000 red blood cells per cubic mm of blood, about 25 trillion altogether. Males have more than females. They circulate in the plasma for about one hundred and twenty days (some authorities estimate a shorter period), and are then destroyed possibly by the liver, spleen, and lymph nodes. Red blood cells contain hemoglobin.

Hemoglobin is a compound of iron and a protein and forms in red blood cells during their development. Hemoglobin unites very readily with oxygen to form oxyhemoglobin. Oxyhemoglobin is bright red, and gives red blood cells, and therefore blood, its red color. Blood appears darker in color when oxygen has been extracted from it.

Functions of red blood cells: these cells by means of their hemoglobin transport oxygen from the lungs to all body tissues and cells through the circulation of blood. They also transport some carbon dioxide from body tissues to the lungs for elimination.

VITAMIN B 12 AND RED BLOOD CELL FORMATION

Red blood cells, as noted, are formed in bone marrow. Vitamin B12, absorbed by blood from the intestine, passes to the bone marrow and assists in the formation of red blood cells.

An intrinsic factor, an enzyme secreted by the stomach, must be present in the intestine for Vitamin B12 to be absorbed and passed to the bone marrow. This intrinsic factor is an enzyme.

A lack of secretion of the intrinsic factor by the stomach interferes with vitamin B12 absorption. When this occurs vitamin B12 is not available in sufficient quantity for the bone marrow to manufacture red blood cells. This may result in the development of pernicious anemia, a type of blood disease.

Normally any excess of vitamin B12 is stored in the liver. Animal livers will therefore contain vitamin B12. Animal liver, fed to any patient will supply extra vitamin B12, some of which will hopefully reach the bone marrow and combat anemia. Vitamin B12 may also be given by injection.

2. WHITE BLOOD CELLS: LEUKOCYTES: LEUCOCYTES

Leukocyte or leucocyte; leukos = white + kytos = a cell - also named white blood corpuscles.

White blood cells, W.B.C., have nuclei. They are less numerous than the red blood cells averaging from 6,000 to 10,000 cells per cubic millimeter of blood. This equals about one for every seven hundred red cells. They are classified according to whether or not they have granules in their cytoplasm as granular or nongranular leucocytes. They are further divided according to their reaction to staining dyes, and also by the shape of their nuclei. A combined stain consisting of an acid stain, eosin, that stains red, and a basic methylene blue stain that stains blue is used to stain them.

(1) **GRANULOCYTES:** these are white blood cells that have granules in the cytoplasm. They are formed in bone marrow. Following repeated division during which the

cells mature they are discharged into bone capillaries. They retain their nuclei. Three varieties of granulocytes are described: neutrophils, eosinophils, and basophils.

a. **Neutrophils** comprise from 60% to 70% of the leucocytes. They have fine granules in their cytoplasm that stain lilac or lavendar color, not red or blue. (neutrophil = neutral + phil = to like). The cell nucleus has from two to five lobes joined together by narrow strands. For this reason neutrophils are also called polymorphonuclear leucocytes, or polys. (poly = many + morpho = structure + nuclear; i.e. many structured nucleus). The nucleus looks like several nuclei joined together.

b. **Eosinophils**: these leucocytes have medium sized granules in their cytoplasm that stain red with an acid dye such as eosin. The nucleus has lobes. These cells make up about 2% to 3% of all leucocytes.

c. **Basophils**: the third group of granulocytes have coarse granules in their cytoplasm that stain blue with a basic dye such as methylene blue. They make up less than 1% of the leucocytes.

(2) AGRANULOCYTES: NONGRANULOCYTES
these leukocytes do not have granules in the cytoplasm. Like granulocytes they retain their nuclei. Two types, the lymphocytes and monocytes are described.

a. **Lymphocytes**: these cells have a single large nucleus that occupies most of the cell and stains a reddish blue color. The cytoplasm surrounding the nucleus is small and stains a faint blue. Lymphocytes are divided into large and small lymphocytes. They comprise about 25% of the leucocytes. Lymphocytes are formed in lymphoid tissues, particularly in the lymph nodes (glands) and spleen.

b. **Monocytes** (mononuclear or transitional): these cells are larger than the other leucocytes and have a single rounded or kidney shaped nucleus, that stains a reddish purple. The cytoplasm stains blue. Monocytes are formed in bone marrow.

Functions of leucocytes:

Leucocytes are mobile, i.e., can move about and may pass out of the capillaries. They do this by passing between adjacent cell margins of the thin capillary walls.

(a) **Leucocytes** defend the body from bacterial infections. They ingest or eat up the bacteria thus destroying them. They may also disintegrate liberating substances that destroy bacteria. They will migrate to the site of a wound or infected area, surround the part and perform their function (phagocytosis). The neutrophils are most active, as are the monocytes.
Phagocytes are cells that can ingest and destroy bacteria, etc. Certain plasma cells, formed by the reticuloendothelial tissues are also phagocytes. (See inflammation in Pathology).
When an acute infection, such as an appendicitis occurs the number of leucocytes in the circulating blood increases. These cells migrate out to the infected area. They form a wall about the lesion and wage a battle to overcome the bacteria present. (If there is an increase in the number of leucocytes in the circulating blood this confirms the suspicion of some inflammatory process in the body).

(b) **Leukocytes** help to repair damaged tissues by a process of phagocytosis, in which dead tissues resulting from injury or bacterial infection are dissolved and removed.

(c) They also aid in the clotting of blood.

(d) Eosinophils are increased in number in asthma and some parasitic diseases.

(e) Lymphocytes are responsible for the development of immunity, the production of antibodies, gamma globulin, etc. In chronic types of infection such as tuberculosis the number of lymphocytes will show an increase.

3. **BLOOD PLATELETS OR THROMBOCYTES**
   F.12-2(3)

These are small irregular fragments of cells that are formed in the bone marrow from much larger cell bodies. They circulate in the blood plasma and are concerned with the clotting of blood. They number from 200,000 to 400,000 per cubic mm and appear as minute stained particles in blood smears.

THE COAGULATION OF BLOOD   F.12-1

The coagulation or clotting of blood consists of the formation of plugs to prevent blood loss and hemorrhage from injured vessels. The bleeding may be externally through the skin, or into body cavities, or into tissues from the rupture of blood vessels, e.g. **Epistaxis** refers to a bleeding nose, a **hematoma** refers to bleeding into some body tissue resulting from torn vessels due to some injury with a bruising or tearing of tissue. If near the surface there may be visible swelling and discoloration of the overlying skin. The color may gradually turn from purple to a greenish yellow, etc., e.g. a black eye. Hemorrhage from a large vessel or persistent or repeated bleeding from a small vessel may be serious.

The factors involved in the process of clotting include:

1. **Fibrinogen**, a blood protein, formed in the liver, circulates in solution in blood plasma.

2. **Prothrombin** is an enzyme formed in the liver, and circulates in solution in the blood plasma.

3. **Calcium** ingested with food also circulates in the blood plasma in an ionized state.

4. **Thromboplastin** or thrombokinase does not circulate in the plasma but is present in tissue cells. Blood platelets may contain a similar substance.

5. **Heparin** or antithrombin is present in the liver and some other organs and dissolved in plasma. It acts to prevent blood from clotting within the blood vessels.

CLOTTING PROCEDURE

Following an injury the injured tissue cells liberate thromboplastin at the site of the injury. Platelets may also give a similar substance.

Thromboplastin thus freed comes into contact with spilled blood and converts the **prothrombin** dissolved in it to **thrombin.**

Thrombin + **calcium** + **fibrinogen** form **fibrin**. Fibrin consists of fine threads precipitated from the dissolved fibrinogen. These form a network of inter-lacing filaments. Blood platelets congregate among the threads, and red blood cells are trapped in the network. A red clot is formed. The fibrin threads with platelets and red blood cells plug the holes and stop bleeding. If no red blood cells are trapped the clot is white.

Fibrinogen
Calcium
Prothrombin } = thrombin
Thromboplastin

Thrombin + fibrinogen + calcium = clot

Clotting of blood within blood vessels or the heart does not normally occur. It may result when a vessel wall is injured or roughened, or when the circulation is slow. Then a **thrombus** may form.

PROTHROMBIN FORMATION

Bile is secreted by the liver and passes through the common bile duct to the small intestine where it is mixed with food.

Vitamin K from food also enters the small intestine. The bile assists in the absorption of vitamin K from the small intestine into the portal vein. It passes to the liver.

Vitamin K in the liver assists in the formation of prothrombin by this organ, to be circulated in solution in the blood.

Obstruction of the common bile duct by a gall stone, tumor, etc., will prevent bile from entering the small intestine. This in turn will interfere with the absorption of vitamin K and prothrombin manufacture. The patient lacking prothrombin may hemorrhage. The patient will become jaundiced, with yellow skin, and sclera. Vitamin K is also formed in the large intestine by bacterial action. Lack of bile may interfere with absorption of the vitamin K here as well.

BLOOD GROUPING

When blood from one human is added to blood of another human, e.g. for transfusion, in some instances no adverse reaction may occur. The bloods are compatible and a transfusion may be given.

In other instances when the blood of two humans is mixed the red blood cells will form clumps, i.e. will agglutinate. The red blood cells then disintegrate and their hemoglobin is set free. As this hemoglobin is conveyed to the kidneys for excretion it will destroy cells of the renal tubules and may cause death to the subject.

The human subject donating blood to another is the **donor**. The subject receiving the transfusion of blood is the **recipient**.

In those cases in which a reaction occurs it may be due to an agglutinin (antibody) in the serum of the donor or recipient, or an agglutinogen (antigen) attached to the red blood cells of the donor or recipient. Human blood falls into four main groups designated as O, A, B, or AB. In the table shown here a plus (+) sign indicates clumping, while a minus (-) sign indicates no clumping.

Corpuscles	Serum			
•	O	A	B	AB
O	-	-	-	-
A	+	-	+	-
B	+	+	-	-
AB	+	+	+	-

Note that the corpuscles of O do not clump when mixed with any serum. O is therefore a universal donor. Group AB serum does not cause clumping of the cells of any group so can receive cells from any group, and is a universal recipient.

In actual practice, because other minor reactions may occur the blood cells of each possible donor are checked directly with the recipient's serum, and the donor's serum is checked directly with the cells of the recipient. This is direct matching.

THE RH FACTOR

RH - rhesus monkey - a species of monkey with a factor similar to that found in the human race.

The RH factor is present in a large percentage of the members of the white race. These persons are termed RH positive. If this factor is not present they are RH negative.

A husband who is RH positive may conceive an RH positive fetus. If the mother is RH negative she will produce antibodies to neutralize the positive RH factor of the fetus. The fetus may absorb these antibodies. This may result in the destruction of fetal red blood cells. At birth a transfusion is done in which the fetal blood is drained off and replaced with blood containing no antibodies against the RH factor.

In a similar manner if a patient with RH negative blood receives a transfusion of blood containing the RH factor the recipient will produce antibodies that will destroy the red cells of the RH positive blood of a second transfusion of RH positive blood is given.

SUMMARY OF FUNCTIONS OF THE BLOOD

1. **Blood plasma**: water for body tissues and cells, to provide a solvent in which digested food products, blood proteins, minerals, gases, vitamins, hormones and waste products are dissolved, for transportation, and a medium in which blood cells may circulate throughout the body.

2. Red blood cells contain hemoglobin for the transportation of oxygen, and some carbon dioxide.

3. **Hemoglobin** is an iron protein compound in red blood cells that unites readily with oxygen forming oxyhemoglobin for the transportation of oxygen to body tissues.

4. **Leukocytes,** circulating in plasma, destroy microorganisms resulting from infection, remove injured and dead tissue following injury or infection, and help to form blood clots.
   Lymphocytes function in immunization.

5. **Blood platelets** function in blood clotting, liberate thromboplastin or similar substances, and clump to plug ruptured vessels.

6. **Coagulation** of blood occurs to arrest hemorrhage by plugging bleeding vessels, and requires thromboplastin, prothrombin, fibrinogen, calcium, blood platelets, and red and white blood cells for this procedure.

S. 156 **SOME PATHOLOGICAL TERMS — BLOOD**

Leucocytosis or leukocytosis - an increase in the number of leucocytes in the circulating blood.

Leucopenia — a decrease in the number of leucocytes in the circulating blood. (Leuco = white + penia = poverty; i.e. a poverty of leucocytes).

Agranulocytosis — an absence or a marked decrease in the number of leucocytes in circulating blood. This may occur in a patient receiving radiation therapy.

Anemia — (1) a decrease in the number of red blood cells in circulating blood;

(2) a decrease in the hemoglobin content of each red blood cell — an iron deficiency (an = without + emia, haima — blood).

Polycythemia — an increase in the number of the red blood cells in circulating blood. (poly = many + cytos = cell + haima = blood).

Agammaglobinemia — the absence of gammaglobulin, containing antibodies.

Hema = Haima - blood; so hemorrhage, hematoma, etc.

S. 157 ANATOMICAL TERMS RELATING TO THE BLOOD

blood
blod (AS)

hemoglobin
white blood cells

Haima; hema	WBC
inorganic salts	leukocytes
blood proteins	leucocytes
albumin	granulocytes
globulin	neutrophils or
fibrinogen	polymorphonuclear
amino acids	basophils
fatty acids	eosinophils
glycerine	nongranulocytes
glucose	lymphocytes, small
hormones	lymphocytes, large
vitamins	monocytes or
antibodies	transitionals or
red blood cells, RBC	mononuclears
erythrocytes	phagocytes
corpuscles	phagocytosis

— THE HEART —

1. **PERICARDIAL SAC**
   1) Parietal pericardium or serous pericardium
   2) Fibrous pericardium; peri = around

2. **STRUCTURE OR COVERINGS**
   1) Visceral pericardium or epicardium; epi = upon
   2) Myocardium — cardiac muscle; mys = muscle
   3) Endocardium; endo = within

3. **CHAMBERS**
   1) Left atrium with auricle
   2) Right atrium with auricle
   3) Left ventricle
   4) Right ventricle

4. **SEPTA**
   1) Interatrial septum
   2) Interventricular septum

5. **SULCI**
   1) Coronary sulcus or groove
   2) Anterior interventricular sulcus (groove)
   3) Posterior interventricular sulcus or groove

6. **OPENINGS WITHIN HEART**
   1) Left atrioventricular orifice (ostium)
   2) Right atrioventricular orifice (ostium)

7. **OPENINGS INTO OR FROM HEART**
   1) Opening (ostium) of superior vena cava ⎤
   2) Opening (ostium) of inferior vena cava ⎬ into right atrium
   3) Opening (ostium) of coronary sinus ⎦
   4) 2 Openings (ostia) of Lt. pulmonary veins ⎤ into left atrium
      2 Openings (ostia) of Rt. pulmonary veins ⎦
   5) Pulmonary orifice (ostium) = opening pulmonary trunk   from right ventricle
   6) Aortic orifice (ostium) = opening aorta   from left ventricle

8. **CARDIAC VALVES**
   1) Left atrioventricular valve, bicuspid valve, mitral valve
   2) Right atrioventricular valve, tricuspid valve — 3 flaps
   3) Pulmonary valve or pulmonary trunk valve, 3 semilunar cusps
   4) Aortic valve or aortic semilunar valve — 3 semilunar cusps

9. **ARTERIES TO HEART**
   1) Left coronary artery & branches
   2) Right coronary artery & branches

10. **VEINS OF HEART**
    1) Coronary sinus   small cardiac vein / middle cardiac vein / great cardiac vein
    2) Other small veins

11. **NERVES TO HEART**
    1) Sympathetic
    2) Parasympathetic — vagus nerves, Rt. & Lt.

13. THE CIRCULATORY SYSTEM

S. 158 **COMPONENTS OF THE CIRCULATORY SYSTEM** F.13-1

The circulatory system or cardiovascular system is the transportation system of the body. There are actually two parts:

the pulmonary — that conveys blood from the heart to the lungs, and back to the heart;

the systemic — that carries blood from the heart to all body tissues and cells (except some parts of the lungs) and back to the heart.

The circulatory system therefore distributes and collects blood. As a result of this circulation the blood, with its cells and dissolved constituents, reaches all body tissues and cells.

The circulatory system is made up of several component parts: (F.13 - 1, 2)

1. The heart
2. Arteries
3. Arterioles
4. Capillaries
5. Venules
6. Veins
7. Lymphatic vessels and nodes
8. Reticuloendithelial tissues.

1. THE HEART is a muscular pump that propels blood either to the lungs or to other body tissues. It will be studied in detail subsequently. F.13-4

2. THE ARTERIES are distributing vessels that carry blood AWAY from the heart. They may be compared to a tree with two trunks, each of which branches and rebranches forming smaller and smaller vessels as they proceed. These branches branch and rebranch, becoming progressively smaller. Some arteries end by dividing into two terminal divisions, often similar in size. Others divide into a series of small vessels. The pattern of branching is quite similar for all members of any one species. In addition, the pattern on both sides of the trunk, head, and both upper and lower limbs is similar bilaterally.

Arteries are sometimes named from their location, e.g. the femoral artery, and sometimes from the organ that is supplied with blood by them, e.g. the renal artery. Some arteries change their name as they proceed, e.g. subclavian, axillary and brachial. The branches of the larger arteries are also named.

The two large trunk arteries are:
- a. the aorta - to all body tissues
- b. the pulmonary trunk - to the lungs.

3. ARTERIOLES are the small final branches of the smallest arteries, and are present in all tissues that are supplied with blood.

4. CAPILLARIES are very minute microscopic hairlike vessels that form networks between the small arterioles

---

F. 13-1   SCHEMATIC DIAGRAM OF CIRCULATORY SYSTEM - TRANSPORTATION

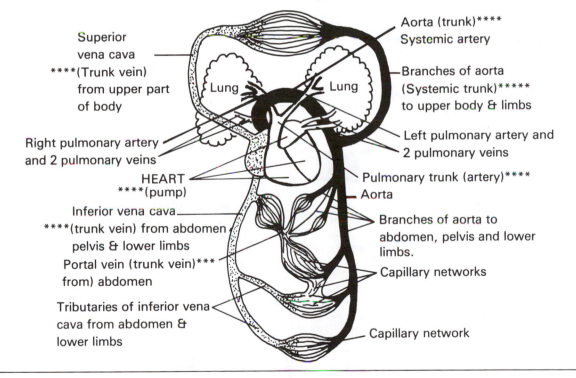

F. 13-2   ARTERY, ARTERIOLES, CAPILLARIES, VENULES, VEIN

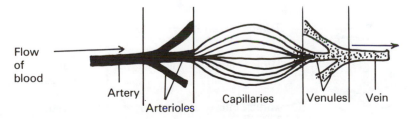

and the venules. Each arteriole divides into many of these minute vessels. At their opposite ends they join to form a venule. The walls of these capillaries have a single layer of flat epithelium with cobblestone cells cemented together to form minute tubes. (L. capillary — a hair, i.e. hairlike).

Functions: exchange of blood and tissue constituents takes place through the capillary walls. Some pass out of the capillaries into tissues, others pass in the opposite direction into the capillaries.

5. **VENULES** are the smallest veins and correspond to the arterioles. They are formed by the union of many capillaries.

6. **VEINS** are collecting vessels that bring blood back to the heart. Veins might be compared to a river with its tributaries. These begin as tiny streams that unite to form larger and larger streams and finally a river. Veins begin as venules that unite to form minute veins. These unite to form larger and larger veins and finally end in the **venous trunks** that empty into the heart.

Veins are described as having tributaries that unite, while arteries branch. Veins have valves that prevent backflow of blood. The limbs have superficial veins close to the skin, and deep veins that accompany arteries. These latter take the same name as the corresponding artery, e.g. femoral artery and femoral vein. The names of many veins are not similar to arteries of the same area, e.g. carotid artery but internal jugular vein. **The trunk veins are:**

a.   Superior vena cava - drains upper half of body.
b.   Inferior vena cava - drain lower half of body.
c.   Four pulmonary veins, two from each lung - drain the lungs, all emptying into the heart.

7. **LYMPHATIC VESSELS AND NODES** — Lymph is a colorless fluid that collects in lymphatic vessels. It is derived from fluid in the spaces between body cells in various tissues — intercellular spaces.

Lymph capillaries — minute vessels which begin as very small vessels. They are not connected to arterioles nor venules.

Lymph vessels are formed by the union of capillaries that unite to form larger and larger vessels. They finally empty into two large trunks.

Lymph nodes (formerly called glands) are connected with the larger lymph vessels.

The trunk lymph vessels are:

a.   Thoracic duct — draining whole body except the upper right part.
b.   Right lymph duct — draining right upper limb, right thorax, right side of head and neck.

The lymphatics form a SECOND COLLECTING SYSTEM.

8. **RETICULOENDOTHELIAL STRUCTURES**
These include bone marrow, lymph nodes, spleen, liver and the thymus gland.

The student may be confused by the terms arterial system, venous system, lymphatic system, reticuloendothelial system, etc. All of these are in fact parts of the circulatory system, and are interdependent.

S. 159   THE STRUCTURE OF BLOOD VESSELS
F.13-3

Before proceeding with a detailed study of the parts of the circulatory system, the microscopic appearance of the walls of blood vessels should be briefly defined. Blood vessels have three coats or coverings: an internal, a middle, and an outer coat.

a. **An internal coat** or layer called the tunica intima or simply the intima, is a layer of flat cobblestone like cells that form a lining membrane, with a connective tissue and elastic tissue base.
b. **A middle coat** or tunica media is a layer of visceral muscle, the fibers encircling the vessel, with considerable elastic tissue as well.
c. **An outer layer** or tunica adventitia is a layer of connective tissue outside of the middle coat.

In medium sized arteries the muscular layer is thick; in very large arteries and trunk vessels it is largely replaced by elastic tissue.

Because of the muscular and connective tissue layers, blood vessels can expand to accommodate an increase in blood flow, and can contract with a decreased flow.

The lumen is the central cavity in any blood vessel.

Veins and lymphatic vessels have similar coats to arteries, but their walls are not as thick. They have valves to prevent a backflow.

Capillaries, as stated previously have very thin walls consisting of a single layer of flat cells.

S. 160   THE HEART — DETAILED STUDY   F.13-4, 8

Heart; (NA), cardia; (AS), heorte
G. Kardia - to L. cardia; adj. cardiac
L. Cor; Ex. precordial; cor bovinum = bull's heart.

The heart is the pump of the circulatory system. It receives blood and distributes it to body tissues. The heart lies in the lower anterior chest, posterior to the sternum and adjacent costal cartilages, and within the mediastinum. It rests upon the upper surface of the diaphragm. Two-thirds of it lie to the left of the median line, one-third to the right. It is about the size of a closed fist. It has an apex and a base.

The apex of the heart is its bluntly pointed end that is directed to the left and anteriorly. In the male, in whom the position of the nipple is fairly constant, the apex lies medial to the nipple.

The base is the broad end that is directed to the right, posteriorly and cranially to the right of the right sternal border.

F. 13-3   CROSS SECTION OF AN ARTERY

Tunica adventitia = outer layer - connective tissue

Tunica media = middle layer (muscle)
(elastic tissue)

Tunica intima = inner layer - endothelium

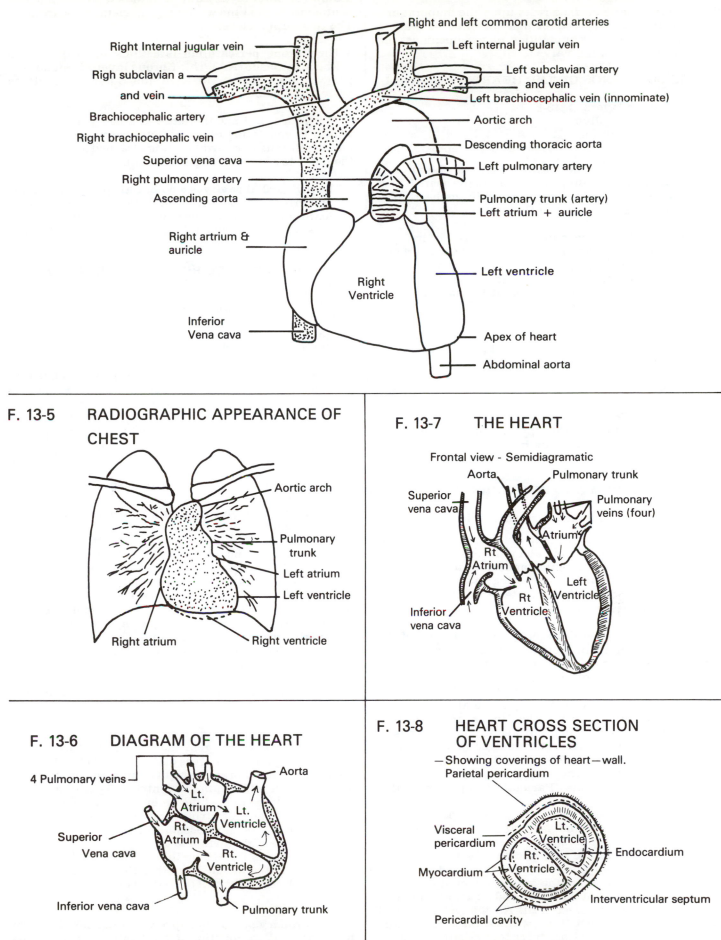

F. 13-4   THE HEART, TRUNK ARTERIES AND TRUNK VEINS

Right and left common carotid arteries

Right Internal jugular vein

Left internal jugular vein

Righ subclavian a and vein

Left subclavian artery and vein

Brachiocephalic artery

Left brachiocephalic vein (innominate)

Right brachiocephalic vein

Aortic arch

Superior vena cava

Descending thoracic aorta

Right pulmonary artery

Left pulmonary artery

Ascending aorta

Pulmonary trunk (artery)

Left atrium + auricle

Right artrium & auricle

Right Ventricle

Left ventricle

Inferior Vena cava

Apex of heart

Abdominal aorta

F. 13-5   RADIOGRAPHIC APPEARANCE OF CHEST

Aortic arch

Pulmonary trunk

Left atrium

Left ventricle

Right atrium

Right ventricle

F. 13-7   THE HEART

Frontal view - Semidiagramatic

Aorta

Pulmonary trunk

Superior vena cava

Pulmonary veins (four)

Atrium

Rt Atrium

Left Ventricle

Inferior vena cava

Rt Ventricle

F. 13-6   DIAGRAM OF THE HEART

4 Pulmonary veins

Aorta

Lt. Atrium

Lt. Ventricle

Superior Vena cava

Rt. Atrium

Rt. Ventricle

Inferior vena cava

Pulmonary trunk

F. 13-8   HEART CROSS SECTION OF VENTRICLES

—Showing coverings of heart—wall.

Parietal pericardium

Visceral pericardium

Lt. Ventricle

Rt. Ventricle

Endocardium

Myocardium

Interventricular septum

Pericardial cavity

131

The sternocostal surface lies posterior to the sternum and costal cartilage

The diaphragmatic surface rests upon the diaphragm.

The heart has three coverings forming its wall, a septum, four chambers or cavities, eleven openings, and four sets of functioning valves.

1. THE COVERINGS OF THE HEART — THE WALL

There are three layers : F.13-8
(1) the endocardium
(2) the myocardium
(3) the pericardium — visceral layer, epicardium

The endocardium; endo = within + cardia = heart.
This layer forms the lining membrane of the heart. It also covers the surfaces of the valves within the heart. It is composed of a single layer of flat cells.

The myocardium; mys = muscle + cardia = heart.
The myocardium is the muscular layer of the heart and consists of a special type of muscle found only in the heart, cardiac muscle. This was described in the section on muscular tissue. The muscle fibers have a single nucleus, cross striations, and branching fibers. The bundles of fibers encircle the atria and ventricles. The muscular layer of the left ventricle is much thicker than that of the other chambers as it must pump blood to the whole of the body.

The Visceral Pericardium or epicardium; peri = around + cardia = heart; also epi = upon + cardia.
This layer forms a thin covering and is applied to the outer surface of the muscular layer. Its smooth surface is directed outwards. This is the inner layer of the serous pericardium described below.

THE PERICARDIUM AND PERICARDIAL SAC
F.13-8
Pericardium = around the heart; adj. pericardial.

The pericardium surrounding the heart consists of two parts; a **serous pericardium**, and a **fibrous pericardium**. The serous pericardium is further divided into a visceral and parietal layer. This serous pericardium consists of a thin serous membrane that secretes a thin watery fluid. See epithelial tissue S.22.

The visceral pericardium (NA), visceral layer of the serous pericardium or epicardium as defined above forms the outer covering of the heart. (Visceral from viscus = organ - part of the organ).

The parietal pericardium (NA, parietal layer of the serous pericardium) forms the lining of a sac that encloses the heart. Its inner surface is smooth and is in contact with the smooth outer surface of the visceral pericardium. The two layers are not attached except at the base of the heart where they surround the cardiac ends of the trunk vessels. (Parietal from paries = wall - forming a wall).
In the developing state the serous pericardium, including the visceral and parietal layers develops as a hollow closed ball of cells. The muscular heart pushes against one side of the ball and finally becomes surrounded by the two layers of the collapsed ball. The student may demonstrate the arrangement by using the closed fist to represent the heart, and a partially inflated toy balloon, as the closed ball. As the fist is pushed against one side of the balloon, while some air is allowed to escape from it, the fist will force the two surfaces of the balloon into contact with each other. The layer that touches the fist is closely applied to it (visceral pericardium). The outer layer while in contact with the inner one is attached to it at the wrist only (parietal pericardium). The two form a closed sac.

The fibrous pericardium is a thick layer of fibrous connective tissue (NA). This fibrous layer surrounds the parietal pericardium to form the outer layer of the pericardial sac.

The pericardial sac therefore is a bag with the heart within it, consisting of the parietal serous pericardium and the fibrous pericardium. It is attached at the base of the heart to the visceral pericardium.

The pericardial cavity is a potential space between the inner surface of the pericardial sac and the visceral pericardium, the outer covering of the heart. It contains a small amount of serous fluid. The heart contracts within this sac.

2. THE FOUR CHAMBERS OR CAVITIES OF THE HEART F.13-6

(1) left atrium; pl. atria; adj. atrial
(2) right atrium
(3) left ventricle; adj. ventricular
(4) right ventricle

The heart is divided into two halves, right and left, by a partition that extends from the base to the apex. There is no communication between these two halves after birth. Each half is further divided into an **atrium** and a **ventricle**. These cavities are named right and left atria, and right and left ventricles. (s. atrium; pl. atria) The atria lie at the basal end, the ventricles towards the apex. F.13-6, 7.

The interatrial septum is that part of the partition between the right and left atria.

The interventricular septum is the partition between the right and left ventricles.

Openings are present between the right atrium and right ventricle, and between the left atrium and left ventricle. Fibrous rings surround these openings. In the (NA) the openings are termed "ostia".

The left atrioventricular opening (ostium) is the opening between the left atrium and left ventricle.

The right atrioventricular opening (ostium) is the opening between the right atrium and right ventricle.

The auricles are two ear shaped pouches, one forming part of each atrium.

The conus arteriosus is the prominent anterior part of the right ventricle where it opens into the pulmonary trunk.

The terms right atrium and right ventricle, left atrium and left ventricle suggest that these cavities occupy either the right or left halves of the heart. Actually as the heart lies in the thorax the right atrium and ventricle lie anterior to the left atrium and ventricle. They might more appropriately be named the anterior atrium and ventricle, and the posterior atrium and ventricle.

The anterior or sternocostal surface of the heart is formed by the right atrium to the right, and almost all of the remainder is formed by the right ventricle. A small part of the left ventricle is visible along the left cardiac border and a small part of the left atrium is visible above it.

The coronary sulcus is a groove on the surface of the heart marking the junction of the atria and the ventricles. It completely encircles the heart.

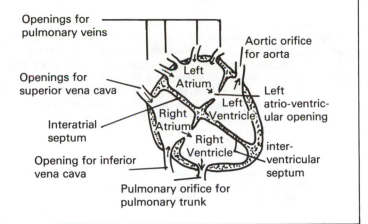

Openings for
pulmonary veins

Aortic orifice
for aorta

Left
Atrium

Openings for
superior vena cava

Left
Ventricle

Left
atrio-ventric-
ular opening

Right
Atrium

Interatrial
septum

Right
Ventricle

inter-
ventricular
septum

Opening for inferior
vena cava

Pulmonary orifice for
pulmonary trunk

F. 13-10   VALVES OF THE HEART

(1) Semilunar Valve — Closed

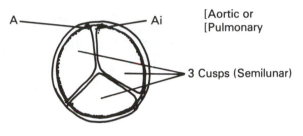

A

Ai

[Aortic or
[Pulmonary

3 Cusps (Semilunar)

(2) Semilunar valve — open

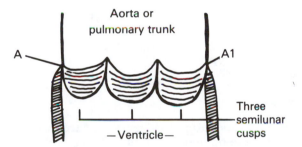

Aorta or
pulmonary trunk

A

A1

Three
semilunar
cusps

— Ventricle —

(3) Right atrioventricular or
tricuspid valve — closed

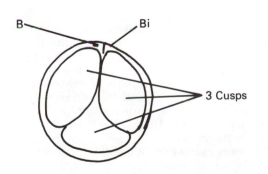

B

Bi

3 Cusps

(4) Right atrioventricular or tricuspid valve — open

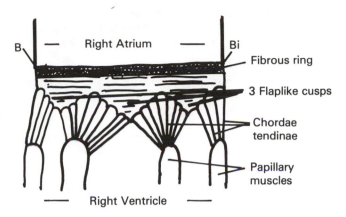

B

Right Atrium

Bi

Fibrous ring

3 Flaplike cusps

Chordae
tendinae

Papillary
muscles

Right Ventricle

(5) Left atrioventricular or bicuspid or mitral valve — closed

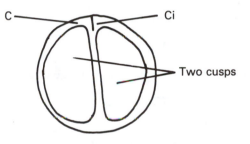

C

Ci

Two cusps

(6) Left atrioventricular or bicuspid or mitral valve — open

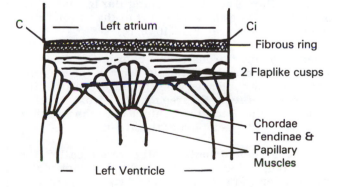

C

Left atrium

Ci

Fibrous ring

2 Flaplike cusps

Chordae
Tendinae &
Papillary
Muscles

Left Ventricle

The **interventricular sulci**, anterior and posterior are the grooves where the ventricles meet each other anteriorly and posteriorly. They extend from the coronary sulcus towards the apex.

These three sulci accommodate the coronary arteries and their main branches.

3. THE OPENINGS OF THE HEART OR OSTIA
F.13-9

There are eleven openings into or within the heart.
(1) and (2) **The right and left atrioventricular** openings (ostia) between the atria and ventricles have been defined.

(3) **The pulmonary opening** (opening of the pulmonary trunk) is located between the right ventricle and the pulmonay trunk.

(4) **The aortic opening** (ostium) is located between the left ventricle and the aorta.

(5) **The opening (ostium) of the inferior vena cava** is located where the inferior vena cava empties into the right atrium.

(6) **The opening (ostium) of the superior vena cava** is located where the superior vena cava empties into the right atrium.

(7) (8) (9) (10) The four openings (ostia), two from the right and two from the left **pulmonary veins** are between the pulmonary veins and the left atrium.

(11) **The opening of the coronary sinus** lies between this sinus and the right atrium. This large vein helps drain blood from the cardiac muscle and it lies in the posterior part of the coronary sulcus.

Note — List the openings in each cardiac chamber.

4. THE VALVES OF THE HEART   F.13-10

The student should note that there is **only one** valve at each opening. Each valve has two or three cusps or flaps. Four of the openings have functioning valves: the two atrioventricular, the pulmonary, and aortic openings.

The left atrioventricular valve (also called the mitral or bicuspid valve) is located at the left anterioventricular opening, between the left atrium and left ventricle. This valve has two cusps or flaps, each somewhat triangular in shape. (Bicuspid; bi = two + cusps), (mitral from mitre - a fold, a bishop's hat, i.e. resembling a bishops hat with its fold). Each flap is somewhat triangular in shape. The base of each flap is attached to part of the fibrous ring surrounding the opening. F.3 - 10 (5, 6)
The other two borders of each flap are free except for attachments by fine fibrous cords (chordae tendinae) to the wall of the ventricle. These flaps close the opening when the ventricle contracts, when blood is forced behind them. The cords prevent the free margins from entering the adjacent atrium. **Backflow of blood from the ventricle is thus prevented from entering the atrium.**

The right atrioventricular valve (also called the tricuspid valve) is located at the right atrioventricular opening between the right atrium and right ventricle. It has three flaps or cusps, hence the name tricuspid, (tri = three). The base of each cusp is attached to part of the circumference of the fibrous ring surrounding the opening, with cords at the free edges as described above. Upon closure the **blood in the ventricle is prevented from flowing back into the right atrium.**

The aortic valve or aortic semilunar valve is located at the opening between the left ventricle and the aorta. It has three semilunar cusps (half-moon shaped). Each cusp is attached by its outer curved margin to the wall of the aorta, with its other free margin in the lumen of the aorta. The student should visualize three shirt pockets sewn around the inner surface of a hollow sleeve, with the top of each pocket free in the lumen. The cusps remain flat against the wall when blood is flowing from the ventricle into the aorta and cause no impedence. When the ventricle dilates to refill the blood tries to flow back into the ventricle from the aorta. The cusps then fill with blood to close the opening and prevent backflow from aorta to ventricle. The pulmonary valve, outlined in the following paragraph is similar in construction and function. F.13-10 (1) (2)

F. 13-10 (1)
The pulmonary valve (valve of the pulmonary trunk) is located at the pulmonary opening between the right ventricle and pulmonary trunk. It is similar to the aortic valve and has three semilunar cusps. When closed it **prevents a backflow of blood** from the pulmonary trunk into the right ventricle. F.13-10 (1, 2)

The inferior vena cava and the coronary sinus also have valves but these are usually not functional due to incomplete formation.

The student should again note that veins and lymphatics also have semilunar valves at intervals along their length.

As stated repeatedly above, valves prevent the flow of blood backwards, but permit it to flow in the proper direction.

5. BLOOD SUPPLY TO THE HEART   F.13-11

The right and left coronary arteries, the first branches of the aorta, supply blood to the heart. These vessels run along the coronary sulcus and the interventricular grooves on the surface of the heart giving off branches 'en route'.

Cardiac veins drain blood from the cardiac capillaries. Some of these open directly into the right atrium.

The coronary sinus is a large vein that lies in the posterior part of the coronary sulcus. It empties into the right atrium. It collects blood from most of the cardiac veins.

6. CONDUCTING APPARATUS AND NERVES OF THE HEART

A special conducting apparatus consisting of a node in the wall of the right atrium initiates the impulses for contraction of the heart. These impulses spread by modified cardiac muscle fibers to the atria and ventricles. The heart is thus capable of contracting without nerve impulses. The sympathetic and parasympathetic do send fibers to the heart consisting of;
a. sympathetic nerve fibers, which when stimulated increase the heart rate.
b. the two vagi nerves (10th cranial) supply fibers also, and stimulation of them slows the heart rate.

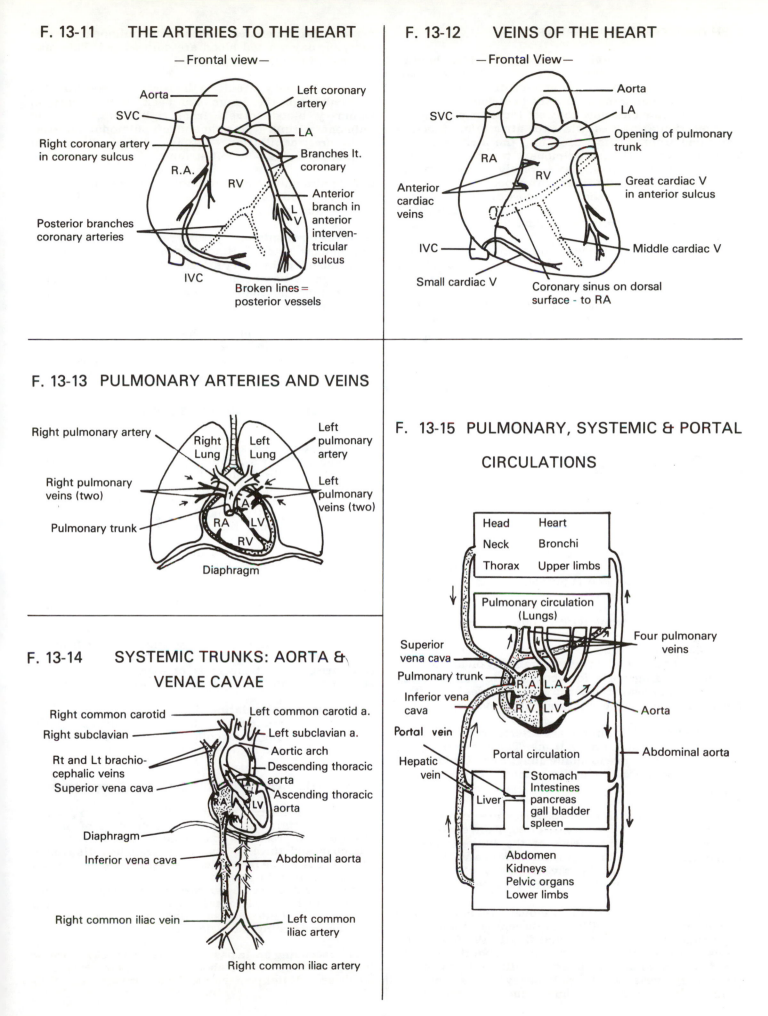

F. 13-11 THE ARTERIES TO THE HEART

— Frontal view —

Aorta
SVC
Right coronary artery in coronary sulcus
Posterior branches coronary arteries
R.A.
RV
IVC
Left coronary artery
LA
Branches lt. coronary
Anterior branch in anterior interventricular sulcus
L V
Broken lines = posterior vessels

F. 13-12 VEINS OF THE HEART

— Frontal View —

SVC
Anterior cardiac veins
IVC
Small cardiac V
RA
RV
Aorta
LA
Opening of pulmonary trunk
Great cardiac V in anterior sulcus
Middle cardiac V
Coronary sinus on dorsal surface - to RA

F. 13-13 PULMONARY ARTERIES AND VEINS

Right pulmonary artery
Right pulmonary veins (two)
Pulmonary trunk
Right Lung
Left Lung
RA
LV
RV
Diaphragm
Left pulmonary artery
Left pulmonary veins (two)

F. 13-15 PULMONARY, SYSTEMIC & PORTAL CIRCULATIONS

Head	Heart
Neck	Bronchi
Thorax	Upper limbs

Pulmonary circulation (Lungs)

Superior vena cava
Pulmonary trunk
Inferior vena cava
Portal vein
Hepatic vein

R.A. L.A.
R.V. L.V.

Four pulmonary veins
Aorta
Abdominal aorta

Portal circulation

Liver
Stomach Intestines pancreas gall bladder spleen

Abdomen Kidneys Pelvic organs Lower limbs

F. 13-14 SYSTEMIC TRUNKS: AORTA & VENAE CAVAE

Right common carotid
Right subclavian
Rt and Lt brachio-cephalic veins
Superior vena cava
Diaphragm
Inferior vena cava
Right common iliac vein
RA
LV
Left common carotid a.
Left subclavian a.
Aortic arch
Descending thoracic aorta
Ascending thoracic aorta
Abdominal aorta
Left common iliac artery
Right common iliac artery

7. PHYSIOLOGY OF THE HEART

The two atria, the receiving chambers of the heart having filled with blood from the trunk veins contract together. They force blood through the atrioventricular openings (these valves being open) into the ventricles.

The filled ventricles now contract together closing the atrioventricular valves, while the atria relax. Blood is prevented from flowing back into the atria. Blood is forced into the aorta and pulmonary trunk.

The two ventricles now relax to receive blood from the atria which have again filled with blood.

The pulmonary and aortic valves close due to blood in the aorta and pulmonary trunk filling their cusps. Blood is thereby prevented from flowing back into the relaxed ventricles.

This cycle is repeated some 72 times a minute, 4,320 times per hour and 103,680 times per day when the subject is at rest.

The closure of the atrioventricular valves produces a sound that is audible when a stethoscope is used. The closure of the semilunar valves also causes a sound, also audible. By stethoscope these sounds are audible as "lub dub, lub dub", etc.

Systole is the contracting phase of the heart. (adj. systolic)

Diastole is the dilation phase of the heart. (adj. diastolic).

S. 161   THE PULMONARY VESSELS AND CIRCULATION

Pulmo = lung; pl. pulmones; adj. pulmonary.   F.13-13

The pulmonary vessels carry deoxygenated blood from the right ventricle to the lungs and return it as oxygenated blood to the heart. The pulmonary vessels include the pulmonary trunk, the right and left pulmonary arteries and their branches, the lung capillaries, and the two right and two left pulmonary veins with their tributaries.

1. The pulmonary trunk (O.T. Pulmonary artery)

The pulmonary trunk arises from the conus arteriosus of the right ventricle. It extends cranially on the left side of the ascending aorta for about two inches inclining posteriorly. Beneath the aortic arch it divides into its two terminal branches, the right and left pulmonary arteries.

2. The right and left pulmonary arteries result from the division of the pulmonary trunk. These arteries pass laterally to the corresponding lung. Each divides into lobar, segmental, and smaller branches to end as lung capillaries.

3. The pulmonary (lung) capillaries are formed from arterioles of the pulmonary arteries. They encircle the hollow cup shaped air sacs of the lungs. Here, an exchange of oxygen into capillaries from alveoli, and of carbon dioxide into the alveoli takes place. These capillaries end by uniting to form venules.

4. The two right and two left pulmonary veins are formed by the union of lung capillaries into venules, small veins, segmental and lobar veins and finally the four trunk veins. These carry oxygenated blood from the lungs to the left atrium of the heart. In illustrations of the pulmonary vessels the pulmonary trunk and two pulmonary arteries are colored blue since they carry venous (deoxygenated) blood. The pulmonary veins carrying oxygenated blood are colored red. This may prove confusing.

5. The pulmonary circuit, is the route of blood from the heart through the lungs and back to the heart. In summary blood travels from right ventricle — to pulmonary trunk — to right and left pulmonary arteries — to lung capillaries — to two right and two left pulmonary veins — to left atrium.

6. The pulmonary circulation therefore is from the heart to the lungs for oxygenation and elimination of carbon dioxide, and from the lungs back to the heart.

Note: the term pulmonary trunk replaces the older name "pulmonary artery". Since the branches were called right and left pulmonary arteries there was always some confusion as to whether pulmonary artery was being used as referring to the trunk artery or to one of the two branches.

The student should note that the vessels supplying nutrients to the lungs, except for the air sacs, are not the pulmonary arteries but the bronchial arteries. See branches of descending aorta.

S. 162   THE SYSTEMIC VESSELS & CIRCULATION   F.13-14

1. The systemic vessels distribute blood from the left ventricle to all parts of the body except the air sacs of the lungs. They collect blood from these same parts and return it to the right atrium. The term systemic as used here refers to the whole body and not to any one system. It might more logically be named the general circulation to distinguish it from the pulmonary circulation. The vessels include:
    the aorta, its branches and capillaries
    the superior and inferior venae cavae and their tributaries.
These blood vessels supply and drain the head, neck, trunk and extremeties, but not the lungs. Each of these vessels will be studied later.

2. The systemic circuit therefore conveys blood from the left ventricle — to the aorta — to its branches — to capillaries — to tributaries of — and finally to the superior and inferior venae cavae — to the right atrium.

3. The systemic circulation distributes oxygenated blood from the left ventricle to the whole body except the alveoli of the lungs, and collects it after deoxygenation, returning it to the right atrium.

S. 163   THE AORTA   (F.13-14, 16 See Chart Aorta)

The aorta is the trunk artery of the systemic or general circuit. It originates at the aortic opening of the left ventricle. It receives oxygenated blood from this chamber and through its many branches **distributes** blood to the whole body except as stated above. Initially the aorta lies in the thorax — the thoracic aorta. Eventually it passes down through the diaphragm into the abdomen — abdominal aorta. The thoracic part is further divided into the ascending aorta, the aortic arch, and the descending thoracic aorta.

1. The ascending aorta passes cranially from the opening in the left ventricle for about two inches to become the aortic arch. It lies posterior to the upper sternum.   F.13-16

2. The arch of the aorta (aortic arch) begins at the sternal angle, as a continuation of the ascending part. It curves to the left and posteriorly to become the descending thoracic aorta.   F.13-16

3. The descending thoracic aorta, a continuation of the arch, descends in the posterior thorax along the left margins of the thoracic vertebrae. Anterior to the twelfth thoracic body it passes through the aortic hiatus (opening) of the diaphragm to become the abdominal aorta.   F.13-16

4. The abdominal aorta, a continuation of the descending thoracic part, passes caudally anterior to the lumbar vertebrae in the posterior abdomen. At the level of the fourth lumbar vertebrae it divides into its two terminal branches, the common iliac arteries.   F.13-16

S. 164   BRANCHES OF THE AORTA   F.13-16

The branches of each part of the aorta are outlined below. Many of the divisions of these branches are named and some have been included. The student should refer to the chart of the branches of the aorta at the end of this chapter. This chart is for reference and not as an exercise in memory training.

I. BRANCHES OF THE ASCENDING AORTA
F.13-11

The ascending aorta has only two branches: the right and left coronary arteries. They arise from the aorta close to its origin and under cover of two of the semilunar aortic cusps. The coronary arteries run along the groove of the coronary sulcus and the interventricular sulci on the surface of the heart. They give off branches to supply blood to the atria and ventricles.

II. BRANCHES OF THE AORTIC ARCH   F.13-11

There are three branches arising from the arch. From right to left they are the brachiocephalic trunk, the left common carotid artery, and the left subclavian artery.

1. The Brachiocephalic Trunk (NA), or INNOMINATE ARTERY (OT); brachium = arm + cephale = head, so artery of the arm and head.
   This artery arises from the right end of the arch and ascends cranially behind the sternum for about two inches. It divides into the two branches: the right common carotid and right subclavian arteries.

(1) **The right common carotid artery** passes up into the neck. Opposite the thyroid cartilage it divides into:
(a) **the right external carotid** artery to the scalp, face, tongue and neck,
(b) **the right internal carotid** artery to the brain.

(2) **The right subclavian artery** passes up posterior to the medial end of the clavicle. (sub = under + clavian = clavicle). Its pulsation can be felt above the clavicle at this point. It then descends behind the clavicle in a groove on the anterior surface of the first rib, and becomes the axillary artery. The subclavian artery supplies branches to the neck, shoulder, thoracic wall and upper limb. (See vessels of the upper limb). One branch, the vertebral artery enters the skull to help supply the brain.

2. THE LEFT COMMON CAROTID ARTERY is the second branch of the aortic arch. Originating to the left of the brachiocephalic trunk, it ascends into the neck and divides into left external and internal carotid arteries, with a distribution similar to the right arteries of the same names. (See vessels of the head).

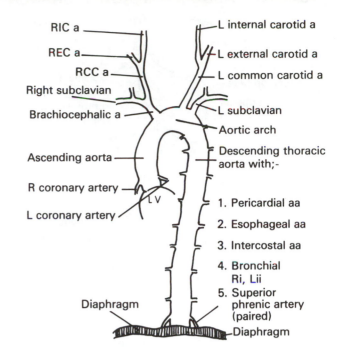

F. 13-16   AORTA — PART — BRANCHES

- RIC a
- REC a
- RCC a
- Right subclavian
- Brachiocephalic a
- Ascending aorta
- R coronary artery
- L coronary artery
- L V
- Diaphragm
- L internal carotid a
- L external carotid a
- L common carotid a
- L subclavian
- Aortic arch
- Descending thoracic aorta with;-
  1. Pericardial aa
  2. Esophageal aa
  3. Intercostal aa
  4. Bronchial Ri, Lii
  5. Superior phrenic artery (paired)
- Diaphragm

3. THE LEFT SUBCLAVIAN ARTERY, the third branch of the aortic arch, arising to the left of the common carotid, has a distribution similar to the right subclavian, including a vertebral branch to the brain.

III. BRANCHES OF THE DESCENDING THORACIC AORTA   F.13-16

As the aorta descends along the left borders of the thoracic vertebrae it gives off branches to the structures in the adjacent mediastinum. These include the **mediastinal, esophageal, pericardial, intercostal and superior phrenic arteries** as well as one **right and two left bronchial** arteries. The bronchial arteries supply blood to the bronchi, and larger bronchioles.

IV. BRANCHES OF THE ABDOMINAL AORTA
F.13-16

There are several paired and unpaired branches of the abdominal aorta. Listed from above, down they are:

The paired branches:
1. **The inferior phrenic** arteries — to the under surface of the diaphragm;

2. **The middle suprarenal** arteries — to the suprarenal glands;

3. **The renal arteries** — right and left, to the kidneys;

4. **The ovarian or testicular arteries,** right and left supply blood to the ovaries or testes. These are branches of the aorta as they were formed while the ovaries or testes were developing in the abdomen, before their descent.

5. **The lumbar arteries,** four pairs, arising opposite the upper four lumbar vertebrae as right and left.

The unpaired branches include arteries to the digestive system and spleen:

1. **The celiac trunk** (OT, celiac axis) divides into three arteries: the gastric, splenic, and common hepatic arteries, to the stomach, spleen, pancreas and liver.

F. 13-16 (2)   ABDOMINAL AORTA BRANCHES

Diaphragm
1 — Inferior phrenic (paired)
Celiac trunk (3 branches)
Lt. gastric a.
Splenic a.
Common hepatic a.
**Superior mesenteric a.
2 — 2 — Middle suprarenal aa (pair)
Rt. renal a.
3 — 3 — Renal aa (paired)
Rt. ovarian or testicular a.
Ovarian or testicular aa (paired)
** Inferior mesenteric a.
4 — 4
Rt. C.I. a.
Lt. common iliac a.
Lt. internal iliac a.
** Middle sacral
Rt. I. I. a.
Rt. E.I. a.
Lt. external iliac a.
Femoral aa

NOTE: 4 Pairs lumbar arteries numbered.
1, 2, 3, 4,

F. 13-17   SUPERIOR VENA CAVA AND TRIBUTARIES

R internal jugular V
L internal jugular V
R subclavian vein
L subclavian vein
R branciocephalic V
L brachiocephalic V
Superior vena cava
Accessory hemiazygos V
Azygos V
Hemiazygos V
Diaphragm
Rt. inf, ph. V.   1
Inferior phrenic

2. **The superior mesenteric artery** — to the small intestine and proximal half of the large intestine. (Superior = upper + mesenteric as it passes in the mesentery to the bowel.)

3. **The inferior mesenteric artery** — to the distal half of the large intestine.

4. **The middle sacral artery** — to the pelvis.

Terminal branches of the abdominal aorta:

At the level of the fourth lumbar vertebra the abdominal aorta divides into its two equal terminal branches, the right and left common iliac arteries.

(1) **The left common iliac artery** descends and opposite the left sacroiliac joint, divides into the left internal iliac and left external iliac arteries.

(a) **The left internal iliac artery** (OT, hypogastric) descends into the pelvis and supplies branches to the pelvic organs and pelvic wall, including the left uterine artery.

(b) **The left external iliac artery** passes around the brim of the pelvis. Under the inguinal ligament it passes into the thigh to become the femoral artery. Its further course is described under the blood vessels of the lower limb.
Note: **The inguinal ligament** extends from the anterior superior spine of the ilium to the pubic bone, running obliquely down and medially in the crease between the abdomen and thigh.

(2) **The right common iliac** artery divides into internal and external iliac arteries with a similar distribution to the left iliac vessels.

S. 165   THE SUPERIOR VENA CAVA AND TRIBUTARIES   F.13-17

The superior vena cava, one of the two trunk veins of the systemic circuit, collects blood from all of the body

above the diaphragm except the air sacs of the lungs. It is formed by tributaries from the head, neck, thorax, shoulders and upper limbs. It lies behind the upper sternum, along its right margin, and descends to open into the right atrium.

1. **The superior vena cava** is formed by the union of the right and left brachiocephalic (innominate) veins in the upper thorax.

2. **The right and left brachiocephalic** (innominate) veins are formed by the union of the corresponding right or left internal jugular and subclavian veins behind the sternoclavicular joint. NOTE: There are two brachiocephalic veins but one artery.

3. **The internal jugular veins**, right and left, begin at the jugular foramina at the base of the skull. They drain the venous sinuses of the dura mater, and some other veins of the head and neck. Each terminates by uniting with the corresponding subclavian vein to form a brachiocephalic vein.

4. **The subclavian veins**, right and left, are continuations of the right and left axillary veins from the upper limbs. They receive tributaries from the head, neck, thorax, and shoulder as well. The external jugular and vertebral veins empty into them. See also blood supply to head and upper limb.

5. **The azygos vein** commences in the right upper posterior abdomen as the ascending lumbar vein. It passes through the diaphragm and ascends to the right of the thoracic vertebrae. It curves around the root of the right lung passing anteriorly to join the superior vena cava. Its tributaries drain blood from the areas supplied with blood by the right sided branches of the descending thoracic aorta. They collect blood from the right side of the mediastinum, by intercostal, esophageal, pericardial, intercostal and right bronchial veins. F.13-17

6. **The hemiazygos vein** and **accessory hemiazygos vein** drain the same areas on the left side. The hemiazygos

vein begins as the left ascending lumbar vein. It passes up through the diaphragm into the chest along the left margins of the upper lumbar and lower thoracic vertebrae. At the 8th or 9th dorsal it crosses over to the right to join the azygos vein.

The accessory hemiazygos vein descends along the left margins of the upper and midthoracic vertebrae and either joins the hemiazygos vein or crosses over to join the azygos vein independently. These latter two veins drain left mediastinal structures and the left bronchial veins.

S. 166   THE INFERIOR VENA CAVA AND TRIBUTARIES   F.13-17

The inferior vena cava is the large trunk vein that collects blood from the inferior half of the body. It is formed by tributaries from the lower limbs, pelvis and abdomen. It ascends in the posterior abdomen in front of the lumbar vertebrae and along the right side of the abdominal aorta. It passes through the vena caval foramen in the diaphragm to enter the thorax and empty into the right atrium.

1. **The inferior vena cava** is formed by the union of the right and left common iliac veins opposite the body of the fifth lumbar vertebrae in the posterior lower abdomen.

2. **The common iliac vein,** right or left is formed by the union of the corresponding right or left internal iliac vein, and external iliac vein opposite the sacroiliac joint.

3. **The internal iliac vein,** right or left (OT - hypogastric vein) is formed by the union of tributaries from the pelvic organs and walls. One of these is the uterine vein. The internal iliac vein collects blood from the pelvic structures and joins the external iliac vein of the same side.

4. **The external iliac vein,** right or left, is a continuation of the femoral vein from the lower limb. It passes around the brim of the pelvis to join the internal iliac vein and form a common iliac vein.

5. **Abdominal tributaries** of the inferior vena cava:
   These drain blood from the abdominal organs;
   — ovarian or testicular veins from these organs
   — renal veins from the kidneys
   — suprarenal veins from these glands
   — phrenic veins from the diaphragm
   — lumbar veins - 4 pairs, right and left.
The veins draining the digestive tract, spleen, and pancreas take a different course and form part of the portal circulation described below.

S. 167   THE PORTAL VESSELS AND CIRCULATION   F.13-18

1. **The portal vein** collects blood from the spleen, pancreas, gallbladder, stomach and intestine. It is formed by the union of the superior mesenteric and splenic veins. It lies in the upper posterior abdomen behind the head of the pancreas. It is about three inches long, and passes to the right to enter the liver at the porta hepatis (L. porta = a gate + hepar = liver; gate to the liver). It divides into right and left brances to the liver lobes. These branches divide into capillaries (termed sinusoids) within the liver. After passing through the liver these capillaries unite to form the hepatic veins that drain into the inferior vena cava, behind the liver.

2. **The superior mesenteric vein** collects blood from the small intestine and part of the colon as far as the left colic flexure. It ascends in the posterior abdomen and joins the splenic vein behind the pancreas to form the portal vein.

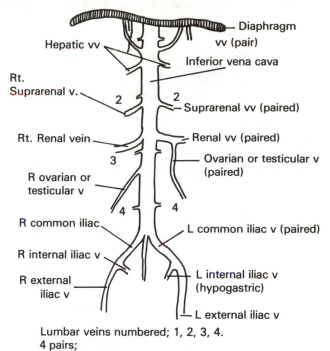

F. 13-17 (2)   INFERIOR VENA CAVA

Hepatic vv
Diaphragm vv (pair)
Inferior vena cava
Rt. Suprarenal v.
2
2
Suprarenal vv (paired)
Rt. Renal vein
Renal vv (paired)
3
Ovarian or testicular v (paired)
R ovarian or testicular v
R common iliac
4
4
L common iliac v (paired)
R internal iliac v
L internal iliac v (hypogastric)
R external iliac v
L external iliac v
Lumbar veins numbered; 1, 2, 3, 4. 4 pairs;

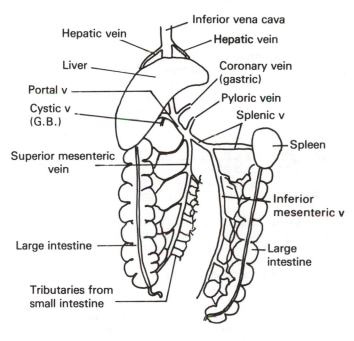

F. 13-18   DIAGRAM PORTAL VEIN & TRIBUTARIES

Inferior vena cava
Hepatic vein
Hepatic vein
Liver
Coronary vein (gastric)
Portal v
Pyloric vein
Cystic v (G.B.)
Splenic v
Spleen
Superior mesenteric vein
Inferior mesenteric v
Large intestine
Large intestine
Tributaries from small intestine

3. **The splenic vein** is formed by veins from the spleen with tributaries from the stomach and pancreas, and by the inferior mesenteric vein. This splenic vein joins the superior mesenteric vein to form the portal vein as stated above. It crosses the abdomen posterior to the stomach.

4. **The inferior mesenteric vein** collects blood from the descending and sigmoid colon and rectum. It runs cranially in the posterior abdomen to join the splenic vein.
A cystic vein from the gallbladder, and coronary and pyloric veins from the stomach also join the portal vein.

In the section on the branches of the abdominal aorta, three unpaired arteries were described as supplying blood to the spleen, pancreas, stomach, intestine, gallbladder and liver. These were:

a. **The celiac trunk** with its three branches:
the left gastric artery
the splenic artery
the common hepatic artery
b. **The superior mesenteric** artery
c. the inferior mesenteric artery

Blood distributed by these vessels, after passing through capillaries in the several structures is collected by the tributaries of the portal vein and conveyed by it to the liver.

5. **The portal circuit** — a poor term as this is not a complete circuit in itself. It includes these unpaired branches of the abdominal aorta; their capillaries; the inferior mesenteric, splenic, and superior mesenteric veins; the portal vein; liver capillaries; hepatic veins; inferior vena cava.

6. **The portal circulation** carries blood containing digested food products to the liver through the portal vein **before** emptying it into the general systemic circulation. The liver therefore gets first chance to extract what it requires.
Note that the portal circulation has two sets of capillaries: those in the intestinal wall formed from arteries here, and the sinusoids, in the liver formed from the portal vein.

BLOOD VESSELS OF THE BRAIN, SCALP, FACE, NECK

Before undertaking this study the branches of the aortic arch and the tributaries of the superior vena cava should be reviewed. The blood vessels to and from the head should be traced to them. Only the larger trunks will be described. Details may be studied in a postgraduate course.

S. 168 BLOOD VESSELS OF THE BRAIN
F.13-19 to 24

1. **ARTERIES** — (Bilaterally similar, right, left)

a. **The internal carotid artery**, a branch of the common carotid, passes up through the neck anterior to the sternomastoid muscle. It enters the cranial cavity by the carotid canal and foramen lacerum. It divides into **anterior and middle cerebral arteries** to supply the cerebrum. F.13-19, 20, 21

b. **The vertebral artery**, a branch of the subclavian, enters the foramen of the transverse process of the sixth cervical vertebra and the five vertebrae above. Passing posterior to the lateral mass of the atlas it enters the skull through the foramen magnum. It joins the vessel of the other side to form the basilar artery. The basilar artery ascends in front of the pons (a part of the brain) and divides into right and left posterior cerebral arteries. It gives off branches to the hindbrain. F.13-19, 22, 23.

c. The circle of Willis (NA, cerebral arterial circle) is an example of an extensive anastomosis of arteries at the base of the brain. Branches of the two internal carotid arteries and branches of the basilar artery unite to form a circle of arteries about the stalk of the pituitary gland. Were one of the branches to become obstructed the other arteries could supply blood to the parts.

F. 13-19 RIGHT CAROTID & VERTEBRAL ARTERIES
—Lateral view—

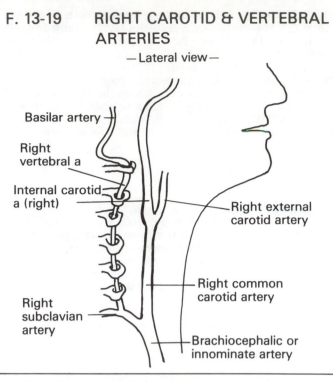

Basilar artery

Right vertebral a

Internal carotid a (right)

Right external carotid artery

Right common carotid artery

Right subclavian artery

Brachiocephalic or innominate artery

2. VEINS OF THE BRAIN
SINUSES OF DURA MATER   F.13-24

The veins within the skull unite to form large venous trunks called the sinuses of the dura mater. They are located in folds of the dura mater. They join together to form the right and left internal jugular veins at the jugular foramina in the base of the cranium.
The superior sagittal sinus (one only) runs from front to back in the median line within the cranium in a shallow groove under the sagittal suture of the skull. It is located in the upper part of the falx cerebri (a fold of the dura mater covering the brain).
The inferior sagittal sinus (one only) runs from front to back in the lower part of the falx cerebri between the two cerebral hemispheres.
The straight sinus is an extension of the inferior sagittal sinus posteriorly to the internal occipital protaberence.
At the internal occipital protuberance, the superior sagittal turns to either the right or left, while the straight sinus turns in the opposite direction. Each follows a groove that passes horizontally, laterally and anteriorly on the inner surface of the occipital bone.

The transverse sinuses are the continuations of the sagittal sinuses on the horizontal grooves of the occipital bone. Each transverse sinus, a right and a left, turns downwards posterior to the petrous part of the temporal bone as the sigmoid sinus.
The sigmoid sinuses, right and left, are the continuations of the transverse sinuses. They leave the skull through the internal jugular foramina close to the foramen magnum as the right and left internal jugular veins. The internal jugular veins pass down through the neck to join the corresponding subclavian vein and form a brachiocephalic vein. (Innominate vein, OT).
Other smaller sinuses within the skull join the larger ones. These have not been named here.

S. 169 VESSELS OF THE SCALP, FACE, NECK

1. **Branches of the external carotid** artery on each side supply branches to these structures. Two of them are sometimes used in counting the pulse, one in front of the opening of the external acoustic meatus, the other as it

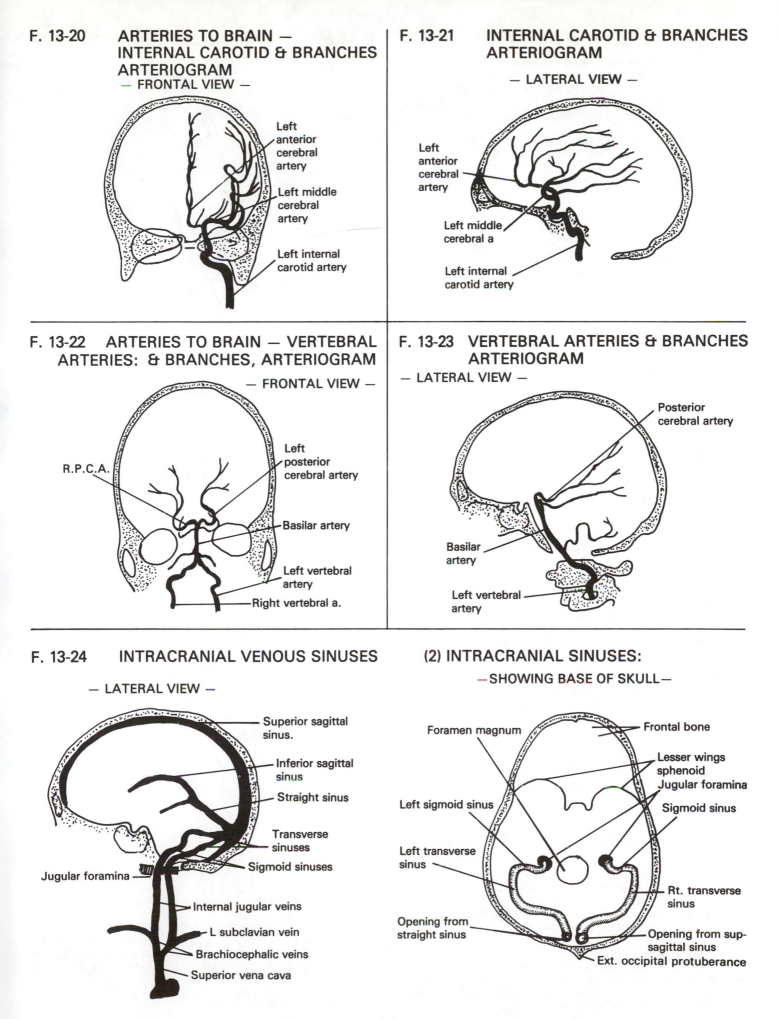

F. 13-20 ARTERIES TO BRAIN —
INTERNAL CAROTID & BRANCHES
ARTERIOGRAM
— FRONTAL VIEW —

Left anterior cerebral artery

Left middle cerebral artery

Left internal carotid artery

F. 13-21 INTERNAL CAROTID & BRANCHES
ARTERIOGRAM
— LATERAL VIEW —

Left anterior cerebral artery

Left middle cerebral a

Left internal carotid artery

F. 13-22 ARTERIES TO BRAIN — VERTEBRAL
ARTERIES: & BRANCHES, ARTERIOGRAM
— FRONTAL VIEW —

R.P.C.A.

Left posterior cerebral artery

Basilar artery

Left vertebral artery

Right vertebral a.

F. 13-23 VERTEBRAL ARTERIES & BRANCHES
ARTERIOGRAM
— LATERAL VIEW —

Posterior cerebral artery

Basilar artery

Left vertebral artery

F. 13-24 INTRACRANIAL VENOUS SINUSES
— LATERAL VIEW —

Superior sagittal sinus.

Inferior sagittal sinus

Straight sinus

Transverse sinuses

Sigmoid sinuses

Jugular foramina

Internal jugular veins

L subclavian vein

Brachiocephalic veins

Superior vena cava

(2) INTRACRANIAL SINUSES:
—SHOWING BASE OF SKULL—

Foramen magnum

Frontal bone

Lesser wings sphenoid

Jugular foramina

Sigmoid sinus

Left sigmoid sinus

Left transverse sinus

Rt. transverse sinus

Opening from straight sinus

Opening from sup-sagittal sinus

Ext. occipital protuberance

141

passes up over the lower margin of the body of the mandible.

The middle meningeal artery is also a branch that enters the cranial cavity through the foramen spinosum and is distributed in small grooves on the inner surface of the lateral aspect of the skull. The grooves are visible in lateral radiographs of the skull.

S. 170   BLOOD VESSELS OF THE UPPER LIMB: THE SUPERIOR MEMBER   F.13-25, 26

The student should again review the branches of the aortic arch, and the tributaries of the superior vena cava.

1. ARTERIES OF THE UPPER LIMB   F.13-25

(1) **The subclavian artery,** on the right side, is a branch of the brachiocephalic; on the left side it is a direct branch of the aortic arch. Each crosses the first rib and is then called the axillary artery.

(2) **The axillary artery,** a continuation of the subclavian crosses the axilla (armpit) to reach the arm and become the brachial artery.

(3) **The brachial artery** descends medial to the humerus to the elbow. It enters the forearm anterior to the elbow joint and divides into the radial and ulnar arteries in front of the neck of the radius.

(4) **The radial artery** descends along the lateral (radial) margin of the anterior surface of the forearm to the wrist. It helps to form the arches of the hand.

(5) **The ulnar artery** descends along the medial (ulnar) margin of the anterior surface of the forearm to the wrist. It helps to form the arches of the hand.

(6) **One dorsal and two palmar arches** are formed from the radial and ulnar arteries. These arches supply metacarpal and digital branches to each hand.

2. VEINS OF THE UPPER LIMB   F.13-26

There are two sets of veins collecting blood from the upper limb, a superficial and a deep set. The superficial veins begin as networks of small veins under the skin of the hands.

The superficial veins
(1) **The cephalic vein** ascends along the anterolateral surface of the forearm and arm, then in front of the upper arm. It joins the axillary vein below the outer part of the clavicle.

(2) **The basilic vein** ascends along the posteromedial surface of the forearm but winds anteriorly below the elbow. It passes up on the medial side of the arm and joins the brachial veins to form the axillary vein.

(3) **The median antebrachial vein** ascends on the anterior surface of the forearm and may join the basilic below the elbow.

These veins as well as the venous arches forming them anastomose freely with each other so that damage to one vessel is not important since other veins can convey their blood. There is considerable variation in the course and position of the larger trunks.

Deep veins of the upper limb
The deep veins begin as small veins in the hand. The veins accompany arteries so have the same names as the arteries. The ulnar, radial and brachial arteries each have two accompanying veins.

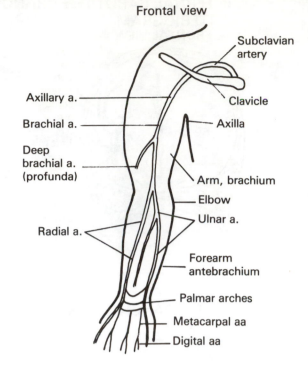

F. 13-25   TRUNK ARTERIES OF UPPER LIMB:
Frontal view

- Subclavian artery
- Axillary a.
- Clavicle
- Brachial a.
- Axilla
- Deep brachial a. (profunda)
- Arm, brachium
- Elbow
- Ulnar a.
- Radial a.
- Forearm antebrachium
- Palmar arches
- Metacarpal aa
- Digital aa

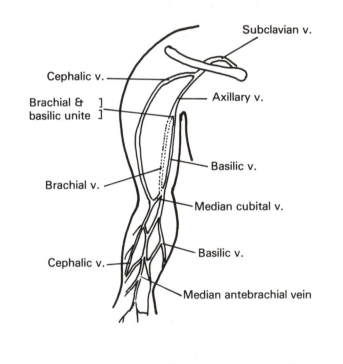

F. 13-26   SUPERFICIAL VEINS UPPER LIMB, & UNION WITH DEEP VEINS

- Subclavian v.
- Cephalic v.
- Axillary v.
- Brachial & basilic unite
- Basilic v.
- Brachial v.
- Median cubital v.
- Basilic v.
- Cephalic v.
- Median antebrachial vein

Two ulnar veins, and two radial veins join to form **two brachial** veins which in turn join the axillary vein, a single vein corresponding to the axillary artery, from 2 brachial + basilic vein.

The axillary vein becomes the subclavian vein as it crosses the first rib in a groove on the upper surface.

The subclavian vein enters the upper thorax and joins the corresponding internal jugular vein to form the brachiocephalic (innominate) vein.

There are connecting veins that pass freely between the superficial and deep veins.

Blood to the upper limb therefore passes through the subclavian, axillary, brachial, radial and ulnar arteries ending in the arches of the hand.

Blood is collected by venous arches, superficial and deep, and passes through the two ulnar, two radial, and two brachial deep veins, or through the superficial veins — cephalic, basilic or median antebrachial veins — to eventually empty into the axillary vein.

Each of the arteries listed above has branches, given off during its course, and each vein receives tributaries. The names of these have been omitted.

S. 171   BLOOD VESSELS OF THE LOWER LIMB: THE INFERIOR MEMBER   F.13-27, 28

The student should first review the terminal branches of the abdominal aorta, and the tributaries that form the inferior vena cava.

1. ARTERIES OF THE LOWER LIMB   F.13-27

(1) **The external iliac artery,** a terminal branch of the common iliac artery passes around the brim of the pelvis. It enters the thigh posterior to the inguinal ligament and becomes the femoral artery.

(2) **The femoral artery,** a continuation of the external iliac artery, descends in the medial part of the thigh. It then curves posteriorly and comes to lie behind the lower femur to become the popliteal artery.
The deep femoral artery (profunda femoris) is a large branch of the femoral and descends in the posterior thigh.

(3) **The popliteal artery,** the continuation of the femoral, descends posterior to the knee in the popliteal space. Below the knee it divides into posterior and anterior tibial arteries.

(4) **The posterior tibial artery** one of the terminal branches of the popliteal, descends posterior to the tibia. It then passes below the medial malleolus of the tibia to reach the sole of the foot. Here it divides into medial and lateral plantar arteries.

(5) **The peroneal artery** is a large branch of the posterior tibial that descends along the fibula, i.e. the lateral side of the calf.

(6) **The anterior tibial,** the other terminal branch of the popliteal artery, passes anteriorly between the upper tibia and fibula and descends in the anterior part of the leg to the ankle where it becomes the dorsal artery of the foot.

(7) **The dorsal artery of the foot,** (dorsalis pedis artery) the continuation of the anterior tibial, runs down in front of the ankle to the foot.

(8) Branches to the foot are formed from the plantar and dorsal arteries.

2. VEINS OF THE LOWER LIMB   F.13-28

As in the upper limb there are two sets of veins draining blood from the lower limb, a superficial and a deep set.

The superficial veins
(1) **The great (NA) or long saphenous** vein lies under the skin in the subcutaneous tissue. It extends from the medial margin of the foot upwards along the medial surface of the leg close to the anterior margin. It passes up along the medial surface of the thigh, then turns to the front and joins the femoral vein below the inguinal

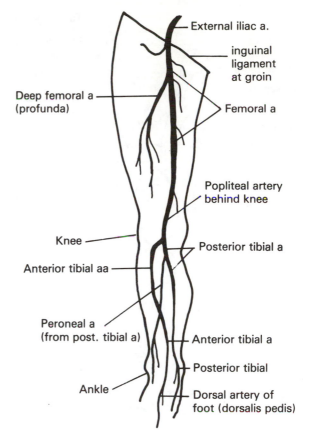

F. 13-27   LARGE ARTERIES, LOWER LIMB
—FRONTAL VIEW—

- External iliac a.
- inguinal ligament at groin
- Femoral a
- Deep femoral a (profunda)
- Popliteal artery behind knee
- Knee
- Posterior tibial a
- Anterior tibial aa
- Peroneal a (from post. tibial a)
- Anterior tibial a
- Posterior tibial
- Ankle
- Dorsal artery of foot (dorsalis pedis)

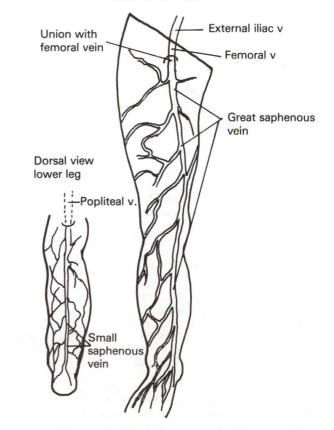

F. 13-28   SUPERFICIAL VEINS LOWER LIMB
—FRONTAL VIEW—

- External iliac v
- Femoral v
- Union with femoral vein
- Great saphenous vein
- Dorsal view lower leg
- Popliteal v.
- Small saphenous vein

F. 13 TRUNK ARTERIES — THORAX & NECK

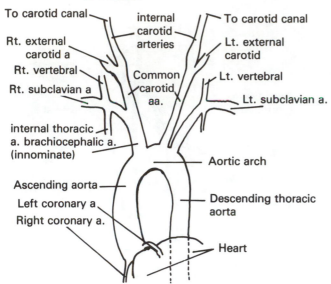

To carotid canal
internal carotid arteries
To carotid canal
Rt. external carotid a
Lt. external carotid
Rt. vertebral
Common carotid aa.
Lt. vertebral
Rt. subclavian a
Lt. subclavian a.
internal thoracic a. brachiocephalic a. (innominate)
Aortic arch
Ascending aorta
Left coronary a.
Right coronary a.
Descending thoracic aorta
Heart

F. 13 TRUNK ARTERIES OF PELVIS

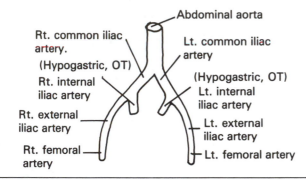

Abdominal aorta
Rt. common iliac artery. (Hypogastric, OT)
Lt. common iliac artery
Rt. internal iliac artery
(Hypogastric, OT)
Lt. internal iliac artery
Rt. external iliac artery
Lt. external iliac artery
Rt. femoral artery
Lt. femoral artery

F. 13 TRUNK VEINS OF THORAX:

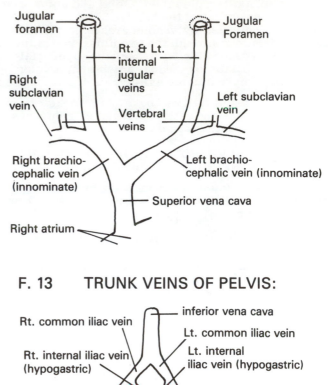

Jugular foramen
Jugular Foramen
Rt. & Lt. internal jugular veins
Right subclavian vein
Vertebral veins
Left subclavian vein
Right brachio-cephalic vein (innominate)
Left brachio-cephalic vein (innominate)
Right atrium
Superior vena cava

F. 13 TRUNK VEINS OF PELVIS:

Rt. common iliac vein
inferior vena cava
Lt. common iliac vein
Rt. internal iliac vein (hypogastric)
Lt. internal iliac vein (hypogastric)
Rt. external iliac vein
Lt. external iliac vein
Rt. femoral vein
Lt. femoral vein

---

ligament. It receives many tributaries. (Saphene, G. = visible)

(2) **The small (NA) or short saphenous** vein commences on the lateral margin of the foot and ascends under the skin up the calf of the leg. It joins the popliteal vein posterior to the knee.

Both of these veins support a long column of blood and are well supplied with semilunar valves to prevent backflow.

The deep veins

The deep veins accompany arteries and each one takes the name of the adjacent artery. Two deep veins accompany each artery below the knee, so that there are 2 posterior tibial, 2 anterior tibial, and 2 peroneal veins. These join to form the popliteal vein posterior to the knee. The popliteal vein becomes the femoral, which becomes the external iliac vein. The external iliac vein joins the internal iliac vein forming the common iliac vein.

The circulation of blood in the lower limb is from the external iliac artery as the source, the femoral, popliteal, anterior and posterior tibial, and peroneal arteries, plantar arteries, dorsal artery of the foot and arches of the foot.

Blood is collected by plantar veins: the dorsal vein of the foot, peroneal, anterior and posterior tibial and popliteal, and femoral vein, and the two saphenous veins, to empty into the external iliac vein.

Many of the arteries have smaller branches and the veins have tributaries, these have been omitted.

Extensive anastomoses are present especially between superficial and deep veins as well as between adjacent superficial or deep veins.

S. 172 THE LYMPHATIC VESSELS, NODES AND ORGANS

The lymphatic structures were defined in S.146, No. 7, dealing with the components of the circulatory system. These lymphatic vessels form a second collecting system, in addition to the veins. Intercellular fluid, including water and waste products, gets back into the blood via lymphatics. In addition considerable fat is absorbed from the intestine by lymphatics. Lymph is the name given to the colorless fluid that circulates in lymphatic vessels.

The lymphatic structures include: F.13-29 to 33

1. Lymph capillaries
2. Lymph vessels (collecting)
3. Lymph ducts — the trunk vessels
4. Lymph nodes (glands)
5. Other organs — tonsils, adenoids, intestinal follicles, spleen, thymus gland.

1. **Lymph capillaries** are minute microscopic thin walled vessels similar to blood capillaries except that they do not originate from arterioles. They form extensive networks in and under the dermis, in the walls of hollow organs, around and within organs and other structures. They have not been demonstrated in the central nervous system and a few other parts.

They are quite numerous in the skin, subcutaneous tissue, and the lining membranes (serous membranes) of the body cavities — the pleura, pericardium, and peritoneum. These lymph capillaries unite to form small lymph vessels.

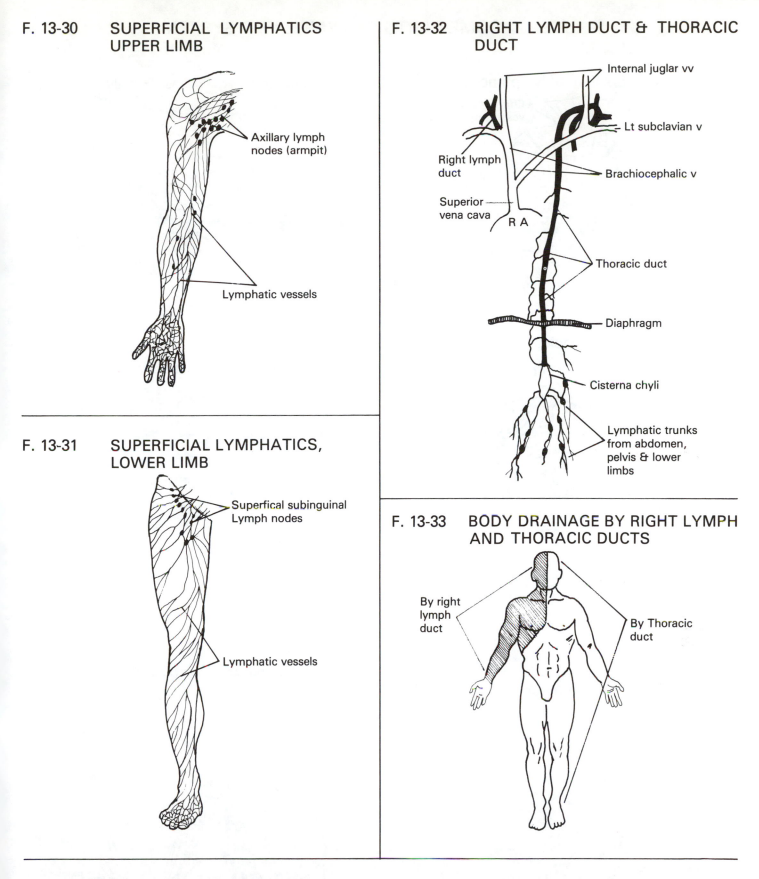

F. 13-30   SUPERFICIAL LYMPHATICS UPPER LIMB

Axillary lymph nodes (armpit)

Lymphatic vessels

F. 13-31   SUPERFICIAL LYMPHATICS, LOWER LIMB

Superfical subinguinal Lymph nodes

Lymphatic vessels

F. 13-32   RIGHT LYMPH DUCT & THORACIC DUCT

Internal juglar vv

Lt subclavian v

Right lymph duct

Brachiocephalic v

Superior vena cava

R A

Thoracic duct

Diaphragm

Cisterna chyli

Lymphatic trunks from abdomen, pelvis & lower limbs

F. 13-33   BODY DRAINAGE BY RIGHT LYMPH AND THORACIC DUCTS

By right lymph duct

By Thoracic duct

2. **Lymph vessels** are tubes formed by the union of lymph capillaries. The smaller vessels unite to form larger and larger vessels which are collecting vessels.

Lymph vessels form a superficial set under the skin and a deep set in the limbs and within the body. The deep vessels accompany the corresponding deep arteries and veins. The superficial vessels join the deeper ones. The large collecting vessels unite to form the thoracic duct or right lymph duct.

3. **Lymphatic ducts or trunks**   F.13-32

(1) **The thoracic duct** is a large lymph trunk that begins in the abdomen by the union of lumbar and intestinal lymphatic vessels. It ascends anterior to the upper vertebrae and passes through the aortic opening in the diaphragm. It ascends in the posterior and superior divisions of the mediastinum to the left side of the neck. It empties into the left brachiocephalic vein at the angle

F. 13-29   SOME PARTS OF LYMPHATIC SYSTEM:

(1) Cross Section of lymph node

Efferent lymph vessels drain node

Lymph follicles (nodules)

Afferent lymph vessels

(4) Anterior chest wall breast & axillary nodes;

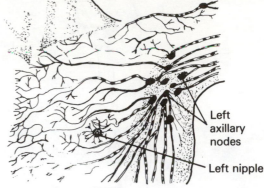

Left axillary nodes

Left nipple

(2) Tonsils & Adenoids

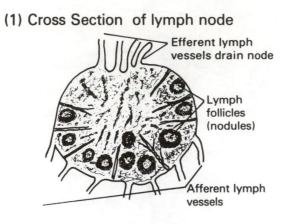

Nasal cavity

Mouth

Tongue

Mandible

Adenoids

Right tonsil

Larynx

Thyroid, cartilage

(5) Tracheal & bronchial nodes

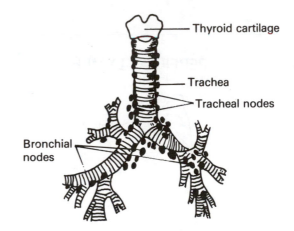

Thyroid cartilage

Trachea

Tracheal nodes

Bronchial nodes

(3) Cervical lymph nodes

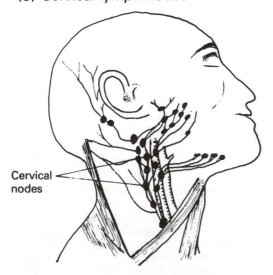

Cervical nodes

(6) Abdominal & pelvic nodes;

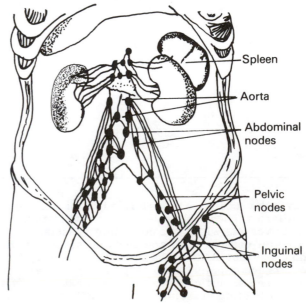

Spleen

Aorta

Abdominal nodes

Pelvic nodes

Inguinal nodes

between the terminations of the left subclavian and internal jugular veins. It drains lymph from the lower limbs, pelvis, abdomen, left half of the thorax, left upper limb and left halves of the head and neck.
The cisterna chyli is the dilated lower end of the thoracic duct in the upper posterior abdomen.

(2) **The right lymph duct** (right thoracic duct) is formed by the union of the right internal jugular, subclavian, and thoracic vessels. It empties into the right brachiocephalic or right subclavian vein. It drains lymph from the right half of the thorax, right upper limb, and right halves of the head and neck.

4. **Lymph nodes:** (OT glands) these are small oval bodies ranging in size from a pin head to a bean. They are distributed along the course of the lymph vessels, sometimes singly but usually in groups or chains. Several lymph vessels empty into each node and larger lymph vessels open out of each one. The contents of each lymph vessel must pass through one or more nodes before reaching the blood stream. There are superficial and deep nodes located at strategic points throughout the body, often close to a large artery or vein. F.13-29 (1 to 6)

Each lymph node has a fibrous capsule with small strands of fibrous tissue (trabeculae) passing into the node. There is also a very fine network of reticuloendothelial fibers present within the node. Many lymphocytes lie in the network. Some of these form clumps of lymphocytes called **lymph follicles**, others are in cords. The spaces between the cords and follicles form channels for the passage of the lymph through the node. F.13-30 (1)

Location of larger groups of lymph nodes
The cervical nodes, superficial and deep lie in the neck close to the sternomastoid muscle. They drain lymph from the head and neck on both sides. F.13-30 (3)

The axillary nodes, in the armpit (axilla) drain the upper limb, chest wall, and breast. F.13-30 (4)

The tracheobronchial nodes are located along the trachea and about the larger bronchi. They drain the lungs, bronchi, heart, and mediastrinal structures. F.13-30 (5)

The aortic nodes lie along the abdominal aorta and drain the testes, or ovaries, uterus, uterine tubes, and abdominal organs. The aortic nodes drain into the thoracic duct in the abdomen. F.13-30 (6)

The internal iliac nodes, adjacent to the internal iliac arteries drain lymph vessels from the pelvis, perineum, genitals and buttocks.

The external iliac nodes lie beside the external iliac blood vessels and drain lymph from the deep vessels of the lower limb, inguinal nodes, and the abdominal wall.

FUNCTIONS OF LYMPHATIC VESSELS AND NODES

The lymphatic vessels form a second collecting system and convey intercellular fluid back into the general circulation. The thoracic and right lymph ducts empty lymph into the trunk veins at the base of the neck.

Lymph nodes act as filters to prevent micro-organisms and other foreign bodies from reaching the blood stream. They destroy bacteria.

Lymph nodes also form new lymphocytes, monocytes, and plasma cells and discharge them into circulating blood.

5. OTHER LYMPHATIC ORGANS   F.13-29 (1, 2)

(1) **The tonsils** (NA) palatine tonsils, one on each side of the opening between the mouth and oral pharynx, contain lymphoid tissue consisting of many follicles (nests of lymphocytes). The opening between the mouth and pharynx is the fauces.

(2) **The adenoids** are located on the roof and posterior wall of the nasopharynx, behind the nasal cavities. They consist of lymphoid tissue. (NA, pharyngeal tonsil).

(3) **Intestinal lymph follicles** are located in the wall of the ileum and colon.

(a) **Solitary lymph follicles** (nodules) are composed of single follicles.

(b) **Aggregated lymph follicles** or Peyer's patches are found in the terminal ileum and contain many lymph follicles. The micro-organisms of typhoid and tuberculosis involve these glands in the bowel.

(4) **The spleen** lies in the upper left anterior abdomen. While it somewhat resembles a kidney in shape its medial or hilar surface is more flattened. The spleen is composed of red blood cells, lymph follicles, lymphocytes, and cords of lymphatic tissue with blood spaces or sinuses between. Like the lymph nodes, strands of fibrous tissue dip into the pulp forming trabeculae, and fine reticuloendothelial networks interlace in the follicles and cords.
Blood vessels of the spleen — The spleen receives blood from the **splenic artery,** a branch of the celiac trunk, a branch of the abdominal aorta. The splenic vein is formed from tributaries within the spleen. This vein joins the superior mesenteric vein to form the portal vein. The inferior mesenteric vein empties into the splenic.

Functions of the Spleen

The spleen forms lymphocytes, monocytes, and the plasma cells.
It destroys worn out red blood cells (phagocytosis).
It is also believed by some to be a reservoir of red blood cells.

(5) **The thymus gland** lies posterior to the upper part of the sternum. There are two lobes enclosed within a fibrous capsule. These lobes have small lobules that contain lymph follicles and reticuloendothelial cells and fibers. As this gland produces lymphocytes it has been reclassified as lymphoid tissue rather than as an endocrine gland.

S. 173   **THE CIRCULATORY SYSTEM IN THE FETUS**

The fetus within the maternal uterus is surrounded by fluid and its lungs do not contain air. Its digestive tract is not functioning. The fetus must obtain oxygen and other requirements and excrete its waste products and carbon dioxide via the mother's blood through the placenta. F.13-34 (1, 2)

The placenta is a flat circular spongy structure about eight inches in diameter and one inch in thickness when fully developed. It develops in the lining of the uterus and consists of maternal and fetal parts. The maternal part has blood spaces in the uterine wall supplied by maternal arterioles and drained by uterine veins. The fetal part of the placenta has capillaries that come into close contact with the maternal blood spaces. Maternal and fetal blood are thus separated by thin membranes so that an exchange of oxygen, carbon dioxide, nutrients and waste may occur. F.13-34 (1)

F. 13-34   THE FETAL CIRCULATION:

(1) BLOOD VESSELS & PLACENTA;

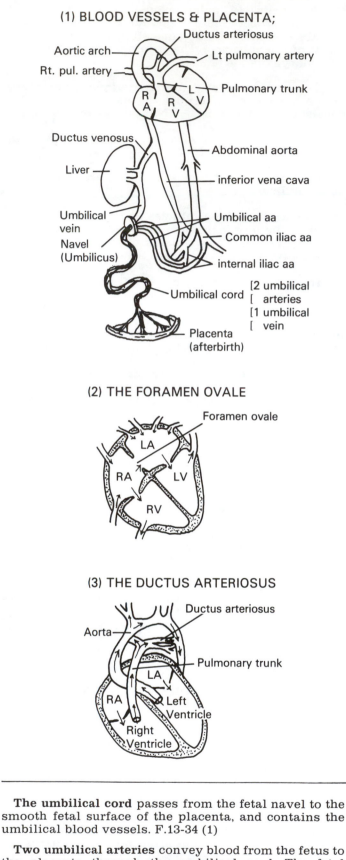

(2) THE FORAMEN OVALE

(3) THE DUCTUS ARTERIOSUS

The umbilical cord passes from the fetal navel to the smooth fetal surface of the placenta, and contains the umbilical blood vessels. F.13-34 (1)

Two umbilical arteries convey blood from the fetus to the placenta through the umbilical cord. The fetal internal iliac arteries, branches of the common iliac vessels, are continued around the inner surface of the pelvis to the anterior abdominal wall and up to the navel. Passing out through the navel they become the two umbilical arteries.

A single umbilical vein carries oxygenated blood with nutrients from the fetal capillaries in the placenta through the umbilical cord to the navel. It passes up the anterior abdominal wall to the liver. One of its branches joins the portal vein, while other branches pass directly into the liver. In this way the liver receives oxygenated blood before it has been passed to any other organs. One branch of the umbilical vein becomes the ductus venosus.

The ductus venosus is a short vein that passes from a branch of the umbilical vein directly to the inferior vena cava, bypassing the liver. F.13-34 (1)

Since the lungs in the fetus do not contain air and do not function as respiratory organs it would be useless for all blood to circulate through the pulmonary circulation. In the fetus the lungs are bypassed by two passages peculiar to fetal life. Only sufficient blood enters the lungs to provide tissue requirements. The foramen ovale and ductus provide the alternate routes.

The foramen ovale is an opening in the interatrial septum between the right and left atria. It is normally present until birth. Blood entering the right atrium from the inferior vena cava is directed across the atrium through the foramen ovale into the left atrium. It then passes into the left ventricle and into the aorta. A valve in the inferior vena cava directs blood through this foramen. The right ventricle and lungs are thus bypassed. The foramen ovale closes after birth. F.13-34 (3)

The ductus arterious is a short vessel that connects the pulmonary trunk to the aorta beyond the origin of the left subclavian artery. Blood entering the right atrium from the superior vena cava is directed to the right ventricle and out through the pulmonary trunk. Instead of entering the right and left pulmonary arteries and passing to the lungs it is carried through the ductus directly into the aorta. The lungs are again bypassed. This vessel should close soon after birth. F.13-34 (3)

After birth the lungs expel the fluid in them. The infant breathes expanding the lungs with air. In a matter of minutes the lungs take on the function of supplying oxygen to the blood that now enters them for oxygenation and excretion of carbon dioxide. Circulation through the umbilical cord and placenta ceases. The lungs now function, the digestive tract must supply nutrients and the kidneys excrete the waste products.

S. 174   THE PULSE — LOCATIONS FOR PALPATION

The pulse is due to alternating dilation and contraction of an artery resulting from contraction of the cardiac ventricles forcing blood out into the aorta and its branches. The pulse rate indicates the rate at which the heart is beating. The pulse should be regular and the rate should be about 72 per minute.

In taking the pulse, two or three fingers should be used. The flat surface in front of the terminal phalanges not the finger tips should be placed over the vessel. Care must be taken not to dig the finger nails into the patient's skin. There are certain points where an artery lies close to the skin. It is here that the pulse is most readily palpable. These are:
1.   Radial artery — anterolateral border of wrist,
2.   Ulnar artery at anteromedial border of wrist,
3.   Brachial artery on medial surface of midarm,
4.   Superficial temporal in front of the ear,
5.   Facial artery in front of angle of mandible,
6.   Abdominal aorta through anterior abdominal wall,
7.   Femoral artery in the groin below inguinal ligament,
8.   Popliteal artery posterior to knee,
9.   Dorsalis pedis artery in front of and below ankle.

S. 175 PRESSURE POINTS — TO ARREST HEMORRHAGE

Bleeding from arteries may be arrested by applying pressure on an artery or the trunk vessel supplying it, above the bleeding point. An attempt should be made to compress the vessel between the hand and an adjacent bone. In bleeding from a vein the pressure should be applied over the bleeding point or below it. The pulse points listed above are selected as the vessels are located close to the skin. In addition to these, in extreme cases, bleeding in the pelvis and lower limbs may be temporarily arrested by direct pressure upon the abdominal aorta. The artery is compressed against the lumbar vertebrae.

S. 176 BLOOD PRESSURE — B.P.

Blood pressure is the pressure of the blood against the containing vessel wall.

Systolic blood pressure is the pressure exerted during the contracting phase of the ventricles of the heart. (Expressed as a number above a line).

Diastolic blood pressure is the pressure exerted during the dilation phase of the ventricles. (Expressed as a number below a line).

(120 = S)
(70 = D)

The blood pressure is elevated in some conditions. A normal blood pressure could be that listed above. (hypertension = high B.P., hypotension = low B.P.)

S. 177 CARDIOVASCULAR CIRCUITS

A circuit implies commencement at one point and a return to that point. In the circulatory system there are no local circuits to an organ and back to the starting point, unless the whole systemic and pulmonary circulations are included. The closest resemblance to a circuit would be the pulmonary circulation or the systemic circulation. In each case blood proceeds out of the heart then back to the heart, BUT NOT TO THE SAME CHAMBER OF THE HEART. It seems more rational therefore to speak of the blood vessels of an organ including the arterial and venous vessels. These will be listed along with the study of each organ as it is defined in this textbook.

SUMMARY OF FUNCTIONS OF CIRCULATORY SYSTEM

1. **General,** distributing and collecting system. A transportation system for the distribution of nutrients, gases, minerals, vitamins, hormones, blood cells, etc., to tissues, and the collection of waste products for excretion from the body.

2. **Heart,** a pump to distribute blood plasma and its components, also receiving chambers, the atria, to accept blood returned to the heart.

3. **Valves** prevent backflow of blood. Atrioventricular valves prevent backflow from ventricles to atria. Semilunar aortic and pulmonary valves prevent backflow from aorta and pulmonary trunk into ventricles.

4. **Arterial trunks,** arteries, and arterioles distribute blood to body tissues. The aorta and branches distribute blood to all of body except alveolar structures, and respiratory bronchioles of lungs. The pulmonary trunk and branches distribute blood to respiratory bronchioles, alveolar sacs and alveoli.

5. **Capillaries,** thin walled minute vessels through which the fluid part of the blood and dissolved substances pass into tissue spaces (intercellular spaces) among cells, and fluid and waste pass back from tissue spaces into the blood.

6. **Venules,** veins, and trunk veins return blood to the heart.

7. **Pulmonary arteries** convey blood from the right ventricle to pulmonary capillaries in lungs for oxygenation and expulsion of carbon dioxide. Pulmonary veins return oxygenated blood to the left atrium.
Pulmonary circuit — right ventricle, pulmonary trunk, pulmonary arteries, pulmonary capillaries, pulmonary veins, left atrium.

8. **Lymphatic structures:**
Lymph capillaries collect colorless fluid, lymph, from tissue spaces.
Lymph vessels convey lymph to lymphatic trunks.
Lymphatic trunks, i.e. right lymph duct and thoracic duct, conduct lymph to brachiocephalic veins, and then into blood stream.
Lymph nodes, along course of lymph vessels, manufacture lymphocytes and filter lymph, screening out microorganisms and cancer cells.

9. **Fetal circulatory structures:**
Foramen ovale, between right and left atria conducts blood entering right atrium from inferior vena cava into left atrium then to left ventricle bypassing the lungs.
Ductus arteriosus, a short vessel between pulmonary trunk and thoracic aorta conducts blood from the right ventricle and pulmonary trunk into the aorta, bypassing the lungs.
Umbilical arteries, two, branches of internal iliacs, convey blood through umbilical cord to the placenta. The placenta, attached to inner surface of the uterus, has maternal and fetal capillaries in close contact for exchange of fluid, gases, nutrients, waste products of fetus, etc. It substitutes for fetal lungs, digestive tract, and urinary system.
One umbilical vein returns oxygenated blood from the placenta through umbilical cord to the fetus. One branch joins the portal vein, another becomes the ductus venosus.
The ductus venosus, a part of the umbilical vein, bypasses the portal vein and liver and joins the inferior vena cava directly.

S. 178 SOME ANOMALIES OF THE HEART AND VESSELS

It was stated in the general discussion of arteries that the pattern of branching is constant for all members of a species, and that there is a similarity in the vessels in both sides of the trunk and both upper or lower limbs. This is generally true but there is some variation from one individual to another in the vessels and also the heart. If the variation does not interfere with function it is probably not important.

A few samples are listed and defined below:

Patent foramen ovale - the opening between the two atria does not close after birth but persists.

Interventricular septal defect - an abnormal opening is present in the interventricular septum between the two ventricles.

Patent ductus - or patent ductus arteriosus - the fetal vessel between the pulmonary trunk and aorta does not close after birth but remains open.

Stenosis of pulmonary orifice - the opening from the right ventricle into the pulmonary trunk is smaller than usual.

Coarctation of the aorta - a narrowing of the lumen of the aorta, usually in the upper descending part.

Right aortic arch - the aortic arch is on the right side rather than on the left.

Transposition - the heart may be on the right side with base directed to the left, and apex close to the right nipple rather than the left. Sometimes the abdominal organs are also reversed.

Other abnormalities, particularly of the large trunks, may occur especially in the superior mediastinum and may cause pressure upon the trachea or esophagus.

S. 179 APPLICATION TO RADIOGRAPHY

The heart lies immediately posterior to the anterior chest wall, with the apex to the left and the base to the right. Radiographs of the chest taken in the posteroanterior projection, with the front of the chest against the cassette will bring the heart close to the film. This will result in a minimal magnification of the cardiac image.

The cardiac image on a processed film of the chest will appear as a single opaque shadow with no separate images of the chambers. Some indication of the size and shape of a chamber is obtainable where a chamber forms one of the borders of the heart. The heart lies obliquely above the diaphragm and is rotated so that the right half lies in front of the left half. The right part of the sternocostal or anterior surface of the heart is formed by the right atrium which also forms the right margin. The remainder of this anterior surface is formed by the right ventricle except at the extreme left where the left ventricle forms a small part, and above this the left atrium forms a smaller part.

In routine P.A. views the right margin will be formed by the right atrium, the diaphragmatic lower border by the right ventricle. The left margin will be formed by the left ventricle below and by the left atrium above.

The posterior surface is formed by the left atrium and left ventricle. With enlargement of these chambers they displace the esophagus backwards. Barium paste in the esophagus will demonstrate this displacement.

Radiographs taken in the lateral and various oblique projections may help to outline the wall of each chamber.

Radiography of the heart and vessels does not form a part of the anatomical study. No study however can be complete without radiographic demonstrations of these structures. As a basic knowledge of the terms used should help the students, these are very briefly defined below.

The chambers of the heart, filled with blood or empty at systole are not visible as separate images. Special procedures consisting of filling them with an opaque medium are therefore utilized in radiography. Blood vessels also do not normally have separate images so must also be filled with a medium.

Cardiography is radiography of the chambers of the heart using an opaque medium to fill them. An injection is made into a vein in the upper or lower limb, or a catheter is inserted through one of these veins into the right atrium and the medium injected.

Arteriography is radiography of an artery and its branches by an injection of an opaque medium into it.

Aortography is radiography of the aorta following the injection of an opaque medium into it.

Venography is radiography of a vein and its tributaries either by direct injection into one of its smaller tributaries, or by injection of the corresponding artery, and radiographs when the veins have filled.

Lymphangiography is radiography of lymphatic vessels, nodes, and ducts, following injection of an opaque medium into small lymph capillaries.

Angiocardiography is a combined examination of the heart, trunk veins and/or trunk arteries. Angion = a vessel or duct; could be an artery, vein, or other small tube or duct.

The student must obtain illustrations of actual radiographs of the normal heart and vessels to correlate the anatomy with radiography.

S. 180 SOME PATHOLOGICAL CONDITIONS

Hypertension is high blood pressure. Tension here is employed as referring to pressure, i.e. the pressure of blood against the vessel wall.
Hypotension is blood pressure below normal.

Blood pressure is influenced by the capillary blood vessels. These function like the nozzle attached to a water hose. By adjusting the size of the opening through which water may pass out, the pressure of the water may be varied. Without the nozzle the water simply passes out with no appreciable force.

A thrombus is a clot of blood that has formed on the inside of a blood vessel or within the heart. In a vessel it may partly or completely block the vessel. Any condition by which the flow of blood is slowed down favors its formation. Lying in bed following an operation especially when the legs are not exercised may induce clotting in the veins of the leg. The condition is called thrombosis.

Phlebitis is an inflammation of the lining membrane of a vein. (Phleb = vein). This results in roughening of the lining membrane and predisposes to thrombus formation termed **thrombo-phlebitis**.

An embolus is a foreign body obstructing a blood vessel. This may be part of a thrombus that has become separated off, or fat cells, bubbles of air, globules of oil, a clump of bacteria, or cancer cells. The embolus lodges where the vessel becomes too small to allow it to proceed. Frequently it occurs where a vessel branches, each branch being too small for further passage. It may come from some distant part of the body. If from a vein it will lodge in a pulmonary artery or branch of it (pulmonary embolus). If from the interior of the heart it may lodge in any branch of the aorta, in the heart, kidney, spleen, or brain, etc. So there may be a coronary embolus, renal embolus, cerebral embolus, etc. The condition is called embolism.

An infarct is a condition produced in an organ as a result of a blocking of an artery supplying it by an embolus or some other obstructing lesion. The blood supply is cut off, the tissues die, and are replaced by fibrous connective tissue, e.g. coronary infarct, pulmonary infarct, etc.

Occlusion of a vessel is its closure due to spasm, embolus, filling in of its lumen due to degeneration of the vessel wall, thrombosis, etc.

Coronary occlusion — blocking of one of the coronary arteries to the heart, or a branch of it. This may be from spasm, embolus, thrombosis, etc.

An aneurysm is a bulging of the wall of a vessel or of one side of the wall due to a weakness from disease of the vessel. It may involve the whole circumference with a fusiform dilatation, or a single side similar to a car tire when the fabric is broken, and the tire bulges. It follows degeneration of a vessel, sometimes from syphilis, etc. An aneurysm may rupture with a fatal hemorrhage resulting. The aorta, cerebral or other vessels may be involved.

Arteriosclerosis and atherosis refer to thickening of a vessel wall with degeneration of the normal coats, and replacement by fibrous tissue, fat, and calcium.

Varicose veins are dilated veins due to weakening of the vessel walls. The veins dilate, the valves cannot close properly and blood may flow in the opposite direction to the normal e.g. in leg.

Hemorrhoids are dilated veins in the anal canal; these frequently bleed.

Esophageal varicosities may occur in the lower part of the esophagus. These may bleed, with the vomiting of blood, or the passage of dark (tarry) stools.

Cardiac hypertrophy - enlargement of the heart from enlargement of its muscle fibers because of an increased work load in pumping blood.

Cardiac dilatation - cardiac enlargement due to stretching of muscle fibers, fibers tired out, and chambers enlarged.

Congestive heart failure (CHF) - heart cannot pump blood through its circuits.

Endocarditis - inflammation of the valves of the heart. They are covered by endocardium.

Myocarditis - inflammation of cardiac muscle.

Pericarditis - inflammation of the pericardium, often with fluid in the pericardial sac - pericardial effusion.

Cardiac infarct - infarction area of dead cardiac muscle, resulting from cutting off of its blood supply, replaced by fibrous tissue in survivors.

Valvular disease - a thickening, deformity, shrinking, and fusion of cusps of cardiac valves from an infection or degeneration, resulting in inability to open or close completely.

Stenosis - a narrowing of an opening or canal, e.g. mitral, aortic, tricuspid or pulmonary stenosis involving the openings of the heart as a result of disease of the cardiac valves.

Incompetence - failure of a valve to close completely from disease, resulting in regurgitation of blood back into the chamber from which it was expelled.

S. 181 ANATOMICAL TERMS DESCRIBING THE CIRCULATORY SYSTEM

Note: a - artery, aa - arteries, v - vein, vv - veins

Heart, cor, cardia
arteries

tunica intima - intima

arterioles
capillaries
venules
veins - phlebs
trunk arteries:
 aorta
 pulmonary trunk
trunk veins:
 superior vena cava
 inferior vena cava
 four pulmonary veins
lymphatic capillaries
lymphatic vessels
lymph ducts;
 lymph nodes (glands)
reticuloendothelial
 structures
atrium, atria, atrial
ventricle, ventricular
interatrial septum
enterventricular
 septum
atrioventricular
 opening
ostium = opening
pulmonary opening
aortic opening
openings of:
 superior vena cava
 inferior vena cava
 coronary sinus
 four pulmonary
 veins
Valves:
 left atrioventric-
 ular or mitral
 or bicuspid
 right atrioventric-
 ular or tricuspid
 aortic semilunar
 pulmonary semi-
 lunar
auricles
conus arteriosus
coronary sulcus
interventricular
 sulci
coronary arteries
coronary sinus
cardiac veins
conducting appara-
tus
systole, adj. sys-
 tolic
diastole, adj. dia-
 stolic

pulmonary blood
 vessels
pulmonary trunk*
pulmonary arteries
pulmonary capil-
 laries
pulmonary veins
pulmonary circu-
 lation
Systemic blood ves-
sels
systemic circu-
 lation aorta *
ascending aorta

tunica media -
 media
tunica adven-
 titia
lumen of blood
 vessel

apex of heart
base of heart
sternocostal
 surface
diaphragmatic
 surface
endocardium
myocardium
visceral peri-
 cardium or

 epicardium
serous peri-
 cardium
parietal peri-
 cardium
fibrous peri-
 cardium
pericardial sac
pericardial
 cavity
external iliac
celiac trunk
 left gastric
 a (artery)
 splenic a
 common hepa-
 tic a
 superior mes-
 enteric a
 middle sac-
 ral a
superior vena
 cava
brachiocepha-
 lic vv (veins)
 internal jug-
 ular vv
subclavian vv
azygos v
 hemiazygos v
 accessory
 hemiazygos

inferior vena
 cava*
common iliac vv
external iliac
 vv
internal iliac
 vv
hypogastric vv
ovarian, tes-
 ticular vv
renal vv
suprarenal vv
phrenic vv
portal v
 superior mes-
 enteric v
 splenic v
 inferior mes-
 enteric v
portal vessels

coronary arteries *
arch of aorta
brachiocephalic a
left common caro-
    tid a
left subclavian a
external carotid a
internal carotid a
vertebral a
descending thoracic
    aorta
mediastinal aa
esophageal aa
pericardial aa
intercostal aa
phrenic aa
bronchial aa
abdominal aorta
inferior phrenic
    aa
middle suprarenal
    aa
renal aa
ovarian, testicu-
    lar aa
lumbar aa
common iliac
internal iliac
brachial vv
axillary v
subclavian v

external iliac a
femoral a
popliteal a

hepatic veins
portal capil-
    laries

blood vessels
    of brain
internal carotid
    aa*
anterior cere-
    bral aa
vertebral aa
basilar a
posterior cere-
    bral aa =
    branch to
    posterior part
circle of Willis

cerebral arte-
    rial circle
venous sinuses
    of the dura
    mater
superior sagi-
    ttal sinus
inferior sagi-
    ttal sinus
transverse sinus
sigmoid sinuses
internal jugular
    vein
cerebral veins
middle menin-
    geal a

posterior tibial a
peroneal a
plantar aabrachial a
anterior tibial a
dorsalis pedis
dorsal a of foot
arches of foot
great saphenous vein
small saphenous vein
veins of foot
dorsalis pedis v
anterior tibial vv
peroneal vv
plantar vv
posterior tibial vv
popliteal v
femoral v
external iliac v*

pulse
pressure points
blood pressure
systolic B.P.
diastolic B.P.

placenta
umbilical cord
umbilical aa
umbilical v
foramen ovale
ductus arteriosis
internal iliac aa
(fetal)
ductus venosus
hepatic vv

subclavian a
axillary a

radial a
ulnar a
palmar arches
dorsal arch
cephalic v
basilic v
median ante-
    brachial v
veins of hand
ulnar vv
radial vv
lymphatics
lymph capil-
    laries
lymph vessels
lymph ducts or
    trunks;
    thoracic duct
    right lymph
    duct
cisterna chyli
lymph nodes
    (glands)
tonsils
adenoids
solitary lymph
    nodes
aggregated lymph
    nodes or Peyers
    patches
spleen
thymus gland

CHART OF BRANCHES OF THE AORTA — FOR REFERENCE ONLY:

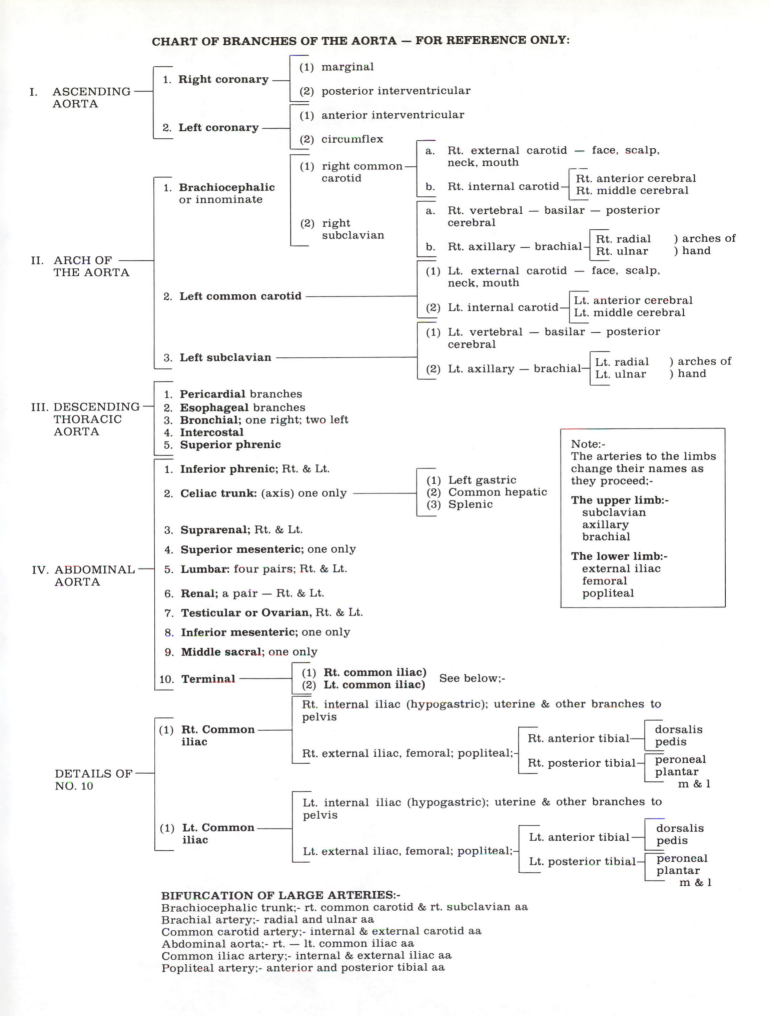

I. ASCENDING AORTA

1. **Right coronary**
   - (1) marginal
   - (2) posterior interventricular
2. **Left coronary**
   - (1) anterior interventricular
   - (2) circumflex

II. ARCH OF THE AORTA

1. **Brachiocephalic** or innominate
   - (1) right common carotid
     - a. Rt. external carotid — face, scalp, neck, mouth
     - b. Rt. internal carotid — Rt. anterior cerebral / Rt. middle cerebral
   - (2) right subclavian
     - a. Rt. vertebral — basilar — posterior cerebral
     - b. Rt. axillary — brachial — Rt. radial) arches of / Rt. ulnar) hand
2. **Left common carotid**
   - (1) Lt. external carotid — face, scalp, neck, mouth
   - (2) Lt. internal carotid — Lt. anterior cerebral / Lt. middle cerebral
3. **Left subclavian**
   - (1) Lt. vertebral — basilar — posterior cerebral
   - (2) Lt. axillary — brachial — Lt. radial) arches of / Lt. ulnar) hand

III. DESCENDING THORACIC AORTA

1. **Pericardial** branches
2. **Esophageal** branches
3. **Bronchial;** one right; two left
4. **Intercostal**
5. **Superior phrenic**

IV. ABDOMINAL AORTA

1. **Inferior phrenic;** Rt. & Lt.
2. **Celiac trunk:** (axis) one only
   - (1) Left gastric
   - (2) Common hepatic
   - (3) Splenic
3. **Suprarenal;** Rt. & Lt.
4. **Superior mesenteric;** one only
5. **Lumbar:** four pairs; Rt. & Lt.
6. **Renal;** a pair — Rt. & Lt.
7. **Testicular or Ovarian,** Rt. & Lt.
8. **Inferior mesenteric;** one only
9. **Middle sacral;** one only
10. **Terminal**
    - (1) **Rt. common iliac)**
    - (2) **Lt. common iliac)** See below;-

> Note:-
> The arteries to the limbs change their names as they proceed;-
>
> **The upper limb:-**
> subclavian
> axillary
> brachial
>
> **The lower limb:-**
> external iliac
> femoral
> popliteal

DETAILS OF NO. 10

- (1) **Rt. Common iliac**
  - Rt. internal iliac (hypogastric); uterine & other branches to pelvis
  - Rt. external iliac, femoral; popliteal;
    - Rt. anterior tibial — dorsalis pedis
    - Rt. posterior tibial — peroneal / plantar / m & l
- (1) **Lt. Common iliac**
  - Lt. internal iliac (hypogastric); uterine & other branches to pelvis
  - Lt. external iliac, femoral; popliteal;
    - Lt. anterior tibial — dorsalis pedis
    - Lt. posterior tibial — peroneal / plantar / m & l

BIFURCATION OF LARGE ARTERIES:-
Brachiocephalic trunk;- rt. common carotid & rt. subclavian aa
Brachial artery;- radial and ulnar aa
Common carotid artery;- internal & external carotid aa
Abdominal aorta;- rt. — lt. common iliac aa
Common iliac artery;- internal & external iliac aa
Popliteal artery;- anterior and posterior tibial aa

NOTES

NOTES

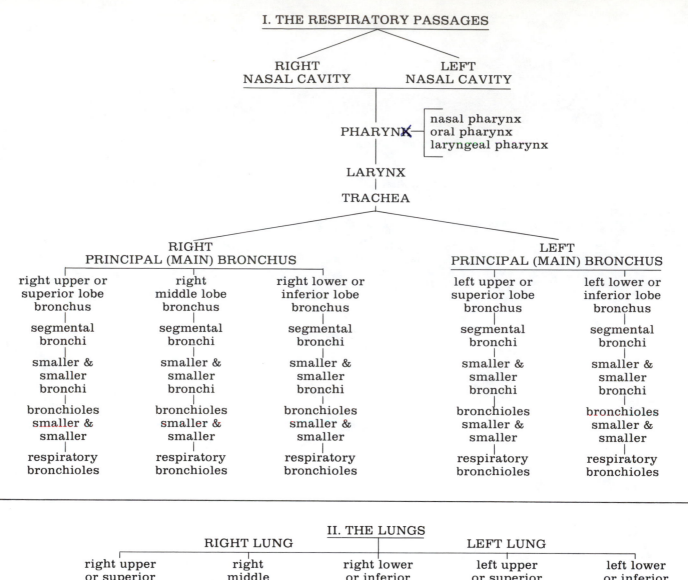

I. THE RESPIRATORY PASSAGES

RIGHT
NASAL CAVITY   LEFT
NASAL CAVITY

PHARYNX — nasal pharynx / oral pharynx / laryngeal pharynx

LARYNX

TRACHEA

RIGHT PRINCIPAL (MAIN) BRONCHUS

right upper or superior lobe bronchus	right middle lobe bronchus	right lower or inferior lobe bronchus
segmental bronchi	segmental bronchi	segmental bronchi
smaller & smaller bronchi	smaller & smaller bronchi	smaller & smaller bronchi
bronchioles smaller & smaller	bronchioles smaller & smaller	bronchioles smaller & smaller
respiratory bronchioles	respiratory bronchioles	respiratory bronchioles

LEFT PRINCIPAL (MAIN) BRONCHUS

left upper or superior lobe bronchus	left lower or inferior lobe bronchus
segmental bronchi	segmental bronchi
smaller & smaller bronchi	smaller & smaller bronchi
bronchioles smaller & smaller	bronchioles smaller & smaller
respiratory bronchioles	respiratory bronchioles

II. THE LUNGS

RIGHT LUNG   **LEFT LUNG**

right upper or superior lobe	right middle lobe	right lower or inferior lobe	left upper or superior lobe with lingula	left lower or inferior lobe
segments upper lobe	segments middle lobe	segments lower lobe	segments upper lobe	segments lower lobe
alveolar sacs	alveolar sacs	alveolar sacs	alveolar sacs	alveolar sacs
alveoli	alveoli	alveoli	alveoli	alveoli

LUNGS — BASIC

apex, apices
base
hilum, hila
root
fissure
lobe
segment
alveolar sac
alveolus
capillary
alveolar
membrane

III. BLOOD VESSELS

pulmonary trunk

right pulmonary a ——— left pulmonary a

3 lobar aa   2 lobar aa

segmental & smaller aa   segmental & smaller aa

arterioles   arterioles

capillaries   capillaries

2 pulmonary vv (Rt)   2 pulmonary vv (Lt)

descending thoracic aorta

1 right bronchial a ——— 2 left bronchial aa

14. THE RESPIRATORY SYSTEM

The respiratory system has been designed to take oxygen into the body, and to get rid of carbon dioxide and some other waste products of bodily activity. Living organisms, whether unicellular or multicellular, breathe. In the human the respiratory system is lodged in the facial cranium, neck, and thorax. It consists of three parts: the respiratory passages, lungs, and pleura, along with their blood vessels.

S. 183   PARTS OF THE RESPIRATORY SYSTEM
I. **THE RESPIRATORY PASSAGES**   F. 14-1, 2
  1. Nasal cavities
  2. Pharynx:
     1) nasal pharynx (NA, nasal part)
     2) oral pharynx (NA, oral part)
     3) laryngeal pharynx (NA, laryngeal part)
  3. Larynx; adj. laryngeal
  4. Trachea; adj. tracheal
  5. Bronchi
     1) Main (principal, NA) bronchi, Rt. & Lt.
     2) Lobar bronchi;
       a. Rt. superior (upper) lobe bronchus
       b. Rt. middle lobe bronchus
       c. Rt. inferior (lower) lobe bronchus
       d. Lt. superior (upper) lobe bronchus
       e. Lt. inferior (lower) lobe bronchus
     3) Segmental, bronchi;
       a. Rt. lung, 10, varies with lobe
       b. Lt. lung, 9, varies with lobe
     4) Smaller and smaller bronchi
  6. Bronchioles = little bronchi
     1) Smaller and smaller bronchioles
     2) Respiratory bronchioles

II. **THE LUNGS** - Right and left;   F. 18-8 to 12
  1. **Components:**
     1) Apex, pl. apices, adj. apical
     2) Base
     3) Hilum (hilus), pl. hila, adj. hilar
     4) Roots - right & left
     5) Surfaces
     6) Fissures
     7) Lobes
     8) Segments & lobules
     9) Alveolar sacs and alveoli
  2. **Details of lobes:**
     Right lung - three lobes
     1) right superior (upper) lobe
     2) right middle lobe
     3) right inferior (lower) lobe
     Left lung - two lobes
     1) left superior (upper) lobe & lingula
     2) left inferior (lower) lobe
  3. **Segments of lobes:**
     Number varies from lobe to lobe
  4. Alveolar sacs and alveoli

III. **THE PLEURA**   F. 14-10
  1. Visceral pleura (NA) pulmonary pleura
  2. Parietal pleura
  3. Pleural cavity

BLOOD VESSELS OF THE LUNGS
1. Pulmonay trunk (Pulmonary artery, (OT)
2. Pulmonary arteries, right and left, lobar, segmental, smaller branches and lung capillaries, origin - Rt. ventricle
3. Pulmonary veins, two right, two left tributaries similar to arteries, termination in left atrium
4. Bronchial arteries, one right, two left
5. Bronchial veins

Note: all bronchi except the main, all bronchioles, and all blood vessels except the large trunks, are within the lungs.

S. 184   DETAILED STUDY - RESPIRATORY PASSAGES

The respiratory passages provide tubes by which air is conveyed into and out of the lungs. The six divisions of these passages are studied in detail below.

1. THE NASAL CAVITIES   F. 14-1, 2, 3

The **nasal cavities** are the two chambers of the nose. The nostrils or nares are the two openings into the nasal cavities. These cavities are separated by a vertical partition, **the nasal septum**. They open posteriorly into the nasal pharynx. They are lined by mucous membrane, with hairs in the anterior part. Openings are present in each nasal cavity from the corresponding frontal, ethmoidal, sphenoidal, and maxillary sinuses, as well as an opening for the nasolacrimal or tear duct. Three curved bony shelves project into each nasal cavity in its lateral wall, the nasal conchae. The lower one, the inferior concha or turbinate, is a separate bone. These shelves partly divide each nasal cavity into four parts, the meatuses. The paranasal sinuses have been described in S. 110. The nasal cavities warm, moisten, and filter inspired air. (air conditioner)

2. THE PHARYNX   F. 14-1

The **pharynx or throat** lies posterior to the nasal cavities, mouth, and larynx. It extends down to the opening into the esophagus. It is about five inches in length, and has three parts: the nasal pharynx, oral pharynx, and laryngeal pharynx. F. 14-1
1) **The nasal pharynx,** nasopharynx, or nasal part (NA), is that part of the pharynx posterior to the nasal cavities, and above and behind the soft palate. The two nasal cavities open into it.
The soft palate projects posteriorly from the hard palate (bony). It forms a partial posterior wall for the mouth. During swallowing it closes off the nasal pharynx preventing regurgitation of fluid or food into the nasal cavities.
The auditory tube, a canal between the nasal pharynx and middle ear, opens into the lateral side of the nasal pharynx. There are two, a left and a right. (OT, Eustachian tube).
The adenoids are located on the posterior wall of the nasal pharynx (NA, pharyngeal tonsil).

2) **The oral pharynx,** oropharynx or oral part (NA), lies posterior to the mouth, between the soft palate above and the epiglottis below. (os = mouth - oral)
The epiglottis is a flat leaflike plate of cartilage with a free upper border that projects back from the anterior wall of the pharynx below the base of the tongue.
The tonsils (NA, palatine tonsils) are masses of lymphoid tissue, one on each side of the opening between the mouth and oral pharynx, the fauces.

3) **the laryngeal pharynx,** laryngopharynx or (NA), laryngeal part extends from the epiglottis down to the opening into the esophagus. The larynx opens off from the anterior surface of the laryngeal pharynx. There are seven openings into the pharynx. Infection of the nasal cavities may extend into the paranasal sinuses. Infection of the oral pharynx, palatine tonsils or adenoids may track out through the auditory tubes to the middle ear.

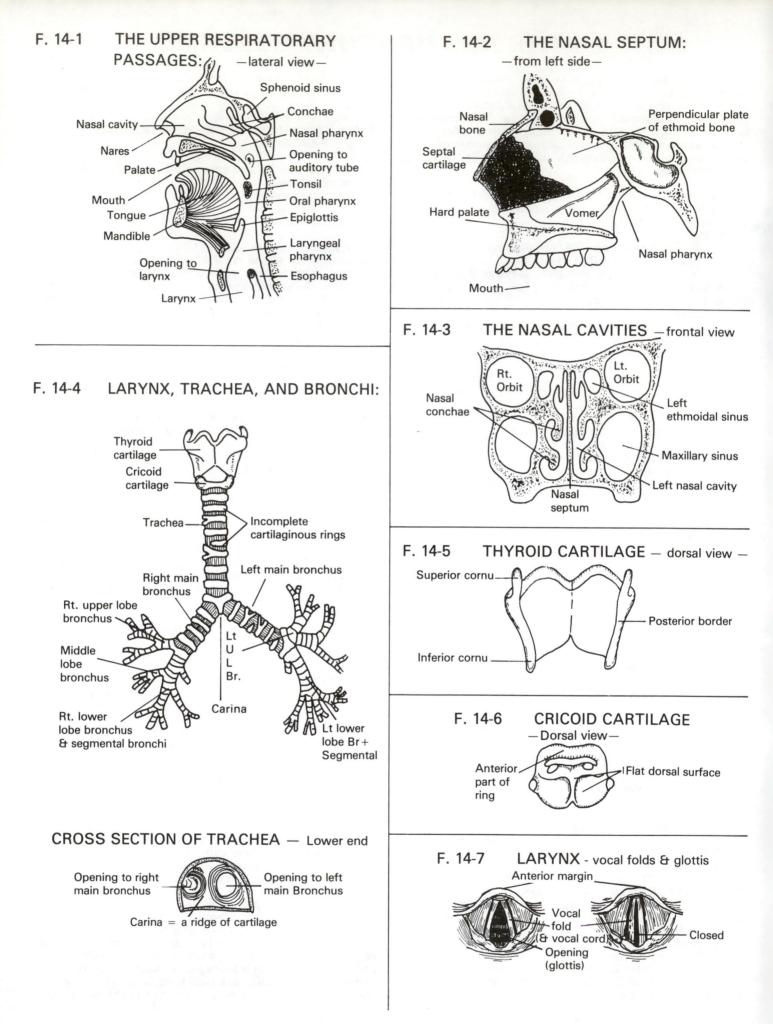

F. 14-1 THE UPPER RESPIRATORARY PASSAGES: —lateral view—

- Sphenoid sinus
- Conchae
- Nasal cavity
- Nasal pharynx
- Nares
- Opening to auditory tube
- Palate
- Tonsil
- Mouth
- Oral pharynx
- Tongue
- Epiglottis
- Mandible
- Laryngeal pharynx
- Opening to larynx
- Esophagus
- Larynx

F. 14-2 THE NASAL SEPTUM: —from left side—

- Nasal bone
- Perpendicular plate of ethmoid bone
- Septal cartilage
- Hard palate
- Vomer
- Nasal pharynx
- Mouth

F. 14-3 THE NASAL CAVITIES —frontal view

- Rt. Orbit
- Lt. Orbit
- Nasal conchae
- Left ethmoidal sinus
- Maxillary sinus
- Left nasal cavity
- Nasal septum

F. 14-4 LARYNX, TRACHEA, AND BRONCHI:

- Thyroid cartilage
- Cricoid cartilage
- Trachea
- Incomplete cartilaginous rings
- Right main bronchus
- Left main bronchus
- Rt. upper lobe bronchus
- Middle lobe bronchus
- Lt U L Br.
- Rt. lower lobe bronchus & segmental bronchi
- Carina
- Lt lower lobe Br + Segmental

F. 14-5 THYROID CARTILAGE — dorsal view —

- Superior cornu
- Posterior border
- Inferior cornu

F. 14-6 CRICOID CARTILAGE —Dorsal view—

- Anterior part of ring
- Flat dorsal surface

CROSS SECTION OF TRACHEA — Lower end

- Opening to right main bronchus
- Opening to left main Bronchus

Carina = a ridge of cartilage

F. 14-7 LARYNX - vocal folds & glottis

- Anterior margin
- Vocal fold (& vocal cord)
- Closed
- Opening (glottis)

causing otitis median with earache. It may extend into the mastoid cells from the middle ear causing mastoiditis.

3. THE LARYNX F. 14-1

The **larynx**, organ of voice or voice box, is a triangular boxlike structure that lies in the anterior part of the neck. Its upper end opens into the laryngeal part of the pharynx. Its lower end is continuous with the trachea. This box is surrounded by nine cartilages, only two of which will be described here.

The thyroid cartilage is composed of two flat plates of cartilage placed vertically and joined together at the midline anteriorly. The posterior ends are separated similar to the blades of a snow plough. The larynx lies between these two plates. The prominence in front where the two plates join is often visible, especially in males and is called the laryngeal prominence or Adam's apple. The upper margin is notched. Each plate or lamina extends back to end in small projections above and below - the cornua.

The cricoid cartilage is shaped like a signet ring, (krikos = a ring + oid). It encircles the larynx below the thyroid cartilage. Its broad flat part lies posteriorly and helps to fill in the gap between the separated plates of the thyroid cartilage. In front the thyroid and cricoid cartilages are bound together by ligaments. Below, the cricoid is attached by ligaments to the superior margin of the first ring of the trachea. F. 14-6

The vocal folds are paired folds of the lining membrane of the larynx that extend medially from each sidewall of the larynx towards the midline. They pass from front to back of the laryngeal cavity. Each contains a **vocal cord** that functions in phonation. F. 14-7

The glottis (rima glottis) is the slitlike opening between the vocal folds. Air must pass in and out through this slit.

4. THE TRACHEA F. 14-3

The **trachea** or windpipe is a hollow tube 11 cm or about 4.5 in. in length extending from the larynx above to the bronchi below. Commencing opposite the sixth cervical vertebra it reaches to the fourth thoracic vertebra. The upper part lies in the neck below the Adam's apple. The remainder lies in the superior mediastinum in front of the esophagus. Like the larynx it is lined by epithelium. Its walls are strengthened by 16 to 20 incomplete rings of cartilage. Each ring encircles the trachea but is open behind similar to the letter "C". These rings prevent collapse of the trachea in the same way as the rib cage prevents collapse of the lungs. Muscle fibers fill in the gaps posteriorly. A large metal coin in the trachea will assume the "on edge" position when viewed from the front since it can push the muscular wall back, whereas the side walls are rigid due to the cartilaginous rings. The trachea divides into two bronchi.

5. THE BRONCHI F. 14-3

The **bronchi** are hollow branched tubes continuous above with the trachea. Bronchi like the trachea have encircling cartilaginous rings as well as circular visceral muscle fibers in their walls, and an epithelial lining. The main bronchi divide into lobar bronchi, these branch into segmental bronchi, and these branch into smaller and smaller bronchi, ending as bronchioles. (s. bronchus, adj. bronchial).

1) **The main bronchi** (NA, principal bronchi) one right and one left, result from the division of the trachea into right and left branches. Each of these enters the corresponding lung and divides into lobar bronchi, one for each lobe of each lung.

The carina is a ridge of cartilage between the openings from the trachea into right and left main bronchi, i.e. at the bifurcation (forking) of the trachea. Carina = a keel, i.e. a ridge. F. 14-4

2) **The lobar bronchi** result from division of the main bronchi into one for each pulmonary lobe. F. 14-4, 10
The **right main bronchus** divides into three lobar bronchi, one for each of its lobes:
a. right superior (upper) lobe bronchus
b. right middle lobe bronchus
c. right inferior (lower) lobe bronchus
The **left main bronchus** divides into two lobar bronchi, one for each of its two lobes:
a. left superior (upper) lobe bronchus
b. left inferior (lower) lobe bronchus

3) **The segmental bronchi**
Each lobar bronchus divides into segmental bronchi, one for each segment of each lung. As the number of segments varies from lobe to lobe the number of segmental bronchi also varies. See bronchopulmonary segments & F. 14-4

4) **Smaller and smaller bronchi**
Due to branching and rebranching of segmental bronchi they become smaller. The smallest ones are called bronchioles.

6. THE BRONCHIOLES F. 14-10

The **bronchioles** are formed by the division of the smallest bronchi. (Bronchiole = a little bronchus). They also branch to form smaller bronchioles. The smallest bronchioles are named respiratory bronchioles, and these lead into the air sacs of the lungs.

Note: the numerous branches of the main bronchi lie within the lung, and form part of it. They are considered separately in this study for the sake of simplicity, and because they do form respiratory passages and are important in radiography.

S. 185 DETAILED STUDY OF THE LUNG
F. 14-8, 12

As. lungen = lung
L. pulmo = lung, pl. pulmones, adj. pulmonary
G. pneumon = lung, adj. pneumonic
G. pneuma = air, adj. pneumatic

The lungs are divided into lobes - segments - alveolar sacs - alveoli.

1. COMPONENTS OF THE LUNGS DEFINED

There are two lungs, a right and a left. Each occupies the corresponding half of the thorax. The lungs are separated by the mediastinum, a partition in the median plane containing all of the thoracic organs except the lungs. See the description at the end of this chapter. Each lung is cone shaped.

The apex is the upper bluntly pointed end, and reaches one inch above the clavicle. pl. apices.

The base, the broad lower end is concave and rests upon the upper convex surface of the diaphragm. Its lateral and posterior edges lie at a lower level than the anterior following the diaphragmatic dome.

The hilum (hilus) is a depression or indention on the medial surface of each lung where the structures enter or leave the lung. pl. hila, adj. hilar.

The roots of the lungs consist of structures entering or leaving the lungs at the hila, the blood vessels, and main bronchi.

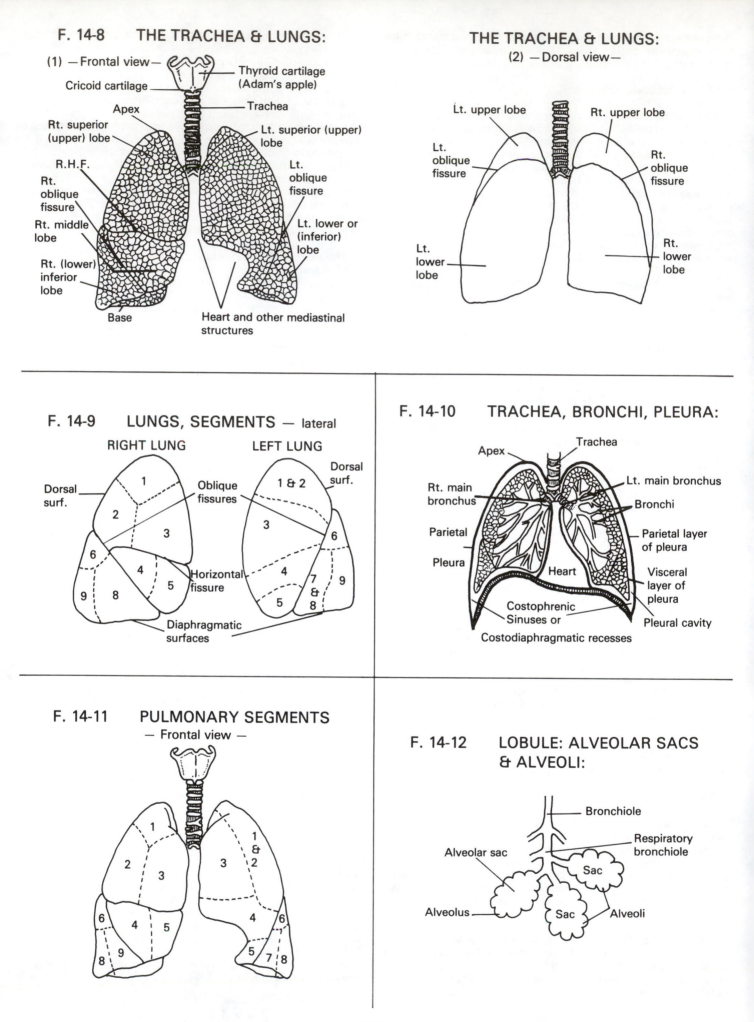

F. 14-8   THE TRACHEA & LUNGS:

(1) —Frontal view—

Thyroid cartilage (Adam's apple)
Cricoid cartilage
Apex
Trachea
Rt. superior (upper) lobe
Lt. superior (upper) lobe
R.H.F.
Rt. oblique fissure
Lt. oblique fissure
Rt. middle lobe
Lt. lower or (inferior) lobe
Rt. (lower) inferior lobe
Base
Heart and other mediastinal structures

THE TRACHEA & LUNGS:

(2) —Dorsal view—

Lt. upper lobe
Rt. upper lobe
Lt. oblique fissure
Rt. oblique fissure
Lt. lower lobe
Rt. lower lobe

F. 14-9   LUNGS, SEGMENTS — lateral

RIGHT LUNG   LEFT LUNG

Dorsal surf.
1
Oblique fissures
Dorsal surf.
1 & 2
2
3
3
6
6
4
5
Horizontal fissure
4
7 & 8
9
9
8
5
7 & 8
9
Diaphragmatic surfaces

F. 14-10   TRACHEA, BRONCHI, PLEURA:

Apex
Trachea
Rt. main bronchus
Lt. main bronchus
Parietal
Bronchi
Pleura
Parietal layer of pleura
Heart
Visceral layer of pleura
Costophrenic Sinuses or
Pleural cavity
Costodiaphragmatic recesses

F. 14-11   PULMONARY SEGMENTS

— Frontal view —

1
1 & 2
2
3
3
6
4
5
4
6
8
9
5
7
8

F. 14-12   LOBULE: ALVEOLAR SACS & ALVEOLI:

Bronchiole
Respiratory bronchiole
Alveolar sac
Sac
Alveolus
Sac
Alveoli

The **costal surface** of the lung is that surface that lies adjacent to the ribs and cartilages.

The **diaphragmatic surface** is that part in contact with the diaphragm.

The **mediastinal surface** is in contact with the mediastinum, the middle partition between the two lungs.

The **fissures** of the lung are narrow grooves or slits that divide the lung into lobes.

The **oblique fissure** can be traced around the chest wall from the fifth thoracic vertebrae behind, obliquely downwards, and forward to the anterior end of the sixth rib. This fissure is present on each lung.

The **horizontal fissure** begins at the axillary border of the right lung. Commencing at the oblique fissure it passes horizontally forward opposite the fourth rib and its cartilage to the anterior margin of the **right** lung. The left lung has no horizontal fissure, having only two lobes.

2. THE LOBES OF THE LUNGS   F. 14-6

The **RIGHT LUNG** is divided into three lobes by the oblique and horizontal fissures: superior, middle, and inferior lobes.

The **right superior (upper) lobe** lies above and in front of the oblique fissure, as well as above the horizontal fissure. It forms the upper anterior part of the right lung, and a part of the posterior surface.

The **right middle lobe** forms a wedge shaped part in the lower medial chest with the horizontal fissure above, and the oblique fissure and the diaphragm below. It is visible on the sternocostal surface of the right lung. There is no left middle lobe.

The **right inferior (lower) lobe** lies below and behind the oblique fissure. It forms most of the posterior part of the right lung.

The **LEFT LUNG** has only two lobes, superior and inferior, as the heart lies in the lower left chest.

The **superior (upper) lobe** lies above and in front of the left oblique fissure. Its lower part is called the lingula and partly overlaps the heart. Lingua = tongue, lingula = little tongue.

The **left inferior (lower) lobe** lies behind and below the left oblique fissure. It forms most of the posterior surface of the left lung.

There is overlapping of the adjacent lobes. Since the oblique fissures pass obliquely down and forward the upper parts of the inferior lobes extend up behind the superior lobes. The right middle lobe lies in front of the lower part of the right inferior lobe. In an anteroposterior radiograph of the chest a shadow on the film could be in the upper or lower lobe or possibly in the middle lobe. Lateral views would indicate whether the shadow is in the anterior or posterior chest, showing the lobe involved. F. 14-8, 9

3. THE PULMONARY SEGMENTS   F. 14-10, 11

The **segments** are the structural units of the lungs, each with its segmental bronchus, segmental artery and vein. A segment can be separated from adjacent segments and may be removed. The number of segments varies from lobe to lobe.

	Right lung	Left lung
Upper lobe	apical	apical posterior
	posterior	anterior
	anterior	superior lingual
		inferior lingual
Middle lobe	lateral	nil
	medial	nil

	Right lung	Left Lung
Lower lobe:	superior	superior
	medial basal	medial basal
	anterior basal	anterior basal
	lateral basal	lateral basal
	posterior basal	posterior basal

In the Nomina Anatomica the superior segment of the lower lobe is subdivided into superior (apical) and subsuperior (subapical) segments. There does not appear to be complete agreement as to the number of segments in the lower lobes, and also there is some variation from one subject to another. The total number of segments therefore varies with the authority quoted. In any event students should not be required to memorize the list since it is not applicable to their work.

4. ALVEOLI AND ALVEOLAR SACS   F. 14-12

An **alveolus** is a cup shaped microscopic structure that with many other similar structures form the wall of an alveolar or air sac. The wall of each alveolus consists of a single layer of flattened cells. Many capillaries surround the external surface of each alveolus. Air in an alveolus is separated from the blood stream by its thin wall and that of a capillary. The exchange of gases takes place between the capillaries and an alveolus. For this reason the alveolus has been named the **functional unit** of the lung. (L. alveolus = a small tub or a cavity).

An **alveolar sac** or air sac is the expanded saclike cavity opening from a small division (alveolar duct) of a respiratory bronchiole. (See bronchioles above). Each alveolar sac is made up of many alveoli. They resemble a series of cups with edges glued together, and with their open ends facing a central cavity. Air from the respiratory bronchioles enters the alveolar sacs through their alveolar ducts.

Lobules of the lungs are described by some authorities. A lobule includes a respiratory bronchiole, its alveolar ducts, their alveolar sacs and alveoli, and the blood vessels of this unit. Each lobule is a small division of a pulmonary segment.

5. SUMMARY OF THE STRUCTURE OF THE LUNGS

The **right lung** has an oblique and a horizontal fissure dividing it into three lobes: a superior, middle, and inferior lobe. The right upper lobe has three segments, the middle lobe two, and the lower lobe five. Each segment is made up of smaller divisions or lobules. The lobules are composed of alveolar sacs (air sacs) with many alveoli forming their walls. The bronchi, and their smaller divisions the bronchioles form passages within the lung.

The **left lung** has a single oblique fissure dividing it into two lobes, a superior and an inferior. The superior lobe has a tonguelike part adjacent to the left cardiac border, called the lingula = a little tongue. There is no left middle lobe. The upper lobe has four segments, the lower lobe five. The left lung like the right has lobules, alveolar sacs and alveoli.

S. 186   THE PLEURA   F. 14-8

pl. pleurae; adj. pleural. The pleura forms the lining membrane of the chest and the covering of each lung. It is a serous membrane similar to the pericardium and peritoneum, and secretes a thin watery fluid.

The **Visceral pleura** (NA, **pulmonary pleura**) covers each lung except at its root. The outer surface is smooth similar to that of the visceral pericardium. It is firmly adherent to the underlying lung. (L. viscus = an organ; visceral — part of an organ).

The **parietal pleura** (NA) lines the chest cavity on each side. The inner surface is smooth and is in contact with the visceral pleura, but is attached about the lung root only.

The **diaphragmatic pleura** is that part of the parietal pleura on the upper surface of the diaphragm.

The **costal pleura** is the parietal pleura that lines the rib cage.

The **mediastinal pleura** is the parietal pleura on each lateral surface of the mediastinum.

The **pleural cavity** is the space between the lung and chest wall, i.e. between the parietal and visceral pleura. With the lung expanded, its normal state, there is no actual space. With air or fluid in it the space becomes apparent.

The **costophrenic sinus** or recess (NA, costo-diaphragmatic recess, an awkward term) is the narrow space where the inner surface of the lower chest wall and the diaphragm are in contact. This is at the lower level of the pleural cavity. This space is shallow anteriorly, lower down laterally, and at the lowest level posteriorly. Any free fluid in the pleural space will gravitate to this sinus, since fluid seeks the lowest possible level.

The pleura forms a closed sac about each lung similar to the pericardial sac surrounding the heart. The lung has been pushed against the sac so that the two layers of the sac form the visceral and parietal pleura and envelop the lung. (Review structure of pericardium).

S. 187 BLOOD VESSELS AND NERVES

Two sets of arteries supply blood to the lungs: the bronchial and pulmonary arteries.

The **BRONCHIAL ARTERIES**, one right and two left, branches of the descending thoracic aorta, send branches into the lungs along with the bronchi. These arteries supply the lungs with blood as far as the respiratory bronchioles, to the bronchi and larger bronchioles. Bronchial veins drain blood from the same areas of the lungs.

The **PULMONARY ARTERIES**, one right and one left, are the two branches of the pulmonary trunk that originate from the right ventricle. Each artery sends branches into the corresponding lung along with the bronchi.

The pulmonary arteries have two functions:
1) They form capillaries about the alveoli for the exchange of oxygen and carbon dioxide,
2) They furnish nutrients to the respiratory bronchioles and alveolar sacs.

The **PULMONARY VEINS**, two right and two left, drain blood from the capillaries about the alveoli, and convey it to the left atrium.

The **vagi and sympathetic nerves** send fibers to the lungs from plexuses about the lung roots.

S. 188 THE PHYSIOLOGY OF RESPIRATION

The diaphragm and intercostals are respiratory muscles, and by their contraction increase the size of the chest cavity.

At inspiration, breathing in, the diaphragm contracts and since its base is fixed to bony parts, the dome must move down. This compresses the abdominal organs and increases the vertical diameter of the chest. The intercostal muscles attached to adjacent ribs contract to elevate the ribs. This results in an increase in the lateral and anteroposterior diameters of the chest. With all the thoracic diameters increased the air within the lungs now occupies a larger area than before.The air pressure within the lungs is decreased below that of atmospheric pressure. Air therefore enters the lungs to equalize the pressure — breathing in.

At expiration, breathing out, the diaphragm and intercostal muscles relax and the chest cavity becomes smaller. Air pressure within the lungs is now greater than that of atmospheric air so air is expelled from the lungs.

This process is repeated about 18 times per minute, and is controlled by impulses from the respiratory center in the hindbrain to the muscles of respiration. Carbon dioxide circulating in the blood stimulates this center.

AIR CAPACITY OF THE LUNGS

Vital capacity — the volume of air that can be expelled (breathed out) from the lungs by a forced expiration, following the greatest possible inspiration, is 3500 to 4800 ml. It includes: tidal air, and inspiratory and expiratory reserves.

Tidal air — the volume of air that is breathed in or out during normal quiet inspiration or expiration: 500 ml.

Inspiratory reserve volume — (complemental) the extra volume of air that can be inspired following a normal inspiration: 1500 to 2000 ml.

Expiratory reserve volume — (supplemental) the volume of air that can be expelled from the lungs following a normal expiration: 1500 ml.

Residual volume — the amount of air that remains in the lungs following a forced expiration, that cannot be expelled is 1200 to 1500 ml.

S. 189 SUMMARY OF FUNCTIONS OF RESPIRATORY SYSTEM

1. **General**
   to supply oxygen to the blood,
   to get rid of accumulated carbon dioxide,
   to get rid of some moisture (H_2O),

   to get rid of some waste products.

2. **Respiratory passages**
   nasal cavities
   pharynx
   larynx
   trachea
   bronchi
   bronchioles

   passages to convey air
   (oxygen) to lungs, and
   carbon dioxide, moisture,
   and waste products
   from the lungs to the
   exterior.

   In addition
   Nasal cavities are air conditioners, and warm, moisten, and filter inspired air. The pharynx conveys fluids, and food to the esophagus as well as air to lungs. The larynx functions in phonation. The vocal cords vibrate in phonation. The soft palate closes the nasal pharynx during swallowing. The epiglottis helps close off the larynx during swallowing of fluid or food.

3. **Lungs** are organs in which:
   Oxygen is transferred into pulmonary capillaries, and hence to the blood. Carbon dioxide, water, and some waste products are transferred from pulmonary capillaries and exhaled.
   Alveoli, the functional units of lung. Alveolar capillary membrane, two membranes thru which gases, water, and wastes must pass, wall of alveolus, and wall of capillary.

These functions are accomplished by the functional units of the lungs — the alveoli in the alveolar sacs and the pulmonary capillaries.

S. 190 ANOMALIES OF THE RESPIRATORY SYSTEM

Agenesis = no lung, lung has not formed, genesis - production; also aplasia is similar.
Hypoplasis = underdevelopment; hypo = below.
Cysts = hollow cavities filled with fluid.
Bronchoesophageal fistula = the main or other bronchus or trachea communicates with the esophagus. The midesophagus may be closed.

The azygos vein curves over apex of lung resulting in a fold of visceral pleura on the apex, a comma shaped shadow on the chest film.

S. 191 SOME PATHOLOGICAL CONDITIONS

Atelectasis — collapse of the whole or part of a lung from obstruction of a bronchus. Air cannot enter so lung collapses. The obstruction may be by a foreign body — tooth, blood clot, vomitus, mucous plug, peanut, growth, etc.
Bronchiectasis — the dilatation of a bronchus.
Bronchitis — inflammation of bronchi
Tracheobronchitis — inflammation of trachea and bronchi.
Laryngitis — inflammation of the larynx.
Pneumonitis — inflammation of the lung.
Bronchopneumonia — inflammation of part of a lobe or segment.
Lobar pneumonia — inflammation of one or more lobes by a specific microorganism.
Virus pneumonia — inflammation due to a virus.
Pleurisy — inflammation of the pleura.
Pleural effusion — fluid in the pleural cavity.
Empyema — pus in the pleural cavity.
Hemothorax — blood in the pleural cavity; to be distinguished from hemithorax which refers to one-half of the chest.
Pneumothorax — air in the pleural cavity.
Encapsulated empyema or effusion — fluid walled off in a part of the pleural cavity.

S. 192 ANATOMICAL PECULARITIES AND RADIOGRAPHY

The scapulae normally overlie the upper posterior lateral thorax, and must be displaced from the lung fields. See a mounted skeleton.
The apices of both lungs extend up about one inch above the clavicles, and must be included.
The costophrenic sinuses (recesses) extend down almost to the costal margins, and should be included.
The heart lies close to the anterior chest wall so there is less distortion of its image when the anterior chest is in contact with the film cassette.
The markings visible in lungs in radiographs are shadows of the arteries, veins, and lymphatics, and not of the bronchi. The bronchi contain air and are not ordinarily visible.
The heart would obscure a density posterior to it in the right or left medial parts of the chest when viewed from the front or back.
The heart might obsure a density in the right or left anterior chest in a lateral view.
The diaphragm is displaced downward with inspiration, but at expiration may rise to the level of the fourth right rib. With inspiration the heart lies more vertically and the vascular markings in the lungs are more widely separated, especially at the bases than at expiration.
The mediastinum is a partition, not a space and is filled with thoracic organs. It may be displaced from its normal midline position.
Bronchography is a procedure employed to outline the trachea and bronchi with an opaque medium instilled into the trachea. Since the bronchi, like the lungs, contain air they are not visible as separate shadows in routine radiographs of the chest.

S. 193 SOME OPERATIVE PROCEDURES

Tracheotomy — an opening is made into the trachea in the neck anteriorly so the air may pass into the lungs, when the passage is obstructed higher up. A tube may be placed in the opening.
Lobectomy — one of the pulmonary lobes is removed.
Pneumonectomy — an entire lung is removed.
Segmental resection — a lung segment is removed.
Thoracoplasty — the ribs are cut through in several places and flattened in order to cause collapse of a lung.
Thoracocentesis — insertion of a hollow needle through the chest wall into the pleural cavity to remove air or fluid.

S. 194 THE MEDIASTINUM   F. 14-13

Mediastinum, (L - middle partition); adj. mediastinal. The mediastinum is a vertical partition located in the median plane of the chest, and separating the two lungs from each other. It divides the thoracic cavity into two parts, a right hemithorax, and a left hemithorax, (hemi = half). It reaches from the neck (thoracic inlet) above to the diaphragm below. It extends from the sternum in front to the thoracic vertebrae behind. It is not a hollow space or a cavity but contains within it all the thoracic organs except the lungs. Its width varies with the size of its contents. It is narrow above but wide below to contain the

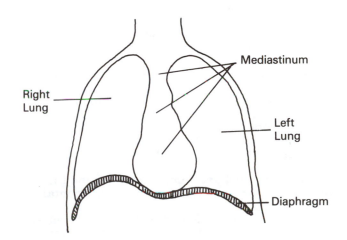

F. 14-13   MEDIASTINUM — (1) Frontal view

Right Lung · Mediastinum · Left Lung · Diaphragm

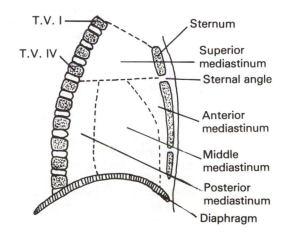

MEDIASTINUM — (2) lateral view

T.V. I · T.V. IV · Sternum · Superior mediastinum · Sternal angle · Anterior mediastinum · Middle mediastinum · Posterior mediastinum · Diaphragm

heart. It is covered on both sides by the parietal pleura (the mediastinal pleura), except at the lung roots where vessels and nerves enter or leave the lungs.

The mediastinum is divided into a superior part, and an inferior part. The inferior part is further divided into anterior, middle, and posterior parts.

1. DIVISIONS OF THE MEDIASTINUM

The superior mediastinum lies above the sternal angle and a plane from this angle to the fourth thoracic vertebra posteriorly.
The anterior mediastinum lies below the superior, directly behind the sternum, and in front of the heart and pericardium.
The middle mediastinum is that part below the superior mediastinum and behind the anterior mediastinum, and containing the heart and its pericardium.
The posterior mediastinum also lies below the superior mediastinum, behind the heart, and in front of all but the upper thoracic vertebrae.

2. CONTENTS

The mediastinum contains several parts of the circulatory system including:
— the heart
— the aorta, ascending part, arch, and descending thoracic part
— pulmonary trunk, pulmonary arteries, Rt. & Lt.
— pulmonary veins, two right, two left
— superior vena cava, and its tributaries
— azygos group of veins
— inferior vena cava - upper end of
— thoracic duct
Other structures:
— trachea
— main bronchi, Rt. & Lt.
— lung roots, and lymph nodes
— esophagus
— thymus gland
— nerves - vagus, phrenic, cardiac, pulmonary

The superior mediastinum, since it extends from front to back in the upper thorax, contains all those structures passing from the neck into the thorax and abdomen. It also contains those structures that pass from the thorax into the neck, or from the abdomen through the thorax into the neck.
The anterior mediastinum, located behind the sternum, and below the superior mediastinum contains some lymph nodes only.
The middle mediastinum, lying between the anterior and posterior mediastina, and below the superior part contains the heart, its pericardium, and blood vessels entering or leaving the heart, as well as the lung roots.
The posterior mediastinum, behind the heart and below the superior mediastinum contains several structures that pass down into the abdomen, or from the abdomen up into the thorax or neck. Some structures pass through the superior mediastinum and one of the other divisions as well.

3. LOCATIONS OF SOME MEDIASTINAL STRUCTURES

Aorta:
{ ascending part — middle mediastinum
arch — in superior mediastinum
descending thoracic — in posterior mediastinum }

Superior Vena cava:
{ upper one-half in superior mediastinum
lower one-half in middle mediastinum }

Trachea:
{ superior mediastinum contains all exept the bifurcation
middle mediastinum contains the bifurcation (some authorities place it in the posterior part) }

Main Bronchi
{ middle mediastinum (some authorities place them in posterior mediastinum) }

Esophagus
{ superior mediastinum contains upper part, behind the trachea
posterior mediastinum contains the lower part }

Thoracic Duct
{ superior mediastinum contains upper part
posterior mediastinum contains the lower part }

S. 195 ANATOMICAL TERMS RESPIRATORY SYSTEM

THE LUNGS:
apex
base
hilum (hilus)
root
costal surface
diaphragmatic surface
mediastinal surface
oblique fissure
horizontal fissure
right lung
left lung
lobes:
Rt. superior (upper)
Rt. middle
Rt. inferior (lower)
Lt. superior (upper)
Lt. inferior (lower)
lingula
bronchopulmonary segments
segments of lungs
lobules
alveolar sac
alveolus, alveoli
alveolar, adj.
functional unit
RESPIRATORY PASSAGES
nasal cavities
conchae
meatuses
paranasal sinuses or accessory nasal ss.
pharynx:
nasal or nasopharynx
soft palate
auditory tube
adenoids or pharyngeal tonsil
oral pharynx or oropharynx
epiglottis
tonsils or palatine tonsils
laryngeal pharynx or laryngopharynx
larynx
thyroid cartilage
cricoid cartilage
vocal folds
vocal cords

glottis or
rima glottis
trachea
cartilaginous rings
carina
bronchus, bronchi
main or principal bronchus
lobar bronchus
segmental bronchus
superior lobe bronchus
middle lobe bronchus
inferior lobe bronchus
small bronchi
bronchiole
respiratory bronchiole
alveolar duct
pleura, pleurae, pleural
pulmonary pleura or
visceral pleura
parietal pleura
diaphragmatic pleura
costal pleura
mediastinal pleura
pleural cavity
costophrenic recess or sinus
costodiaphragmatic recess or sinus
bronchial arteries 1 Rt. 2 Lt.
pulmonary trunk
pulmonary aa. Rt & Lt
pulmonary capillaries
pulmonary veins, 2 Rt, and 2 Lt.
vagus nerve, vagi
phrenic nerves
inspiration
expiration
diaphragm (phren & phrenic)
intercostal muscles
vital capacity
excretion of CO_2
oxygenation
mediastinum
mediastina, mediastinal
superior mediastinum
anterior mediastinum
middle mediastinum
posterior mediastinum
contents of mediastinum

NOTES

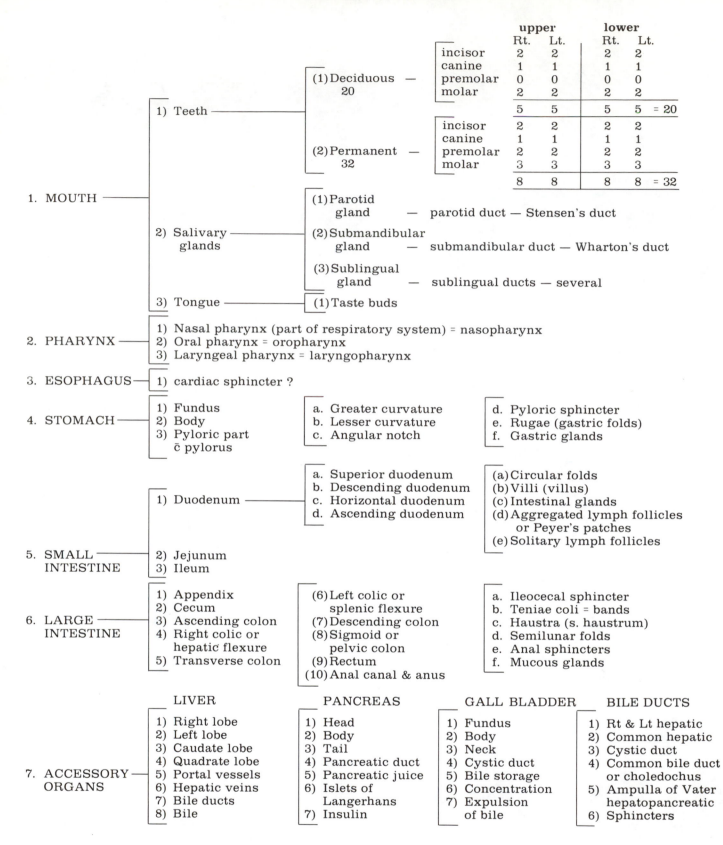

1. MOUTH

1) Teeth
  (1) Deciduous — 20

	upper Rt.	upper Lt.	lower Rt.	lower Lt.	
incisor	2	2	2	2	
canine	1	1	1	1	
premolar	0	0	0	0	
molar	2	2	2	2	
	5	5	5	5	= 20

  (2) Permanent — 32

	upper Rt.	upper Lt.	lower Rt.	lower Lt.	
incisor	2	2	2	2	
canine	1	1	1	1	
premolar	2	2	2	2	
molar	3	3	3	3	
	8	8	8	8	= 32

2) Salivary glands
  (1) Parotid gland — parotid duct — Stensen's duct
  (2) Submandibular gland — submandibular duct — Wharton's duct
  (3) Sublingual gland — sublingual ducts — several

3) Tongue
  (1) Taste buds

2. PHARYNX
1) Nasal pharynx (part of respiratory system) = nasopharynx
2) Oral pharynx = oropharynx
3) Laryngeal pharynx = laryngopharynx

3. ESOPHAGUS
1) cardiac sphincter ?

4. STOMACH
1) Fundus
2) Body
3) Pyloric part c̄ pylorus

a. Greater curvature
b. Lesser curvature
c. Angular notch
d. Pyloric sphincter
e. Rugae (gastric folds)
f. Gastric glands

5. SMALL INTESTINE
1) Duodenum
  a. Superior duodenum
  b. Descending duodenum
  c. Horizontal duodenum
  d. Ascending duodenum

  (a) Circular folds
  (b) Villi (villus)
  (c) Intestinal glands
  (d) Aggregated lymph follicles or Peyer's patches
  (e) Solitary lymph follicles

2) Jejunum
3) Ileum

6. LARGE INTESTINE
1) Appendix
2) Cecum
3) Ascending colon
4) Right colic or hepatic flexure
5) Transverse colon
(6) Left colic or splenic flexure
(7) Descending colon
(8) Sigmoid or pelvic colon
(9) Rectum
(10) Anal canal & anus

a. Ileocecal sphincter
b. Teniae coli = bands
c. Haustra (s. haustrum)
d. Semilunar folds
e. Anal sphincters
f. Mucous glands

7. ACCESSORY ORGANS

LIVER
1) Right lobe
2) Left lobe
3) Caudate lobe
4) Quadrate lobe
5) Portal vessels
6) Hepatic veins
7) Bile ducts
8) Bile

PANCREAS
1) Head
2) Body
3) Tail
4) Pancreatic duct
5) Pancreatic juice
6) Islets of Langerhans
7) Insulin

GALL BLADDER
1) Fundus
2) Body
3) Neck
4) Cystic duct
5) Bile storage
6) Concentration
7) Expulsion of bile

BILE DUCTS
1) Rt & Lt hepatic
2) Common hepatic
3) Cystic duct
4) Common bile duct or choledochus
5) Ampulla of Vater hepatopancreatic
6) Sphincters

15. THE DIGESTIVE SYSTEM

S. 196 DIVISIONS OF THE DIGESTIVE SYSTEM

I. THE DIGESTIVE TRACT, TUBE or ALIMENTARY TRACT
(1) Mouth
(2) Pharynx
(3) Esophagus
(4) Stomach
(5) Small intestine
(6) Large intestine

2. ABDOMINAL ACCESSORY ORGANS OF DIGESTION

(1) Pancreas
(2) Liver
(3) Gall bladder & bile ducts

S. 197 GENERAL FUNCTION — FOODS

The digestive system has been developed to provide a means by which food is taken into the body and broken down into simple molecules that can be absorbed by the blood. The blood carries these molecules to all body tissues. Food is necessary to provide energy, to promote growth, to replace tissues worn out by day to day wear and tear, and for the manufacture of secretions that the body requires in order to carry on its functions. Food for humans includes animal tissues (especially meat), dairy products, and some plants. There are six types of substances required to meet bodily requirements:
1. Carbohydrates
2. Fats or lipids
3. Proteins
4. Mineral salts
5. Vitamins
6. Water

1. Carbohydrates includes starches and various sugars. They vary from simple to complex compounds but all contain carbon, hydrogen, and oxygen. Glucose, fructose and galactose, the simple sugars, contain 6 carbon atoms, and 6 molecules of water, i.e. 6 carbon, 12 hydrogen and 6 oxygen atoms. They are called **monsaccharides**. Sucrose (cane sugar), lactose (milk sugar) and maltose (malt sugar) have two units of simple sugar and are called **disaccharides**. Starches, cellulose, glycogen and dextrins are called polysaccharides. They contain several or many molecules of simple sugar. Carbohydrates are found in fruits, vegetables, and cereals. They are burned up in the body to provide energy. The excess is converted into fat or is stored as glycogen in the liver, muscles and some other tissues.

2. Fats are simple lipids from plant and animal foods. They are composed of a molecule of glycerine and three molecules of a fatty acid. Like carbohydrates they are utilized to provide energy, or are stored in the fat depots of the body. They contain carbon, hydrogen, and oxygen.

3. Proteins are present in animal tissues, such as lean meat (muscle), as well as in many plants, notably beans, peas, nuts, and some cereals. There are many different proteins. Some are simple, others are complex, being made up of thousands of atoms. Each kind of animal or plant has its own proteins.

Protein molecules are composed of smaller molecules of amino acids. These are compounds of atoms of carbon, hydrogen, oxygen, and nitrogen, the latter in its (NH^2) form. There are some 25 different amino acids, and these in various multiples and combinations make up animal and plant proteins. There are ten amino acids that the body must have to function properly - the essential amino acids. Not all of these are in any single food, and some proteins such as gelatin contain very few of them. Therefore a variety of protein foods is necessary to supply the requirements.

Amino acids from ingested proteins are utilized to produce new proteins in the body for growth, and to replace those lost by the wear and tear of living.

If an excess of protein food is eaten this is not stored up in the body as such, but may be converted into fat or glucose and stored. In this case the nitrogen part of this excess is converted by the liver into urea, and is excreted by the kidneys in urine. No other food can replace protein in a diet.

4. Mineral salts include sodium, potassium, magnesium as chlorides, and calcium and phosphorus as tricalcium phosphate. Some iron, copper, zinc, iodine and fluorine are also necessary. These are supplied in food or water.

5. Vitamins are contained in plant and animal foods and are necessary for the normal functioning of the body.

6. Water is taken in as fluids, but also makes up a considerable portion of foods.

S. 198 TERMS APPLIED TO DIGESTIVE FUNCTIONS

The digestive system is responsible for the ingestion, digestion, and absorption of food.

1. Ingestion is the taking into the mouth of the food defined above, and their passage into the stomach. It includes **mastication** (chewing), and **deglutition** (swallowing). (L. ingestere = a pouring in).

2. Digestion is the breaking up of complex food molecules into small simple molecules that can pass through living membranes. (L. digere = to break apart). The process is mechanical and chemical. Mechanical digestion involves the chewing, mixing, churning, and liquifying procedures. Chemical digestion is accomplished by enzymes that are secreted in gastric juice, pancreatic juice, and intestinal juice. Enzymes are protein. Each one is specific in that it acts upon one type of food only, and accomplishes a single process in the break down of any food. Enzymes are named with a suffix "ase", for example — protease, amylase, lipase, disaccharase, etc. Some older names for enzymes have been retained and are still used — pepsin, erepsin, trypsin concerned with protein digestion, and ptyalin used in carbohydrate digestion. Enzymes break up the three foods:
a. proteins to amino acids
b. fats, (lipids) to fatty acids and glycerine
c. carbohydrates to glucose, fructose, and galactose.

There are many other enzymes in the human body besides those concerned with digestion.

3. Absorption, with reference to the digestive tract, is the passage of digested food products — amino acids, fatty acids, glycerine, glucose, fructose, and galactose — into blood or lymph capillaries. It also includes the passage of mineral salts, vitamins, and water.

The glucose, amino acids, and some glycerine and fatty acids enter **blood** capillaries to pass to the liver through the portal system of veins. The liver therefore may extract its requirements before these products enter the inferior vena cava and general circulation. Some glycerine and some fatty acids however are absorbed by lymph capillaries (lacteals) in the villi and pass through the thoracic duct to the left subclavian vein, bypassing the liver.

Absorption is a term that is not only used in connection with the uptake of digested products from the intestine, but may refer to the passing of dissolved substances from blood capillaries into intercellular spaces and from these to any body cells.

S. 199 FACTORS DETERMINING POSITION OF THE ABDOMINAL ORGANS

While the abdominal part of the digestive system, and many other structures lie in the same general area in all human subjects, their position is also influenced and altered by the factors listed below:

1. **Habitus** or body build; stocky individuals have their organs at a higher level than the slim, narrow shouldered person.
2. **Weight**; the organs in a thin person will lie at a lower level than in the **obese** subject.
3. **Position** of subject; in the upright position the organs descend, and lie at a lower level than when the subject is lying down.
4. **Phase of respiration**; with inspiration, especially a deep inspiration the diaphragm descends and pushes the abdominal organs down. They ascend with expiration.
5. **State of fullness**; an organ such as the stomach when full lies at a lower level than when empty.

S. 200 SURFACE DIVISIONS OF THE ABDOMEN

In order to accurately define the normal or abnormal position of abdominal organs the anterior surface of the abdomen has been divided into:
1. Four quadrants, or into
2. Nine regions
by drawing horizontal and vertical lines.

1. DIVISION INTO FOUR QUADRANTS F. 15-1

(1) A line is drawn transversely across the abdomen at the level of the umbilicus (navel).
(2) A second line is drawn longitudinally to correspond to the median line of the body, and intersecting the first line at the umbilicus. These two lines divide the abodmen into four quadrants:

right upper quadrant — R U Q or U R Q
right lower quadrant — R L Q or L R Q
left upper quadrant — L U Q or U L Q
left lower quadrant — L L Q

2. DIVISION INTO NINE REGIONS F. 15-1

(1) A line is drawn transversely across the abdomen at the level of the transpyloric plane at the tips of the ninth costal cartilages (difficult to locate).
(2) A second line is drawn transversely across the abdomen at the level of the iliac crests.
(3) Vertical lines are drawn on each side of the abdomen from points midway between the anterior spines of the iliac bones and the symphysis pubis, up to the thorax. These four lines divide the abdomen into nine regions. These are named from right to left, as:

Upper
{ (1) right hypochondriac region
(2) epigastric region
(3) left hypochondriac region

Middle
{ (4) right lumbar region
(5) umbilical region
(6) left lumbar region

Lower
{ (7) right iliac region
(8) hypogastric region
(9) left iliac region

In the Nomina Anatomica list lumbar is called lateral region; iliac is called inguinal; and hypogastric is called pubic (not significant).

The division of the abdomen into nine regions is the one published in many anatomical texts. It has several disadvantages. The location of the landmarks used to draw the required lines is different in various texts and difficult to locate. The regions outlined are too small to pinpoint the location of an organ that may vary in position in a single subject to some degree, and in different individuals to a considerable degree.

The division of the abdomen into four quadrants although not as popular in texts is by far the most practical and is used by most medical practitioners. Except for the term epigastrium the names of the regions are not seen on requisitions. During the study of the various structures in the abdomen the student should learn the position of each, at least as to the quadrant in which each organ lies. It was pointed out in the section dealing with the abdominal and pelvic cavities that there is no partition separating them. It was suggested that perhaps a better term might be abdomino-pelvic cavity. An organ may lie in the abdomen or pelvis, or partly in both, and may change its position with changes in the body position, etc.

Even the four quadrants have a limited application. They are only usable when the subject is lying on the back — supine. When lying face down — prone — they are not visible.

Of more value in radiography are the bony landmarks that may be seen or palpated (felt) in either the supine or prone positions. The iliac crests are examples, and are invaluable in radiography.

S. 201 DEFINITIONS OF SOME BASIC TERMS

A lumen is the cavity within any hollow organ. (L. lumen = a light, a window, to let light in).

The peritoneum is the membrane that **lines** the abdominal and pelvic cavities, and **covers** most of the abdominal and pelvic organs. It is a serous membrane consisting of a layer of flat epithelial cells on a basement membrane. Serous fluid in small amounts is secreted by it. The peritoneum is a closed sac with no communication with the outside, except in the female where the uterine tubes open into the pelvic cavity. F. 15-2

The parietal peritoneum is the lining of the abdominal and pelvic cavities. It also forms the mesentery of the intestine. (Paries = a wall).

The visceral peritoneum forms the outer covering membrane of most abdominal and pelvic organs, particularly their anterior surfaces. (Viscus = an organ). See pericardium & pleura.

The peritoneal cavity is the potential space between the visceral peritoneum covering the organs and the parietal peritoneum. Actually these surfaces are in contact but not attached to each other. See pericardial & pleural.

The mesentery is a double layer of peritoneum that extends from the parietal peritoneum on the posterior abdominal wall to the small intestine. Arteries, veins, lymphatics and nerves pass between the two layers to reach the bowel from the posterior abdominal wall.

Note: assuming that during early development the small intestine lies behind the peritoneum in front of the posterior abdominal wall, the small intestine pushes forward, pushing the parietal peritoneum in front of it to form the outer layer covering the intestine. This displaced peritoneum behind the intestine now forms the double layer from abdominal wall to intestine. These two layers fall together to form the mesentery. (Meso = middle + enteron = intestine), so middle or small intestine.

168

F. 15-1　DIVISIONS OF THE ABDOMEN:

(1) FOUR QUADRANTS　(2) NINE REGIONS

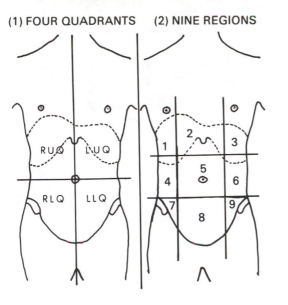

F. 15-3　PERISTALSIS IN THE STOMACH:

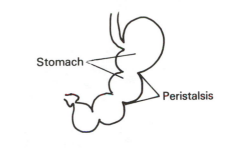

F. 15-2　DIAGRAM OF PERITONEUM

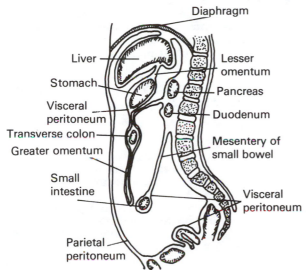

(2)　CROSS SECTION OF ABDOMEN:
— the peritoneum —

Mesocolon　Mesocolon

Colon　Descending colon

Visceral peritoneum　Mesentery

Small intestine　Parietal peritoneum

Anterior abdominal wall

The mesocolon is a similar double layer of the parietal peritoneum between some parts of the large intestine and the posterior abdominal wall. Some parts of the large intestine are not displaced forward to any degree, such as the ascending and descending colon. They are covered by peritoneum but no definite mesentery exists. The cecum, transverse, and sigmoid parts may have a definite mesocolon.

The omentum is a double layer of peritoneum that extends between two organs, e.g. stomach and spleen.

The greater omentum is a fold of peritoneum that extends from the lower margin of the stomach to the transverse colon, and is draped down like an apron below the transverse colon.

The lesser omentum is a fold of peritoneum between the stomach and liver.

A sphincter is a thickened ring of the circular layer of visceral muscle that surrounds the opening of a hollow organ. It acts as a valve to keep the opening closed, e.g. stomach.

Peristalsis is a contraction wave that passes along the wall of a hollow organ, and alternates with a wave of dilatation. As the circular layer contracts the longitudinal layer relaxes ahead of it and bowel contents are propelled along the canal. (See radiograph of a barium filled stomach). F. 15-3

Segmentation is a contraction of a small segment of bowel that divides the contents into two parts. This is followed by relaxation and a further contraction in an adjacent segment.

Stenosis is the narrowing of the lumen of a hollow organ. It may involve a short or long segment.

Atresia refers to the absence of a lumen or canal in a hollow organ that normally has one. (a = without + tresia = an opening).

S. 202　STRUCTURE OF WALLS OF DIGESTIVE TRACT　F. 15-4

Four coats or coverings form the walls of the parts of the digestive tract. As they are somewhat similar in all parts a general description is given here. Variations from this pattern will be pointed out as each area is studied. These coats include:

1. a serous layer
2. a muscular layer
3. a submucosal layer
4. a mucous coat — mucosa

1. **The serous layer** is the visceral layer of peritoneum that covers most abdominal and pelvic organs. It does not cover the posterior wall of the organs that lie against the posterior abdominal wall. See peritoneum above & S. 22 on epithelium.

The esophagus or gullet, lying in the thorax has **no** serous (peritoneal) covering. Instead, its outer **layer** consists of a layer of connective tissue.

F. 15-4 CROSS SECTION OF SMALL INTESTINE

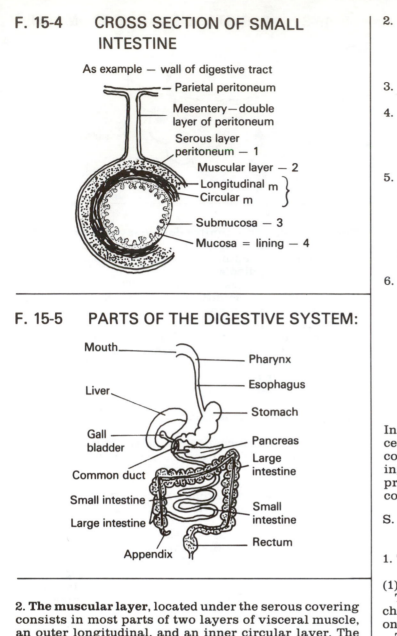

As example — wall of digestive tract

- Parietal peritoneum
- Mesentery—double layer of peritoneum
- Serous layer peritoneum — 1
- Muscular layer — 2
  - Longitudinal m
  - Circular m
- Submucosa — 3
- Mucosa = lining — 4

F. 15-5 PARTS OF THE DIGESTIVE SYSTEM:

Mouth
Pharynx
Esophagus
Liver
Stomach
Gall bladder
Pancreas
Common duct
Large intestine
Small intestine
Large intestine
Small intestine
Rectum
Appendix

2. The muscular layer, located under the serous covering consists in most parts of two layers of visceral muscle, an outer longitudinal, and an inner circular layer. The longitudinal layer has fibers (cells) extending lengthwise, while the fibers of the circular layer encircle the organ. These layers are responsible for peristalsis and segmentation.

3. The submucous coat or submucosa is a layer of loose (areolar) connective tissue. It carries blood vessels, lymphatics and nerves to the lining membrane. Do not be confused by the term submucosa; here it refers to the fact that this layer is under the mucous layer, "sub", when examined from the cavity of the organ.

4. The mucous coat or mucosa, or lining membrane of an organ is a layer of epithelium adjacent to the lumen or cavity. It is a single layer of columnar cells in many organs with mucous cells and other glands. The type of gland varies with the organ. Frequently the lining membrane forms folds that are named differently in each organ. These will be defined as each part is studied.

S. 203 PARTS OF EACH DIVISION

1. **Mouth**
   (1) vestibule & cavity of mouth
   (2) teeth
   (3) tongue
   (4) salivary — (parotid)
       glands   (submandibular)
                 (sublingual)

2. **Pharynx**
   (1) nasal pharynx, (NA, nasal part)
   (2) oral pharynx, (NA, oral part)
   (3) laryngeal pharynx (NA, laryngeal part)

3. **Esophagus**

4. **Stomach**
   (1) fundus
   (2) body
   (3) pyloric part
   (4) pylorus

5. **Small Intestine**
   (1) duodenum; (superior part (cap) (descending part (horizontal part (ascending part
   (2) jejunum
   (3) ileum

6. **Large Intestine**
   (1) cecum and appendix
   (2) ascending colon
   (3) right colic flexure (hepatic)
   (4) transverse colon
   (5) left colic flexure (splenic)
   (6) descending colon
   (7) sigmoid colon (pelvic)
   (8) rectum
   (9) anal canal
   (10) anus

In anatomical texts the large intestine is divided into: cecum, colon, rectum, and anal canal; included in the colon are — ascending colon, transverse colon, descending colon, and sigmoid colon. In radiology and surgical practice the whole intestine is often included in the term colon.

S. 205 DETAILED STUDY OF THE PARTS OF THE DIGESTIVE TRACT

1. **THE MOUTH** F. 15-6, 7, 8

(1) **Vestibule and oral cavity.**

The **vestibule** of the mouth is the space between the cheeks and lips on its outer side and the teeth and gums on its inner side.

The **cavity** of the mouth is the hollow space, the receiving chamber of the digestive tract. It opens at the lips upon the face, (NA, rima oris) and behind by the isthmus of the fauces into the oral pharynx. Its walls are formed by the teeth, its roof by the hard and soft palate, and its inferior limit by the tongue and floor of the mouth in front of the tongue.

The hard palate, studied with the facial bones, consists of the palatine processes of the maxillae and the palatine bones. These structures also form the floor of the nasal cavities.

The soft palate contains no bone but is a fold of epithelium that hangs down like an apron from the posterior margin of the hard palate. During swallowing it prevents food from entering the nasal pharynx and nasal cavities.

The fauces is the space between the mouth and the oral pharynx. It is sometimes referred to as the isthmus of the fauces. It is formed by the soft palate, the two palatine arches, and the tongue.

The palatoglossal arch (palate + glossus = the tongue) extends from the soft palate to the base of the tongue on either side. It lies anterior to the palatopharyngeal arch.

The palatopharyngeal arch (palate + pharynx) reaches from the soft palate to the side wall of the pharynx on either side. It is posterior to the other arch. The space between the two arches is occupied by the palatine tonsils.

F. 15-6 NASAL CAVITY, MOUTH, AND PHARYNX — Viewed from the left side —

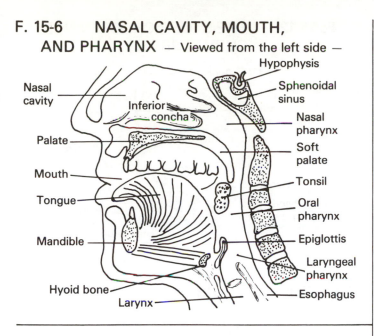

F. 15-7 CAVITY OF THE MOUTH:

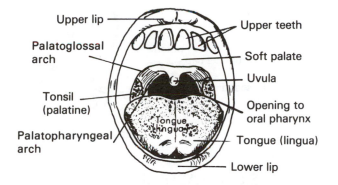

F. 15-8 FLOOR OF THE MOUTH
— Mouth open —

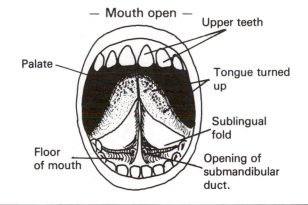

Students should examine these structures in the open mouth of a classmate or their own in a mirror to demonstrate their locations.

(2) **The teeth:** these were studied along with the facial bones, and should be reviewed. In the digestive process the teeth bite (incise) food, and chew (masticate) it to produce smaller particles more accessible to the digestive enzymes.

(3) **The tongue:** (L. lingua, adj. lingual) this is a muscular organ covered by epithelium with small projections (papillae) on the upper surface. These papillae have taste buds. Besides tasting, the tongue has a function in speech. The tongue forces food between the upper and lower teeth to be chewed. Finally it forms the portion of

chewed food into a ball or bolus and flips it back through the fauces into the oral pharynx during the act of swallowing or deglutition.

(4) **The salivary glands:** F. 15-9

There are three pairs of salivary glands, and all of them empty through ducts into the mouth. They are: right and left parotid, right and left submandibular, and right and left sublingual. They are alveolar glands and secrete saliva, a watery mucous fluid containing the enzyme **ptyalin.** This digests carbohydrates. (G. sialon = saliva, hence sialography — radiography of the salivary glands.)

The parotid gland lies on the side of the face in front of and below the opening into the ear. It extends downwards from the zygomatic arch above to behind the ramus of the mandible, on each side.
The parotid duct (NA), formerly called Stensen's duct passes anteriorly from the parotid gland in the cheek. It opens into the mouth through a slitlike opening opposite the second upper molar tooth.

The submandibular gland lies under the floor of the mouth, one on each side. It lies against a depression on the medial surface of the body of the mandible anterior to the angle, but extends down below the inferior border of the mandible. It was formerly named the submaxillary gland.
The submandibular duct (NA) formerly called Wharton's duct passes forward in a fold of the mucous membrane on the floor of the mouth. It opens at a small elevation beside the frenulum of the tongue close to the midline.

The sublingual gland also lies in the floor of the mouth, one on either side of the median line and anterior to the submandibular gland. (sub = under + lingua = tongue)
The sublingual ducts, of which there are several open upon the surface of the fold of the mucous membrane on the floor of the mouth by several openings on each side. The duct of the submandibular gland lies in this same fold.

2. THE PHARYNX   F. 15-10

The pharynx or throat has been studied along with the respiratory passages.— S. 173. It is 12.5 cm (5 in.) long. The oral pharynx and laryngeal pharynx have a dual function in that they convey food and fluid to the esophagus and air to the lungs. The oral pharynx lies behind the mouth with the fauces forming an opening between the two. During swallowing (deglutition) the nasal pharynx is closed off from the oral pharynx to prevent the food from entering the nasal cavities. The glottis (opening to the larynx) is also closed off to prevent food entering the larynx.

3. THE ESOPHAGUS   F. 15-10

The esophagus or gullet is a tube about 25 cm (10 in.) in length and extends from the laryngeal part of the pharynx to the stomach. It is anterior to the thoracic vertebrae but posterior to the trachea (windpipe). Its upper part lies in the superior mediastinum above the heart. Its lower part lies in the posterior mediastinum behind the heart. It passes down into the abdomen through the esophageal hiatus or opening in the diaphragm. It opens into the medial side of the stomach at the cardiac orifice (NA, ostium). The esophagus since it lies in the thorax has no serous covering (peritoneum). It has an outer covering of fibrous connective tissue instead. Food passes down through the esophagus by peristalsis and gravity. The fluids pass down by gravity unless the subject is positioned horizontally when peristalsis becomes active.

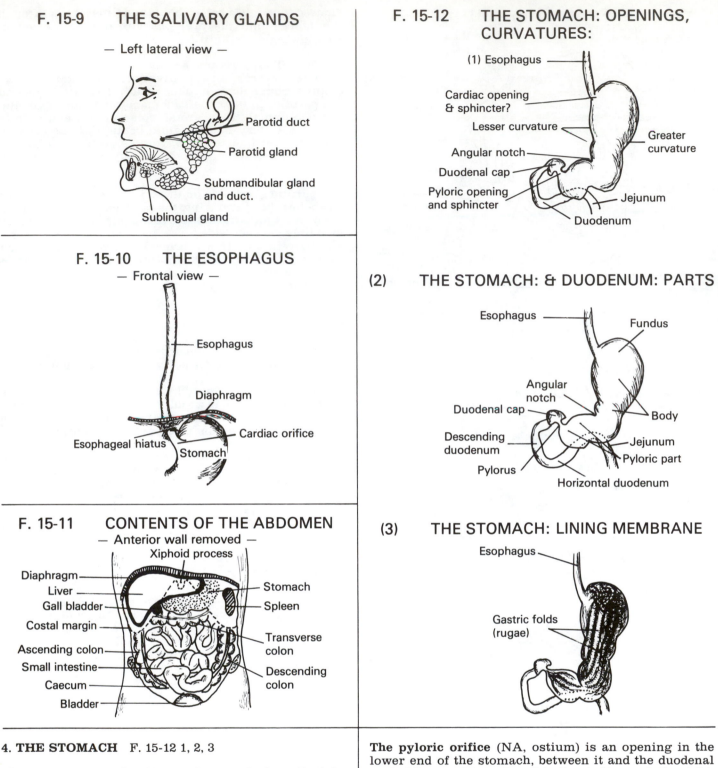

F. 15-9 THE SALIVARY GLANDS
— Left lateral view —

- Parotid duct
- Parotid gland
- Submandibular gland and duct.
- Sublingual gland

F. 15-10 THE ESOPHAGUS
— Frontal view —

- Esophagus
- Diaphragm
- Cardiac orifice
- Esophageal hiatus
- Stomach

F. 15-11 CONTENTS OF THE ABDOMEN
— Anterior wall removed —

- Xiphoid process
- Diaphragm
- Liver
- Gall bladder
- Costal margin
- Ascending colon
- Small intestine
- Caecum
- Bladder
- Stomach
- Spleen
- Transverse colon
- Descending colon

F. 15-12 THE STOMACH: OPENINGS, CURVATURES:

- (1) Esophagus
- Cardiac opening & sphincter?
- Lesser curvature
- Angular notch
- Duodenal cap
- Pyloric opening and sphincter
- Greater curvature
- Jejunum
- Duodenum

(2) THE STOMACH: & DUODENUM: PARTS

- Esophagus
- Fundus
- Angular notch
- Duodenal cap
- Descending duodenum
- Pylorus
- Body
- Jejunum
- Pyloric part
- Horizontal duodenum

(3) THE STOMACH: LINING MEMBRANE

- Esophagus
- Gastric folds (rugae)

---

4. THE STOMACH F. 15-12 1, 2, 3

(G. gaster = stomach, adj. gastric; ventriculus = Latin).

The stomach is a collapsible reservoir for food. It lies vertically or obliquely in the upper left abdomen, to the left of the liver and the median line of the body. When empty it is usually "J" shaped with its walls collapsed except for the upper end that usually contains some air. The stomach has:

two openings — cardiac and pyloric
two sphincters — cardiac and pyloric
an angular notch
two curvatures — lesser and greater
three parts — fundus, body, pyloric part
rugae or gastric folds with pylorus gastric glands.

The cardiac orifice (NA, ostium) is an opening between the lower end of the esophagus and the upper stomach. It is at the right upper medial margin of the stomach and is termed cardiac orifice because it is close to the heart.

The pyloric orifice (NA, ostium) is an opening in the lower end of the stomach, between it and the duodenal part of the small intestine. It is directed towards the right. See also pylorus.

The cardiac sphincter is located at the cardiac end of the stomach. It prevents regurgitation (backflow) of the stomach contents into the esophagus. The presence of an actual sphincter is disputed, but the arrangement of muscle appears to act as one.

The pyloric sphincter is located at the pyloric end of the stomach in the pylorus. It keeps the opening closed except for relaxation to allow the stomach contents to enter the duodenum. The stomach curves to the right as it approaches the pylorus. This results in curved right and left borders:

The lesser curvature is the short right curved border that extends from the cardiac to the pyloric openings of the stomach.

The **greater curvature** is the much longer left curved border of the stomach.

The **angular notch** (incisura angularis) is a notch on the lower part of the lesser curvature of the stomach where it bends sharply to the right.

The stomach has three parts:

The **fundus** of the stomach is that part that lies above the cardiac opening.

The **body** of the stomach is that part between the cardiac opening and the angular notch.

The **pyloric part** (NA) is that part between the angular notch and the pyloric opening. This part is directed to the right rather than down.

The **pylorus** is a constricted area between the pyloric part of the stomach and the first or superior part of the duodenum. It contains the pyloric opening of the stomach and the pyloric sphincter.

The stomach has the usual **four coverings**. The outer layer or serous covering is the visceral peritoneum. There are three muscular layers, and submucosal and mucosal layers.

Rugae or gastric folds are folds of the lining membrane that usually run longitudinally. They disappear when the stomach is filled with food or fluids.

Gastric glands are simple tubular glands in the mucosal layer of the stomach. They secrete gastric juice including hydrochloric acid, pepsin, milk curdling rennin, and perhaps a fat splitting enzyme. The glands in the pylorus differ in the type of secretion. The enzymes of gastric juice act upon ingested foods but do not complete the digestive process. Nothing is absorbed through the stomach except alcohol which is rapidly absorbed from an empty stomach.

5. THE SMALL INTESTINE   F. 15-

The small intestine is called small because its lumen is smaller than that of the large intestine. Its walls are much thinner but it is over four times as long. It is about 7 m (23 ft.) in length. It is a single continuous tube but in order to accommodate to the abdomen it is coiled upon itself repeatedly. It extends from the pyloric end of the stomach to an opening in the large intestine, the ileocecal opening in the right lower abdomen. It is supported by the mesentery, by which blood and lymph vessels and nerves reach it. The small intestine has three parts, with no actual division between but so called to pinpoint anatomical and pathological descriptions more accurately - the duodenum, jejunum, and ileum.

The **duodenum**, the first 25cm (10 in.) of the small intestine form a loop in the upper right abdomen. It first passes up and to the right from the pyloric opening in the stomach, then downwards, then to the left across the midline, and then upwards to join the jejunum posterior to the stomach. The head of the pancreas lies within this loop. (L. duodenum = 12, the width of 12 fingers when placed side by side). The duodenum has four parts: superior, descending, horizontal, and ascending. F. 15-12 (1, 2, 3)

The **superior duodenum**, the first part passes up and to the right from the pyloric opening of the stomach. It includes the duodenal cap or bulb. It joins the descending part.

The **descending duodenum** passes downwards from the superior part, and ends by joining the horizontal part. It lies to the right of the median line. The common bile duct opens into it from the back.

The **horizontal duodenum** extends to the left from the descending part. It crosses the median line to join the ascending part.

The **ascending duodenum** passes up from the horizontal part, and joins the jejunum posterior to the stomach, close to the angular notch. F. 15-11

F. 15-13   SMALL INTESTINE:
Longitudinal section

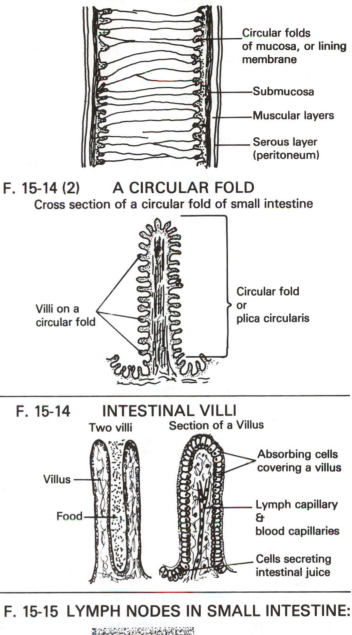

- Circular folds of mucosa, or lining membrane
- Submucosa
- Muscular layers
- Serous layer (peritoneum)

F. 15-14 (2)   A CIRCULAR FOLD
Cross section of a circular fold of small intestine

Villi on a circular fold

Circular fold or plica circularis

F. 15-14   INTESTINAL VILLI
Two villi   Section of a Villus

Villus

Food

- Absorbing cells covering a villus
- Lymph capillary & blood capillaries
- Cells secreting intestinal juice

F. 15-15 LYMPH NODES IN SMALL INTESTINE:

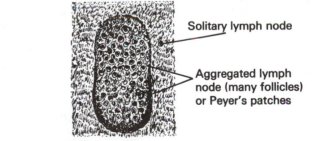

Solitary lymph node

Aggregated lymph node (many follicles) or Peyer's patches

The **jejunum**, the second part of the small intestine, forms about two-fifths of the remaining part of the **small intestine** and is about 2.7m (9 ft.) in length. It extends from the ascending duodenum behind the stomach to the ileum. It lies in the upper abdomen, but is quite mobile. (L. jejunum = empty, adj. jejunal) F. 15-11

The **ileum**, the third part of the small intestine forms the remaining three-fifths of the small intestine and is about 3.9m (13ft.) long. It extends from the jejunum to the junction with the large intestine, at the cecum, either in the right lower abdomen or in the pelvis. F. 15-11

The ileocecal opening (NA ostium) is the opening between the terminal part of the ileum and the cecum. It is usually located in the lower right abdomen or pelvis.
The ileocecal sphincter or valve is a thickened layer of circular muscle at the ileocecal opening. Frequently this sphincter does not prevent backward flow from the cecum as barium instilled per rectum usually enters the ileum. (Compare the spelling ileum with ilium, part of the hip bone).

CHARACTERISTIC FEATURES OF SMALL INTESTINE:

Circular folds or plicae circulares are folds in the lining membrane of the small intestine that encircle its lumen. They give this part of the bowel a characteristic accordionlike appearance when the intestine is opened or distended with barium or air. These are found in the small intestine only. F. 15-13
Villi are very minute microscopic fingerlike projections of the mucosal lining of the small intestine that cover the circular folds and the hollows among them. They are also found only in the small intestine. Each fingerlike process has a covering of columnar epithelium with blood and lymph capillaries running lengthwise within it. These lymph capillaries are called lacteals. The villi increase the absorptive surface of the small intestine in contact with food. It is by these capillaries that digested food is absorbed into the blood. L. villus — shaggy hair, pl. villi. F. 15-14. Students might compare the villi with shag carpeting.
Aggregated lymph follicles (NA), also called Peyer's patches are collections of lymphatic tissue in the mucosa of the small intestine. They are most numerous in the terminal ileum. They become involved in infections such as typhoid fever and tuberculosis of the intestine. F. 15-15
Solitary lymph follicles (nodules) are single follicles of lymphatic tissue, and are also found in the lining of the small intestine.
Intestinal glands lie in the mucosa in the intervals between the bases of the villi. They are tubular glands and secrete intestinal juice containing digestive enzymes. This juice enters the intestine to mix with food and complete the digestion. As stated before the digested products are absorbed into blood or lymph capillaries.

6. THE LARGE INTESTINE   F. 15-16, 17, 18

The large intestine is the final division of the digestive tube. It is about 1.5m (5 ft.) in length. The transverse diameter of its lumen is greater than that of the small intestine, but it is only about one-quarter as long. It begins in the lower right abdomen at the ileocecal opening and ends at the anus. It forms an inverted "U" shaped structure that passes up from the right lower abdomen or pelvis in the right lateral abdomen. It then crosses the upper abdomen to the left side and descends in the left lateral abdomen to the pelvis. For descriptive purposes it is divided into nine parts plus the anus. It must be emphasized that it is a single continuous tube. These divisions are:

    (1) cecum and appendix
    (2) ascending colon
    (3) right colic flexure
    (4) transverse colon
    (5) left colic flexure
    (6) descending colon
    (7) sigmoid colon
    (8) rectum
    (9) anal canal and anus (opening)

The term colon is not considered by anatomists to be synonymous with the large intestine but includes only those parts of which "colon" forms part of the name. The large intestine by these authorities consists of the cecum, colon (with ascending, transverse, descending, and sigmoid,) the rectum, and anal canal. Surgeons and radiologists frequently apply the term "colon" to the entire large intestine.

The cecum is that part of the large intestine that forms a pouch below the ileocecal opening. Its caudal or inferior end is rounded and is closed (blind) except where the appendix is attached. (L. cecum = blind). The cecum is usually located in the right lower abdomen or pelvis, but its position varies. F. 15-16

The appendix (vermiform appendix) is an appendage of the cecum. It is a hollow pencil-like structure that is attached to the blind end of the cecum. As the cavity within the appendix is continuous with that of the cecum feces may pass into the appendix and back out into the cecum. The appendix may be from two to six inches in length. It usually hangs down from the cecum, but may lie in front of, behind, medial to or lateral to the cecum. As it is attached to the cecum it will be found wherever the cecum is located. There is only one appendix. It should not be referred to as "they" or "them". (appendix = appendage, something hanging down). F. 15-16, 17
Vermiform signifies wormlike, hence vermiform appendix.
The ileocecal opening and valve or sphincter were defined along with the ileum.

The ascending colon passes upwards (cranially) from the cecum to the inferior surface of the liver along the right lateral abdomen. It ends at the right colic flexure.
The right colic flexure (NA) or hepatic flexure is a bend of the colon to the left under the right liver margin. It lies in the right upper quadrant or right hypochondriac region.
The transverse colon passes across the upper abdomen from the right colic flexure to the spleen in the upper left abdomen. Frequently the medial part hangs down to a varying degree.
The left colic flexure (NA) or spenic flexure is a bend downwards (caudally) in the colon where the transverse colon ends, at the spleen.
The descending colon extends down from the left colic flexure to the brim of the pelvis in the left lateral abdomen.
The sigmoid colon (NA) or pelvic colon is the "S" shaped curved part of the distal colon that extends from the descending colon at the pelvic brim to the rectum. It is sometime quite long and may form an incomplete loop with a long mesocolon. It may extend across to the right lower abdomen. Its position is variable. (G. sigma — Greek letter S + oid = like, hence shaped like the Greek letter sigma).
The rectum begins anterior to the sacrum at about the third sacral segment. It passes down within the pelvis to end at the anal canal. When viewed from the front it appears to be a straight tube, hence the name rectum (from rectus = straight). When viewed from the side it is seen to follow the curve of the sacrum as it lies in the concavity of this bone. The rectum is from four to five inches in length.
The anal canal is the distal 2.5 to 4cm (1 to 1.5 inches) of the digestive tube. It is continuous with the rectum above and ends below at the anus. The anal canal is directed posteriorly and down from the rectum so that its lumen forms an angle with that of the rectum. A tube being inserted into the anal canal should be directed towards the umbilicus (patient in lateral position) for the first 4cm (1.5 in.). Then it should be directed posteriorly to follow the rectal curve. Careless insertion, particulary with a rigid nozzle might perforate the bowel.
The internal anal sphincter encircles the anal canal. It is due to a thickening of the circular layer of visceral muscle. It keeps the lumen of the canal closed.

F. 15-16 THE LARGE INTESTINE:

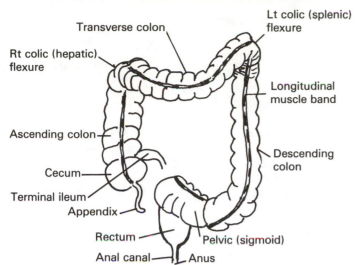

Transverse colon
Lt colic (splenic) flexure
Rt colic (hepatic) flexure
Longitudinal muscle band
Ascending colon
Descending colon
Cecum
Terminal ileum
Appendix
Rectum
Pelvic (sigmoid)
Anal canal
Anus

F. 15-17 CECUM: APPENDIX: ASCENDING COLON:

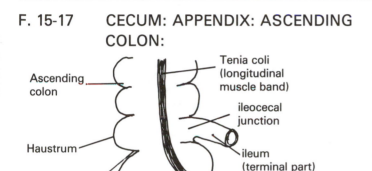

Ascending colon
Tenia coli (longitudinal muscle band)
ileocecal junction
Haustrum
Semilunar fold
ileum (terminal part)
Cecum
Appendix

(2) CECUM, ILEOCECAL SPHINCTER & SEMILUNAR FOLDS

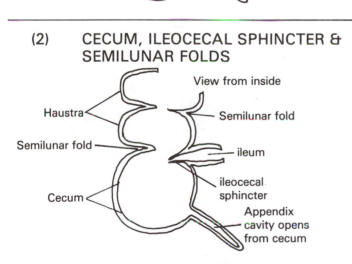

View from inside
Haustra
Semilunar fold
Semilunar fold
ileum
Cecum
ileocecal sphincter
Appendix cavity opens from cecum

F. 15-18 CROSS SECTION ABDOMEN:

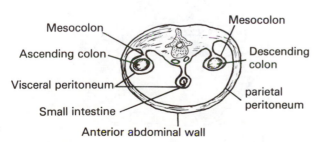

Mesocolon
Mesocolon
Ascending colon
Descending colon
Visceral peritoneum
parietal peritoneum
Small intestine
Anterior abdominal wall

The external anal sphincter is a second safeguard of skeletal muscle that surrounds the canal.

The anus is the opening at the lower end of the anal canal. While in most subjects this opening is present at birth the embryonic membrane that closes it during development may persist — an **imperforate anus**. The infant cannot have a bowel movement. A group of veins surrounds the anus. These may enlarge to form hemorrhoids.

Mesocolon. The attachment of the various parts of the large intestine to the posterior abdominal or pelvic wall is by peritoneum, similar to the mesentery of the small intestine. Here it has been named the mesocolon. In some areas the two layers of the fold of peritoneum come together behind the colon similar to mesentery. The colon in these areas may be quite movable. The cecum, transverse and sigmoid parts may follow this pattern. The ascending and descending parts lie closer to the posterior abdominal wall so the peritoneum may not form a mesocolon. Blood and lymph vessels and nerves reach the large intestine between the two layers of peritoneum.

CHARACTERISTIC FEATURES OF LARGE INTESTINE

Teniae coli (s. tenia coli). These are three bands of muscle fibers that pass lengthwise along the length of the large intestine. They replace the layer of longitudinal muscle found in other parts of this tract. They are spaced at even intervals around the bowel. The muscle fibers are shorter than are the other layers. Therefore they cause a puckering of the colon. This appearance is seen in the large intestine only. (L. tenia = a band or tape + coli = of the colon). The rectum does not have this pattern. F. 15-16, 17

Haustra (S. haustrum). These are saclike pouches in the wall of the large intestine resulting from the puckering due to the teniae coli. F. 15-16, 17

Semilunar folds are the folds visible on the inner surface of the large intestine that pass part way around the intestine between haustra. They do not completely encircle the bowel. They help to make the large intestine distinguishable from the small bowel.

Mucous glands in the lining membrane secrete mucus into the large intestine.

S. 206 ABDOMINAL ACCESSORY DIGESTIVE ORGANS

In addition to the parts of the digestive tract (tube) described above three other organs form abdominal accessory organs of digestion. They are the pancreas, liver, gall bladder and bile ducts.

1. THE PANCREAS   F. 15-19

G. pankreas = sweetbread; adj. pancreatic.

The pancreas is a long tapering gland that lies transversely in the upper posterior abdomen behind the stomach. Therefore it is located in the upper right and left quadrants. It has a head, a body, and a tail.

The head of the pancreas is its large, right, bluntly rounded end that lies in the curve of the duodenum.

The body of the pancreas is the long tapering part extending to the left behind the stomach.

The tail of the pancreas is its pointed left end that reaches the adjacent splenic border.

A neck is described by some authorities as being between the head and body.

The pancreas is made up of many lobes, which in turn consist of many lobules. Secreting cells in the lobules are

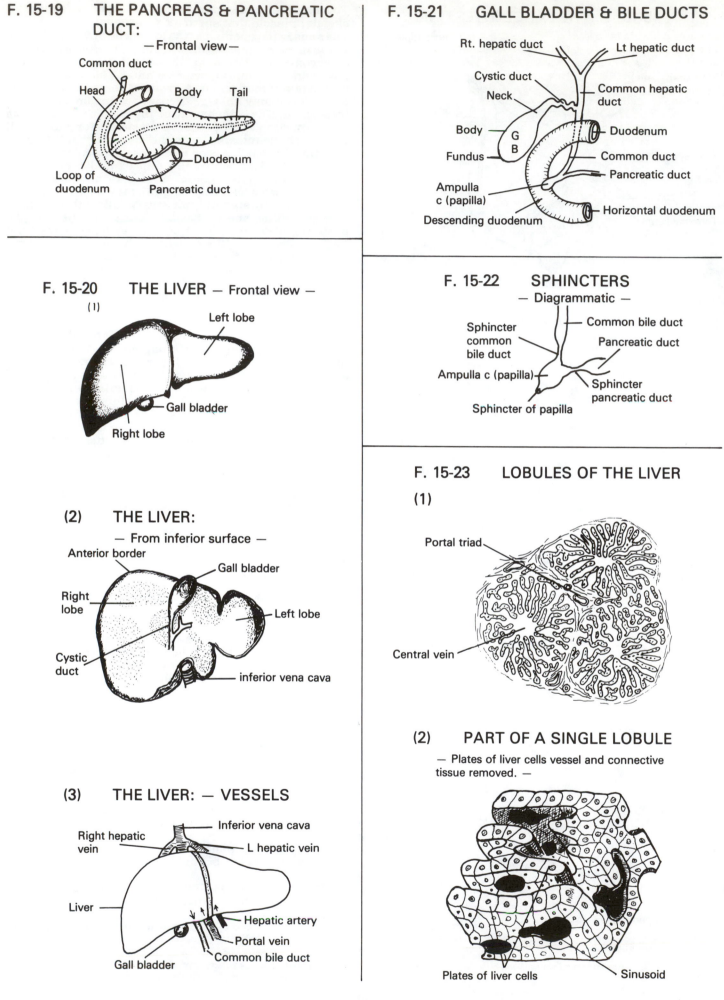

F. 15-19 THE PANCREAS & PANCREATIC DUCT:

— Frontal view —

Common duct
Head
Body
Tail
Loop of duodenum
Duodenum
Pancreatic duct

F. 15-20 THE LIVER — Frontal view —

(1)

Left lobe
Gall bladder
Right lobe

(2) THE LIVER:

— From inferior surface —

Anterior border
Gall bladder
Right lobe
Left lobe
Cystic duct
inferior vena cava

(3) THE LIVER: — VESSELS

Right hepatic vein
Inferior vena cava
L hepatic vein
Liver
Hepatic artery
Portal vein
Common bile duct
Gall bladder

F. 15-21 GALL BLADDER & BILE DUCTS

Rt. hepatic duct
Lt hepatic duct
Cystic duct
Neck
Common hepatic duct
Body
G B
Duodenum
Fundus
Common duct
Pancreatic duct
Ampulla c (papilla)
Descending duodenum
Horizontal duodenum

F. 15-22 SPHINCTERS

— Diagrammatic —

Sphincter common bile duct
Common bile duct
Pancreatic duct
Ampulla c (papilla)
Sphincter pancreatic duct
Sphincter of papilla

F. 15-23 LOBULES OF THE LIVER

(1)

Portal triad
Central vein

(2) PART OF A SINGLE LOBULE

— Plates of liver cells vessel and connective tissue removed. —

Plates of liver cells
Sinusoid

connected with minute ducts that unite to form larger ducts. These empty into the pancreatic duct. The lobules are comparable to clusters of grapes with their stems.

The pancreatic duct is a hollow tube that extends from the tail through the body and head to join the common bile duct just before the latter opens into the duodenum. Occasionally the pancreatic duct has a separate opening into the duodenum, and sometimes there is a second or accessory pancreatic duct.

Pancreatic juice, a mixture of digestive enzymes, is secreted by the pancreatic lobules and passes through the pancreatic duct to the small intestine. Here it mixes with and digests food.

MIXED GLAND — the pancreas is a dual or mixed gland consisting of an exocrine and an endocrine part. The exocrine part secreting pancreatic juice has been briefly described above.

The Islands (Islets) of Langerhans are nests of cells scattered throughout the pancreas and form the endocrine part. They have no ducts, and do not drain into the pancreatic duct. Their secretion, **insulin**, is a hormone that is absorbed directly into blood capillaries. Insulin is necessary for the utilization of glucose by the body cells. Its absence results in a disease called diabetes. See Endocrine glands. Ch. 20. (insulin, from insula = an island).

Enlargement of the pancreas from cysts or tumors may cause pressure upon the duodenum, or stomach or bile ducts.

2. THE LIVER   F. 15-20 (1, 2, 3)

G. Hepar = liver, adj. hepatic; e.g. hepatic v

The liver is the largest solid organ in the body, and may weigh three pounds. It lies for the most part in the upper right abdomen, but extends to the left of the midline, so is in the upper right and left abdominal quadrants. When examined from the front it is roughly triangular in shape, with its pointed end to the left. Its upper surface is convex corresponding to the dome of the diaphragm under which it lies. The inferior (caudal) surface is concave and passes from the right costal margin obliquely upwards and to the left. This surface has depressions that are in contact with the right colic flexure, right kidney and gall bladder. The liver is attached to the under surface of the diaphragm by ligaments, and all but its posterior surface is covered by peritoneum.

The porta hepatis (gate of liver) or transverse fissure is a slit on the inferior surface of the right lobe by which the portal vein, hepatic artery, lymphatics, and nerves enter the liver substance. The hepatic ducts leave the liver here.

The portal vein, draining the stomach, intestine, pancreas and spleen enters the liver. It breaks up into smaller branches, and these end in sinusoids or minute blood spaces within the liver.

The hepatic artery, a branch of the celiac trunk, also enters the liver beside the portal vein. It divides into capillaries that accompany the sinusoids of the portal vein.

Three hepatic veins, beginning as minute central veins in each liver lobule, collect blood from the sinusoids and hepatic capillaries, and empty into the inferior vena cava behind the liver.

LIVER STRUCTURE   F. 15-

The liver, similar to the lung is composed of incomplete lobes, a large right, a smaller left, and small quadrate and caudate lobes. Each lobe is made up of many lobules. Each lobule is a minute polygonal (many sided) unit surrounded by connective tissue; with a small vein in its center; the central or intralobular vein. The central veins leaving the lobules unite to form the three hepatic veins.

Each lobule consists of plates of liver cells. Each plate is one cell thick, with cells placed one upon another like a brick wall. Each plate runs from the margin of a lobule to the central vein at the center of the lobule. The plates have openings and may be curved. Between adjacent plates are spaces in which minute branches of the hepatic artery, and the sinusoids from the portal vein, run towards the central vein. The liver cells are therefore in contact with blood vessels. They extract required substances from the blood. They manufacture and store up new substances from these, and secrete their products back into the blood capillaries as required.

Minute bile ducts called canaliculi lie within each lobule with one tier of liver cells above and one below. These ducts extend from the center of the lobule to its outer border and end by joining to form larger bile ducts. These in turn join together and form the right and left hepatic ducts. **Bile** secreted by liver cells is thus conveyed out of the liver to the duodenum.

3. GALL BLADDER AND BILE DUCTS   F. 15-21, 22

(1) **The gall bladder** is a hollow pear shaped organ that lies against a depression on the inferior surface of the liver, in the upper right anterior abdomen. It may be from 5 to 10.8 cm or 2 to 4.5 inches in length. It has serous, muscular and mucosal coverings, so is capable of contracting and becoming smaller. Its location is somewhat variable, so that while it often lies posterior to the ninth costal cartilage it may be higher or lower. Its position is also influenced by the body build, position of the patient, state of fullness, and the stage of respiration. The gall bladder has a fundus (or rounded end), a body, and a neck (its constricted end). The gall bladder opens into the cystic duct. (Cholecyst = gall bladder; (chole = bile, + kystis = bladder); adj. cholecystic; (NA), vesica fellea, never used.

(2) **The right and left hepatic ducts** collect bile from the lobes of the liver and convey it to the common hepatic duct outside of the liver.

(3) **The common hepatic duct**, formed by the union of the right and left hepatic ducts, joins the cystic duct.

(4) **The cystic duct** is a hollow tube that passes from the gall bladder to join the common hepatic duct.

(5) **The common bile duct**, choledochus or bile duct, is formed in the upper right posterior abdomen by the union of the common hepatic and the cystic ducts. It descends posterior to the descending duodenum, and opens into this part. It is usually joined by the pancreatic duct. The short part of the duct after this union is called the **hepatopancreatic ampulla** (ampulla of Vater).

The hepatopancreatic sphincter (the sphincter of Oddi) is present at the opening into the duodenum. The sphincter may extend to the distal ends of the common bile and pancreatic ducts. When viewed from the inside of the duodenum a nipplelike prominence, the duodenal papilla, may be seen where the duct opens into the duodenum. (chole = bile + dochus = a duct, hence bile duct).

Since the gall bladder and bile ducts have visceral muscle in their walls they are capable of contracting to force bile into the duodenum. The student should reconstrnct the pattern of the bile ducts that carry bile into the duodenum to mix with food. Bile secreted by the liver passes through the right and left hepatic and common hepatic ducts, and cystic duct to the gall bladder, then back through the cystic duct to the common bile duct, and on into the descending duodenum.

Bile secreted by the liver passes into the gall bladder. The gall bladder does not manufacture bile. Capillaries in the wall of the gall bladder extract water from the bile within it. More bile enters the gall bladder and more water is extracted. This process is repeated many times and the bile in the gall bladder becomes thick or concentrated. Bile is stored in the gall bladder, and does not pass into the duodenum until the sphincters between the bile ducts and duodenum relax. This occurs when food entering the duodenum from the stomach causes contraction of the gall bladder and bile ducts with relaxation of the sphincters. Bile then enters the small intestine and mixes with food. (See cholecystokinin, S. 208)

S. 207 BLOOD VESSELS OF THE ABDOMINAL DIGESTIVE ORGANS

The student should refer to S. 167, the portal vein, celiac trunk, and portal circulation.

1. **Arteries**

The abdominal digestive organs are supplied with blood by three unpaired branches of the abdominal aorta.

(1) **Celiac trunk:**
   a. left gastric — to esophagus and stomach,
   b. splenic — to pancreas and spleen
   c. common hepatic — to liver, gall bladder, pancreas, and duodenum.

(2) **Superior mesenteric artery** — small intestine, and proximal half of large intestine.

(3) **Inferior mesenteric artery** — to distal half of large intestine.

2. **Veins**

The veins draining these organs form the portal vein, liver capillaries (sinusoids) and hepatic veins, which then drain into the inferior vena cava.

(1) **the inferior mesenteric vein**, from the distal half of the large intestine, joins the splenic vein.

(2) **Superior mesenteric vein**, from proximal half of large intestine and small intestine.

(3) **Splenic vein**, from the spleen, pancreas and stomach, also joined by inferior mesenteric vein.

(4) **Portal vein**, formed by the union of the superior mesenteric and splenic veins enters the liver and divides into many capillaries (sinusoids). These unite and form the **hepatic veins**, which empty into the inferior vena cava.

A **cystic vein** from the gall bladder, and coronary and pyloric veins from the stomach also join the portal vein. **Note:** blood from the digestive organs and spleen passes through the liver before entering the general circulation.

S. 208 HORMONES SECRETED BY DIGESTIVE ORGANS

Several hormones are manufactured and secreted by the stomach and intestine. They are absorbed directly by blood capillaries and stimulate other structures to secrete or perform other functions. A few of these are defined:

1. **Gastrin**, a hormone secreted by pyloric glands, absorbed by blood, carried to the glands of the body and fundus of the stomach, to stimulate the secretion of gastric juice by these glands. Certain foods in the stomach stimulate the secretion of gastrin, e.g. meat extracts, soups etc. It should be noted that stimulation of the vagus nerve has an effect similar to gastrin.

2. **Secretin and pancreozymin** hormones secreted by the mucosa of the small intestine, carried by blood to the pancreas and liver, stimulate the secretion of pancreatic juice and bile. Food entering the duodenum from the stomach stimulates the secretion of secretin & pancreozymin. Vagus nerve stimulation has similar effects upon the liver and pancreas.

3. **Cholecystokinin**, a hormone secreted by the mucosa of

the small intestine and carried by blood to the gall bladder and bile ducts, stimulates the gall bladder to contract and to empty its bile into the duodenum. Food, particularly fats entering the duodenum stimulates the secretion of cholecystokinin. These physiological facts are utilized in cholecystography.

S. 209 FUNCTIONS OF THE DIGESTIVE SYSTEM

General - to take in food, fluids, minerals, and vitamins, and to prepare them for absorption by blood and lymph capillaries. (Ingestion, mastication, deglutition, digestion). See S. 198

The mouth is concerned with the intake, chewing and swallowing of food, and the mixing of it with saliva.
*The teeth bite, and chew (masticate) food, tearing it apart to produce smaller pieces.
*The tongue pushes food between the upper and lower teeth for chewing and grinding, and then forms a bolus or ball for swallowing.
*The salivary glands secrete a watery saliva containing an enzyme ptyalin, also called salivary amylase, that begins the digestion of carbohydrates. (amylose — starch).

The pharynx is a passage that receives food from the mouth, and passes it on into the esophagus by the act of swallowing (deglutition).

The esophagus is a passage to convey food through the chest from the pharynx to the stomach.

The stomach is a reservoir for food, allowing a considerable quantity to be ingested at a meal.
*It is a mixing bowl, and churns food to produce a semiliquid mixture called chyme.
*Gastric juice is secreted into the stomach by gastric glands in the lining membrane. This includes pepsinogen, hydrochloric acid, and rennin.
*Pepsinogen and hydrochloric acid combine to form **pepsin**, a digestive enzyme (gastric protease) that begins the digestion of proteins.
*Rennin is an enzyme that curdles milk in the stomach.
*An intrinsic factor, an enzyme that promotes the absorption of vitamin B12 from the intestine, for the manufacture of R.B.C. by the bone marrow, is also secreted.
*Gastrin, a hormone, is secreted by the pyloric glands in the stomach. It stimulates the fundus and body of the stomach to secrete gastric juice. It is doubtful if any fat splitting enzyme is secreted by the stomach.
*The stomach, by peristalsis propels chyme into the duodenum at intervals, the pyloric sphincter relaxes to allow this flow.

The small intestine. Most of the digestion and absorption of food takes place in the small intestine.
Three secretions are conveyed to the small intestine and mix with food (chyme) from the stomach.
Bile, secreted by the liver reaches the descending duodenum thru the bile duct. It mixes with the food to emulsify fats (breaks fat globules into smaller globules) so that enzymes may come into more intimate contact to digest them.
Pancreatic juice, secreted by the pancreas is carried by the pancreatic duct to the duodenum, to mix with food.
Intestinal juice, secreted by glands in the lining membrane of the small intestine, is secreted directly into the intestine, where it also mixes with food.

The pancreas secretes **pancreatic juice** that contains pancreatic enzymes.
*Trypsinogen, secreted by the pancreas is inactive until activated by enterokinase, an enzyme secreted by the

intestine, forming the enzyme trypsin (a pancreatic protease). It digests proteins to proteoses, and peptones then to peptids.

*Pancreatic amylase, secreted by the pancreas, digests carbohydrates.

*Pancreatic lipase, secreted by the pancreas digests fat.

Note: during the digestion of proteins they are split first into proteoses, which in turn are divided into peptones, these are split into peptids, and these split into amino acids.

The pancreas also manufactures a hormone, insulin, that is absorbed by blood capillaries, and is concerned with the metabolism of sugar. It is secreted by cell nests, the Islands of Langerhans.

The small intestine secretes intestinal juice, also termed succus entericus, by glands in its lining membrane. This mixes with food within the intestine. Intestinal juice contains enzymes:

*Erepsin, secreted by the small intestine is an enzyme that breaks down the peptids produced by trypsin into amino acids for absorption.

Carbohydrate digestants are also secreted into the small intestine:

*Sucrase splits sucrose (cane sugar) into glucose, and fructose — monosaccharides.

*Maltase splits malt sugar, from barley and other grains.

*Lactase splits milk sugar into glucose and galactose. Enterokinase, an enzyme, combines with trypsinogen to form trypsin. (See pancreas).

The large intestine does not secrete digestive enzymes.

*Mucus is secreted by goblet cells, and glands in the lining membrane. This lubricates the passage.

*Bacteria in the large intestine produce some vitamins that the body requires, such as vitamin K, required by the liver in the production of prothrombin. Bacteria also produce gases notably hydrogen sulphide.

*Water is absorbed from the large intestine. The undigested food residue, the feces, then becomes semisolid. It is stored in the large intestine, and is evacuated at intervals.

Peristalsis, and an increased intraabdominal pressure from straining, plus relaxation of the anal sphincters are responsible for this action.

The liver has several very important functions.

*It forms and secretes bile, containing bile salts and bile pigments. Bile as stated emulsifies fats in the small intestine.

*Glucose is converted into glycogen by the liver and is stored in the liver. It is reconverted into glucose as required.

*Blood proteins, fibrinogen, albumin, globulin, as well as prothrombin, and heparin, are manufactured by the liver.

*Excessive amino acids are broken up by liver cells into carbohydrates, and urea. The urea contains the nitrogen part of the amino acid and is excreted by the kidneys.

*It breaks down worn out red blood cells, producing bile pigments.

*It stores some vitamins, including vitamin B12

*It renders inactive toxic products of bacterial action formed in the large intestine.

*It forms reticuloendothelial cells — phagocytes.

The gall bladder stores bile, concentrates bile, and expels it as required for digestion.

*Bile enters the gall bladder by the cystic duct.

*Bile is concentrated by capillaries in its wall absorbing water, and making it thicker.

*The gall bladder contracts to expel bile back through the cystic duct into the common bile duct, and duodenum to emulsify fats.

THE GALL BLADDER DOES NOT MANUFACTURE (SECRETE) BILE

The bile ducts store some bile. They convey bile from the liver to the gall bladder, and into the descending duodenum.

The sphincter in the common duct relaxes to permit this to occur.

Bile, following its secretion into the duodenum of the small intestine:

*emulsifies fat, i.e. forms minute fat globules to provide for close contact with fat splitting enzymes.

*promotes the absorption of vitamin K from the intestine, for prothrombin formation in the liver.

ENZYMES, HORMONES, AND OTHER SECRETIONS LISTED

Salivary glands — saliva:
    salivary amylase or ptyalin, enzyme for starch

Stomach — gastric juice, etc.
    pepsinogen + hydrochloric acid = pepsin
    rennin — enzyme for milk
    gastrin — a hormone
    intrinsic factor — vitamin B12 absorption

Small intestine — intestinal juice etc.
    erepsin — a protease enzyme — proteins
    sucrase, maltase, lactase — enzymes for sugars
    enterokinase — to combine with trypsinogen
    secretin and pancreozymin ⎫
    cholecystokinin   ⎬   hormones

Large intestine —
    mucus
    bacterial products

Pancreas — pancreatic juice, etc.
    trypsinogen + enterokinase = trypsin, protease
    pancreatic amylase — carbohydrates
    pancreatic lipase — fats
    insulin — a hormone for sugar metabolism

S. 210   CONGENITAL ANOMALIES DIGESTIVE ORGANS

Cleft palate — an opening in the roof of the mouth, so that fluids or food pass into the nasal cavities from the mouth.

Atresia of esophagus — the esophagus is closed at some point, and may communicate with the trachea.

Hypertrophic pyloric stenosis — a narrowing of the pyloric part of the stomach causing some obstruction to the passage of food into the duodenum; it may occur in infants.

Hernia or rupture — protrusion of an organ either through a weak part of the wall, or where a wall has not formed. Some are congenital, and some from straining when lifting, etc.

Hiatal hernia — through the esophageal opening (hiatus) of the diaphragm.

Umbilical hernia — at the umbilicus or navel.

Inguinal hernia — in the groin.

Imperforate anus — absence of an opening at the anus.

Interposition — the right colic flexure lies between the liver and diaphragm.

Transposition — situs inversus — organs lie in the opposite side of the body to their normal position. One or several organs may be involved.

Meckel's diverticulum — a diverticulum of the ileum close to the ileocecal junction. It may sometimes be demonstrated by the ingestion of barium and by taking films at intervals, if it fills with the barium.

S. 211   SOME PATHOLOGICAL TERMS

Itis — an inflammation — a suffix.
appendicitis — inflammation of the appendix
cholecystitis — of the gall bladder
colitis — of the large intestine
duodenitis — of the duodenum
diverticulitis — of diverticula
enteritis — of the intestine

179

esophagitis — of the esophagus
hepatitis — of the liver
medianstinitis — of the mediastinum
pancreatitis — of the pancreas
peridiverticulitis — around diverticula
peritonitis — of the peritoneum
regional enteritis — of a part of the intestine, often the ileum
appendiceal abcess — an abscess of the appendix.

Calculi: (s. calculus) stones, that may form in any part of the biliary system, appendix, etc., but most frequently in the gall bladder. They may be calcified and opaque. They may have an outer layer of calcium and cast ring-like images, or may not be calcified and so not visible, except when the gall bladder is filled with an opaque medium. Then they present as dark translucent densities on the processed film. This type will not be visible in scout films of the abdomen.

Cholelithiasis: calculi in the gall bladder.

Diverticula: (s. diverticulum) pouchlike protrusions of the wall of an organ. They may occur in any organ of the digestive tract, and in some other organs. They are frequently present in the large intestine of older subjects.

A pulsion diverticulum — a pushing outwards of the small saclike part, e.g. esophagus.

A traction diverticulum — one caused by pulling of the wall from outside, by scar tissue about lymph nodes, etc. e.g. esophagus.

A fissure: a narrow slit or crack such as may occur about the anus.

Fistula: an opening between two organs where no opening normally exists, e.g. gastrocolic fistula between the stomach and colon.

Intussusception: a telescoping of one part of the intestine into the part beyond it, as if one part were swallowed by another, or sucked into it, e.g. the terminal ileum into the cecum.

Mesenteric thrombosis: a thrombus in one of the mesenteric arteries or their branches, resulting in death of the intestine supplied with blood by that vessel.

Volvulus: a twisting of a loop of intestine upon itself, e.g. the sigmoid colon.

Jaundice: a deposition of bile in body tissue visible in the skin, nails, eyeballs, etc., as a yellow tinting. This may be due to obstruction to the flow of bile into the duodenum from a calcus, tumor, etc., or to the rapid destruction of red blood cells (hemolytic jaundice), or to inability of the liver to secrete bile as in hepatitis (inflammation of the liver).

S. 212   TERMS USED TO DEFINE OPERATIONS

Cholecystectomy = removal of the gall bladder
Cholecystotomy = a cutting into gall bladder
Cholecystduodenostomy = making an opening between the gall bladder and duodenum, and sewing the cut edges together, in order to allow bile to enter the small intestine when the common bile duct is obstructed.
Enterocolostomy = an opening between the small and large intestines in order to bypass some obstruction.
Gastrectomy = removal of the stomach.
Partial gastrectomy = removal of part of the stomach.
Gastroenterostomy = an opening made between the stomach and small intestine, usually into the upper jejunum to allow food to bypass the duodenum.
Colostomy = an opening made into some part of the large intestine to bypass some obstruction.
Colectomy = removal of the colon, frequently a right or left hemicolectomy (one-half).

APPLICATION TO RADIOGRAPHY.

The following facts can be demonstrated.

(1) Hollow organs such as the parts of the digestive tract and bile system will not be outlined as separate images on x-ray film, if they are empty.
(2) In normal adults there is usually some air in the stomach, duodenum and colon, but none in the jejunum, ileum nor esophagus.
(3) In infants, small children, and old subjects there is usually air in the jejunum and ileum as well.
(4) In bowel obstruction, air will accumulate in the small bowel or colon or both, and will outline the bowel. This fact is utilized and films are taken.
(5) Most parts of the digestive system lie closer to the anterior than the posterior abdominal wall. The films are usually taken in the postero-anterior position. This may be awkward in colon examinations.

The **fundus** of the stomach fills with barium when the patient lies supine, as it lies close to the posterior abdominal wall. Films are taken in this position. The **body** and **pylorus**, lying closer to the anterior abdominal wall, fill best with the patient prone; so this position is utilized.

The **duodenal cap** or bulb often lies directly behind the pylorus and can be seen with the patient oblique.

The **colon** examination is usually done with the patient in the supine position. It is thus easier to handle the patient, enema tube, and tip. It is simple to rotate the patient to oblique positions to view those parts of the colon that overlap, especially the flexures and sigmoid.

Opaque media. Since the empty organs do not show on films, it is necessary to use some opaque medium to outline them. Barium sulphate is the medium in common use to visualize the parts of the digestive tract. It is mixed with water and given by mouth or as an enema. In esophageal examinations thin barium passes down too rapidly; so a thick barium paste is employed.

Gall bladder. This organ usually lies in the upper right abdomen opposite the ninth costal cartilage and close to the anterior abdominal wall. Films are therefore taken in the prone position. If empty, the gall bladder is not visible. A special examination is done to outline it.

In examinations of the gall bladder or bile ducts a medium containing iodine is used. The various media are usually referred to as dyes — not strictly correct.

Cholecystography is the procedure used to visualize the gall bladder. It consists of giving the opaque medium (dye) by mouth and taking films at proper intervals.

A cholecystogram is a film or series of films outlining the gall bladder.

Some knowledge of the function of the gall bladder is necessary to understand what instructions should be given the patient regarding diet, the time to take the dye, and when to report to the department for films.

The gall bladder dye containing iodine is given to the patient after a fat-free supper. This medium (dye) is absorbed from the small intestine and passes through the portal vein to the liver. It is excreted by the liver along with bile, into the bile ducts. Some of the bile and dye enter the gall bladder. Water is absorbed from the bile in the gall bladder so that the bile and dye become more concentrated — thick. More bile enters the gall bladder, more water is absorbed, and by morning there is a sufficient concentration of dye in the bile in the gall bladder to render the gall bladder opaque to x-rays. Films are taken before the patient eats breakfast. Sometimes a fat meal is then given. Fat in the duodenum stimulates the gall bladder and bile ducts to contract and empty. Further films taken after the fat meal should show a smaller gall bladder or no shadow if the gall bladder has emptied. These films may also outline the bile duct.

In cholecystitis the gall bladder will not concentrate bile, and films will show a faint image or no image. Other

F. 15-24   T-TUBE CHOLANGIOGRAM

(1) T-Tube in position in common duct;

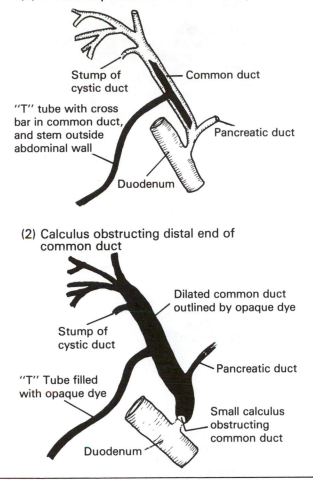

Stump of cystic duct

Common duct

"T" tube with cross bar in common duct, and stem outside abdominal wall

Pancreatic duct

Duodenum

(2) Calculus obstructing distal end of common duct

Dilated common duct outlined by opaque dye

Stump of cystic duct

Pancreatic duct

"T" Tube filled with opaque dye

Small calculus obstructing common duct

Duodenum

factors may be responsible. The patient may have vomited the dye. If there is obstruction at the pylorus or duodenum, dye will not pass into the small bowel and will not be absorbed. In diseases of the liver such as hepatitis the liver will not excrete bile or dye. If a fat meal is taken in the morning before films, the gall bladder will empty. Under any of these conditions the gall bladder may not outline. In section 138 gallstones were discussed. The student must remember that nonopaque calculi will not be visualized if the gall bladder is not outlined.

Cholangiography (chole, bile; angeion, a vessel).
This is a procedure used to outline the bile ducts, particularly the common bile duct.

(1) **Pre-operative.** If, weeks or months after a gall bladder operation, symptoms suggest obstruction of the common duct from calculi, etc., or if the patient vomits gall bladder dye when given by mouth for a gall bladder examination, an intravenous cholangiogram is done. Opaque dye, containing iodine, may be injected intravenously. The dye will pass through the heart to the liver and should be excreted into the bile ducts and gall bladder if present. Films are taken at intervals following the injection.

(2) **During operation.** After the abdomen is opened for removal of the gall bladder, opaque dye may be injected directly into the common duct, by means of a syringe and needle. This may outline calculi or obstruction in the common duct.

(3) **Post-operative.** If calculi are found in the common duct at operation, these are removed. The crossbar of a rubber T tube is then passed into the common duct. One limb is directed upward towards the liver; the other end is passed down towards the duodenum. The

stem of the T tube is brought out through the skin. Bile may pass through the cross bar to the duodenum. If the common duct is still obstructed, bile can drain off through the stem of the tube. Opaque dye may be injected through the stem and films taken to determine if obstruction is present and if calculi are still blocking the duct.

S. 213   ANATOMICAL TERMS OF DIGESTIVE SYSTEM

Foods
mineral salts
vitamins
water
carbohydrates
  monosaccharides
    glucose
    fructose
    galactose
  disaccharides
    sucrose (cane sugar)
    lactose (milk sugar)
    maltose (malt sugar)
  polysaccharides
    starches
    cellulose
    glycogen
    dextrins
fats — lipids
  glycerine
  fatty acids
proteins
  amino acids
Ingestion
  mastication
  deglutition
digestion
  mechanical
  chemical
digestive enzymes
  proteases
  lipases
  amylases
  disacchreases
  pepsin (old)
  erepsin "
  trypsin "
  ptyalin "
absorption
  blood capillaries
  lymph capillaries or lacteals
position of organs
  body build (habitus)
  weight
  position
  respiration
  fullness

Divisions of abdomen

four quadrants
  right upper
  right lower
  left upper
  left lower
Nine regions
  rt. hypochondriac r
  epigastric region
  lt. hypochondriac r
  rt. lumbar region
  or rt. lateral r
  umbilical region
  lt. lumbar region
  or lt. lateral r

{ rt. iliac region
or rt. inguinal r
hypogastric region
or pubic region
lt. iliac region
or lt. inguinal r

General definitions
lumen
peritoneum
visceral peritoneum
parietal peritoneum
peritoneal cavity
mesentery
mesocolon
omentum
greater omentum
lesser omentum
sphincter
peristalsis
segmentation
stenosis
atresia

Coverings digestive organs
serous membrane or
visceral peritoneum
muscular layer;
    longitudinal
    circular
submucous coat
mucosal layer or
    mucosa
    folds
    glands

Divisions digestive tract
mouth
pharynx
stomach
small intestine
large intestine

Mouth
cavity of mouth
vestibule
cavity proper
hard palate
soft palate
glossopalatine
    arch
glossopharyngeal
    arch
fauces
teeth
tongue
salivary glands;
    parotid & parotid
    duct = Stensen's duct
    submandibular & submandibular duct
      = Wharton's duct
    sublingual glands & sublingual ducts

Pharynx
oral pharynx or
    oropharynx
laryngeal pharynx or
    laryngopharynx

Esophagus
Stomach or gaster (G)
adj. gastric
ventriculus (L)
cardiac orifice (ostium)
pyloric orifice (ostium)
cardiac sphincter?
pyloric sphincter
angular notch or
    incisura angularis
lesser curvature
greater curvature

parts:
fundus
body
pyloric part
pylorus
rugae (gastric folds)
gastric glands
hydrochloric acid
pepsin
rennin
gastric lipase

Small intestine
parts:
duodenum
adj. duodenal
{ superior part (cap)
descending part
horizontal part
ascending part
jejunum, adj. jejunal
ileum, adj. ileac
ileocecal opening
ileocecal sphincter (or valve)
terminal ileum

general:
circular folds or
plicae circulares
villi, s. villus
aggregated lymph
    follicles (Peyer's patches)
solitary lymph follicles
intestinal glands
intestinal juice

Large intestine
parts:
cecum & appendix
ascending colon
rt. colic or hepatic flexure
lt. colic or splenic flexure
descending colon
simoid colon or
    pelvic colon
rectum
anal canal
anus
internal anal
    sphincter
external anal
    sphincter
longitudinal
    muscle bands
or teniae coli
haustra
semilunar folds

Accessory abdominal
digestive organs
pancreas
liver
gall bladder
bile ducts

Pancreas
parts;
head
body
tail
lobules
pancreatic duct
accessory pancreatic duct
pancreatic juice
Islands of Langerhans
Insulin

Liver
hepar (G) hepatic

parts;
- right lobe
- left lobe
- two small lobes
  - caudate lobe
  - quadrate lobe
- segments
- lobules
- small bile ducts
- liver capillaries
- sinusoids
- bile, G. chole
- rt. hepatic duct
- lt. hepatic duct

Gall bladder
- G. cholecyst
- adj. cholecystic

parts:
- fundus
- body
- neck

Bile ducts
- rt. hepatic duct
- lt. hepatic duct
- common hepatic duct
- cystic duct
- common bile duct
  - or choledochus

or bile duct
- sphincter of common bile duct
- pancreatic duct
- hepatopancreatic ampulla or
- ampulla of Vater
- sphincter of ampulla
- or Sphincter of Oddi

Blood vessels of the
digestive system

arteries
- celiac trunk (OT axis)
  - left gastric artery
  - splenic artery
  - common hepatic a
- superior mesenteric a
- inferior mesenteric a

veins
- portal vein formed by
- superior mesenteric
- & splenic veins
- inferior mesenteric
  - a tributary of the
  - splenic vein
- cystic vein
- pyloric vein
- coronary vein

NOTE:enteron = intestine, adj. is enteric

183

NOTES

NOTES

kidney	**macroscopic anatomy**	number | pedicle size | cortex shape | medulla location | renal pelvis capsule-fibrous | major calyces peritoneum | minor calyces extremities | renal pyramids hilum | renal papillae sinus | renal columns	
	microscopic anatomy	nephron — glomerulus = capillaries	afferent arterioles efferent arterioles
		renal tubules	glomerular capsule (Bowman's) proximal convoluted tubule descending limb of loop loop of Henle ascending limb of loop distal convoluted tubule **renal corpuscle = glomerulus + glomerular capsule
		collecting tubules — renal pyramids renal papillae	
	blood vessels	renal artery | afferent arter- anterior div | ioles posterior div | glomeruli lobar aa | efferent arterioles interlobar aa | peritubular capil. arcuate aa | renal veins & interlobular aa | tributaries	

ureter	number location parts — abdominal part pelvic part origin termination length	structure	fibrous outer layer muscular layer mucous layer, mucosa
		openings peristalsis	from renal pelvis; ureteral opening to bladder;

bladder	location — male female	sphincter	vesical (bladder)

	structure — peritoneum or
	     serous layer
	muscular
	mucosa, mucous
	  layer
		trigone
		capacity
		function	

	openings — left ureteral right ureteral internal urethral		

urethra	location — male female	sphincter —— urethral sphincter	
	length — male female	parts in male — prostatic part; membranous part; penile or spongy part;	
	openings — internal urethral external urethral	functions — male female	

urine —— quantity —— specific gravity — acidity (ph) — composition

16. THE URINARY SYSTEM

S. 214 The urinary system is recognized as the excretory system that gets rid of waste products produced in the body. Three other systems also excrete some waste products:
1. The skin — excretes water and salts
2. The lungs — excrete carbon dioxide and water
3. The large intestine — excretes undigested food residue as feces
4. The kidneys — as described in this chapter

Genitourinary or **urogenital** are terms that are used to include both the urinary and reproductive organs. In the male the urethra within the penis forms part of both the urinary and reproductive systems. The other organs have no common function. In the female the organs of these two systems are entirely separate. In the Handbook the urinary and reproductive systems are studied separately.

The urinary system, female or male, has the following parts: (F. 16-1)
1. Right and left kidneys — excretory organs
2. Right and left ureters — pipe lines to bladder
3. Urinary bladder — a reservoir for urine
4. Urethra — a waste pipe to the outside

S. 215 DETAILED STUDY OF PARTS OF URINARY SYSTEM

1. THE KIDNEYS: MACROSCOPIC ANATOMY
F. 16-1, 2, 3

L. ren = kidney; adj. renal; G. nephros = kidney; adj. nephritic

There are two kidneys, a right and a left. They are bean shaped organs that lie in the upper posterior abdomen, one on either side of the vertebral column. They are located behind the peritoneum, and their anterior surfaces are covered by this membrane. Each kidney is about 11.25 cm or 4.5 inches in vertical length. The upper extremity lies opposite the body of the twelfth thoracic vertebra and the lower extremity may reach the third lumbar vertebra. The right kidney lies at a slightly lower level than the left, since it is displaced down by the liver.

Basic Definitions;

A renal capsule, a covering of fibrous connective tissue encloses each kidney, with perinephritic fat surrounding this covering. F. 16-2

The hilum (hilus) of the kidney is a depression or indentation on its medial border where the structures composing the pedicle join the kidney. F. 16-2

The renal sinus is a vertical cleft or space at the hilum through which the parts of the pedicle enter or leave the kidney.

The renal pedicle includes the renal artery, renal vein, renal pelvis, lymphatics and nerves that join the kidney at the hilum.

The renal segments, five in number, are subdivisions of the kidney. F. 16-2

The superior (upper) extremity of the kidney is its upper pole or end. F. 16-2

The inferior (lower) extremity is the lower pole or end. F. 16-2

F. 16-1   PARTS OF THE URINARY SYSTEM:

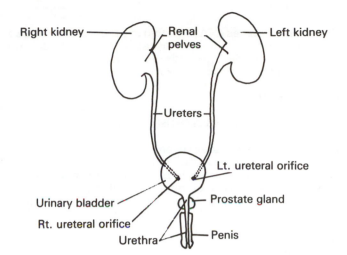

Right kidney — Renal pelves — Left kidney — Ureters — Lt. ureteral orifice — Urinary bladder — Prostate gland — Rt. ureteral orifice — Urethra — Penis

F. 16-2   KIDNEY - LONGITUDINAL FRONTAL VIEW:

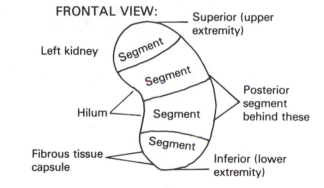

Left kidney — Superior (upper extremity) — Segment — Segment — Posterior segment behind these — Hilum — Segment — Fibrous tissue capsule — Segment — Inferior (lower extremity)

2. THE URETERS   F. 16-1

There are two ureters, a right and a left. They are hollow tubes 25 to 30 cm, or 10 to 12 inches in length that extend from a kidney to the urinary bladder. Each consists of abdominal and pelvic parts. The abdominal part passes down on the inner surface of the posterior abdominal wall behind the peritoneum. It crosses the pelvic brim and descends along the lateral wall of the pelvis (pelvic part) to the pelvic floor. It joins the bladder on the lateral and posterior border, opening by the ureteral orifice into the bladder. The lower part of each ureter lies obliquely to the bladder. As the bladder fills with urine it presses upon the lower end of the ureter closing it to any back flow of urine into the ureter. The ureter has three coats, an outer fibrous layer, a layer of visceral muscle, and a lining of epithelium.

The ureter conveys urine from the kidney to the urinary bladder by peristalsis of its muscular layer and by gravity.

The ureteral openings (NA, ostia) are the two openings between the lower ends of the ureters and the urinary bladder. F. 16-5

The renal pelvis F. 16-3 is the upper expanded funnel-shaped end of each ureter. It lies partly outside of the kidney, and partly within the renal sinus in the kidney. It is continuous with the ureter. L. pelvis = a basin, pl. pelves, adj. pelvic. G. pyelos = a basin or pelvis, hence pyelogram, pyelitis — each referring to the renal pelvis.

The **major calyces** (NA, calices) are the two or three short tubes into which the renal pelvis divides. They are located in the renal sinus within the kidney. Major = greater; larger. F. 16-3, 4 G. calyx = the cup of a flower, pl. calyces, adj. calyceal; (NA, calix, calices, & caliceal). F. 16-3, 4

The **minor calyces** (NA, calices) are the small cup shaped tubes into which each major calyx divides. Each major calyx may divide into several minor calyces. The cup shaped end of a minor calyx receives the bluntly pointed end of a renal pyramid, the papilla. Minor = lesser; smaller. F. 16-3, 4

3. THE URINARY BLADDER   F. 16-1, 5, 6, 7

AS. blaedre = a bladder or bag;
L. vesica = a bladder; adj. vesical;
G. kystis = cyst = a bladder or sac, any bladder, must specify as — urinary bladder, gall bladder, etc.
    The urinary bladder is a sac, a reservoir for urine, and lies in the pelvis, protected by the pelvic bones. It lies posterior to the symphysis, and anterior to the rectum. In the female when it is empty it lies anterior to and below the uterus, and in front of the upper part of the vagina. Its upper surface is covered by peritoneum. It has an outer fibrous covering, a middle muscular coat, and an epithelial lining. The muscular layer consists of three layers similar to the stomach. When the bladder is empty it is collapsed and its upper surface flat. It may be distended to hold 500 ml of urine, but 250 ml will cause considerable distension and discomfort.

Openings of the urinary bladder (three) — F. 16-1, 5

1. **The internal urethral orifice** (NA, ostium) is an opening into the urethra. It is located at the lowest part of the bladder, at its neck.
The urethral sphincter surrounds the upper part of the urethra below the opening from the bladder. Some authorities define a bladder sphincter as well, at the bladder neck.
The ureteral openings (NA, ostia) right and left are located at the lower ends of the ureters where they join the bladder. They lie at the lateral margins of the posterior surface of the bladder.
The trigone is a flat smooth triangular area on the lining epithelium of the bladder at its base or posterior part, between lines joining the ureteral openings and internal urethral opening, i.e. the three openings in the bladder.
    As the amount of urine in the bladder increases it exerts pressure upon the bladder wall stimulating the sensory nerve fibers. These transmit impulses to the brain. If the time and place are suitable the subject voids, the sphincter(s) relax(es), and the muscular wall contracts to expel urine.
    The bladder functions as a reservoir for urine, and contracts to expel it when required to do so.

4. THE URETHRA   F. 16-5, 6, 7

The urethra is the single passage from the bladder to the outside of the body through which urine is excreted. In both female and male subjects it extends from the opening of the bladder, the internal urethral orifice, to an external opening.
The external urethral orifice, opening or ostium, is an opening at the outer end of the urethra. It has been incorrectly named the external urethral meatus. Meatus indicates a canal.

The urethral sphincter, as defined above under the urinary bladder, surrounds the urethra below the bladder.

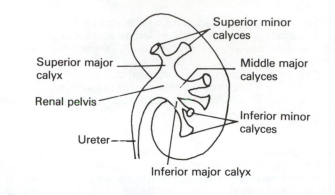

F. 16-3   RENAL PELVIS, AND CALYCES

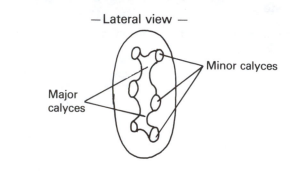

F. 16-4   RENAL CALYCES -

— Lateral view —

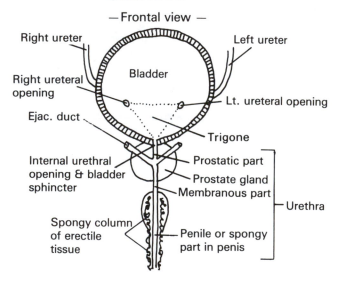

F. 16-5   URINARY BLADDER & MALE URETHRA:

— Frontal view —

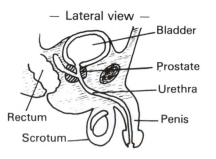

F. 16-6   MALE BLADDER & URETHRA

— Lateral view —

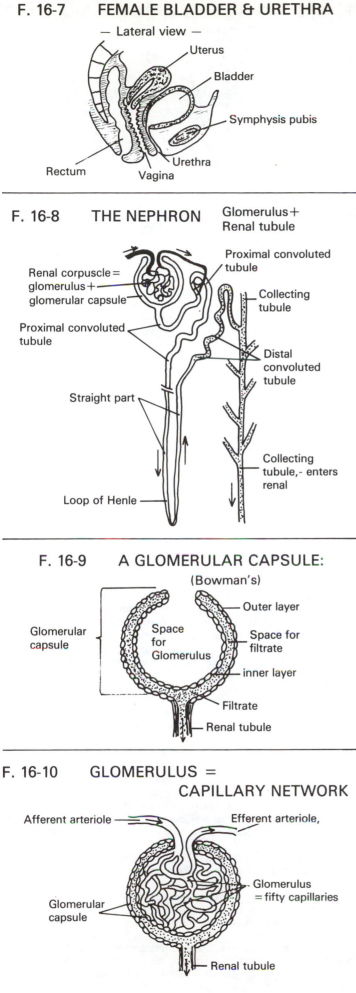

F. 16-7 FEMALE BLADDER & URETHRA

— Lateral view —

- Uterus
- Bladder
- Symphysis pubis
- Urethra
- Vagina
- Rectum

F. 16-8 THE NEPHRON

Glomerulus + Renal tubule

- Renal corpuscle = glomerulus + glomerular capsule
- Proximal convoluted tubule
- Proximal convoluted tubule
- Collecting tubule
- Distal convoluted tubule
- Straight part
- Collecting tubule, - enters renal
- Loop of Henle

F. 16-9 A GLOMERULAR CAPSULE:

(Bowman's)

- Outer layer
- Glomerular capsule
- Space for Glomerulus
- Space for filtrate
- inner layer
- Filtrate
- Renal tubule

F. 16-10 GLOMERULUS = CAPILLARY NETWORK

- Afferent arteriole
- Efferent arteriole,
- Glomerulus = fifty capillaries
- Glomerular capsule
- Renal tubule

The **female urethra** is about 4 cm or 1.5 inches in length. It passes down and forward from the bladder neck to its opening on the perineum. The symphysis is in front and the vagina behind it. The external urethral opening is a small opening in the midline in front of the vaginal opening. F. 16-7

The **male urethra** is about 20 cm or 8 inches in length, much longer than that of the female. It passes through the prostate gland below the bladder, also through the membranes covering the perineum, and the entire length of the male penis. The three parts of the male urethra are defined as: (F. 16-5, 6)

The **prostatic part** or prostatic urethra, passes through the prostate, and is about 3 cm or 1.2 inches in length. The prostatic and ejaculatory ducts open into this part.

The **membranous part** or membranous urethra runs from the inferior surface of the prostate through the membranes of the perineum.

The **spongy part**, or **penile part**, previously called the cavernous urethra extends through the length of the penis, in the corpus spongiosum, one of the cylinders of erectile tissue in the penis. The urethra has a lining membrane of epithelium.

The urethra in the male conveys urine from the bladder, but also serves as a passage for spermatozoa that are expelled into the vagina during intercourse — hence a genitourinary organ.

5. BLOOD VESSELS OF THE URINARY SYSTEM
F. 16-15, 16

The **right and left renal arteries,** paired branches of the abdominal aorta, supply blood to the kidneys.

The **right and left renal veins** drain blood from the kidneys and empty into the inferior vena cava.

The ureters are supplied by branches of vessels adjacent to them.

Vesical arteries, branches of the internal iliac arteries supply the bladder, and vesical veins drain blood into the internal iliac veins.

Details concerning the renal arteries and veins are included following a description of the microsopic structure of the kidneys.

S. 216 MICROSCOPIC ANATOMY OF THE KIDNEY F. 16-8, 9, 10

The **nephron** is the structural and functional unit of the kidney. There are about one and one quarter million of these microscopic units in each kidney. A nephron consists of two parts:
1. A glomerulus — i.e. capillary network
2. A renal tubule, made up of —
   (1) A glomerular capsule — covering
   (2) A proximal convoluted part of tubule
   (3) A straight tubule with loop of Henle
   (4) A distal convoluted part or tubule

Each distal convoluted tubule joins with others to form a collecting tubule.

A **glomerulus** is a network of about 50 blood capillaries that lie within a cup shaped glomerular capsule (See below). An arteriole from a branch of the renal artery enters the open end of this cup-shaped capsule and divides into the tuft of blood capillaries. These unite at their opposite ends to form a second arteriole which leaves the capsule. (Glomus = a ball or skein + ulus = little, hence a small ball or skein of capillaries). F. 15-10

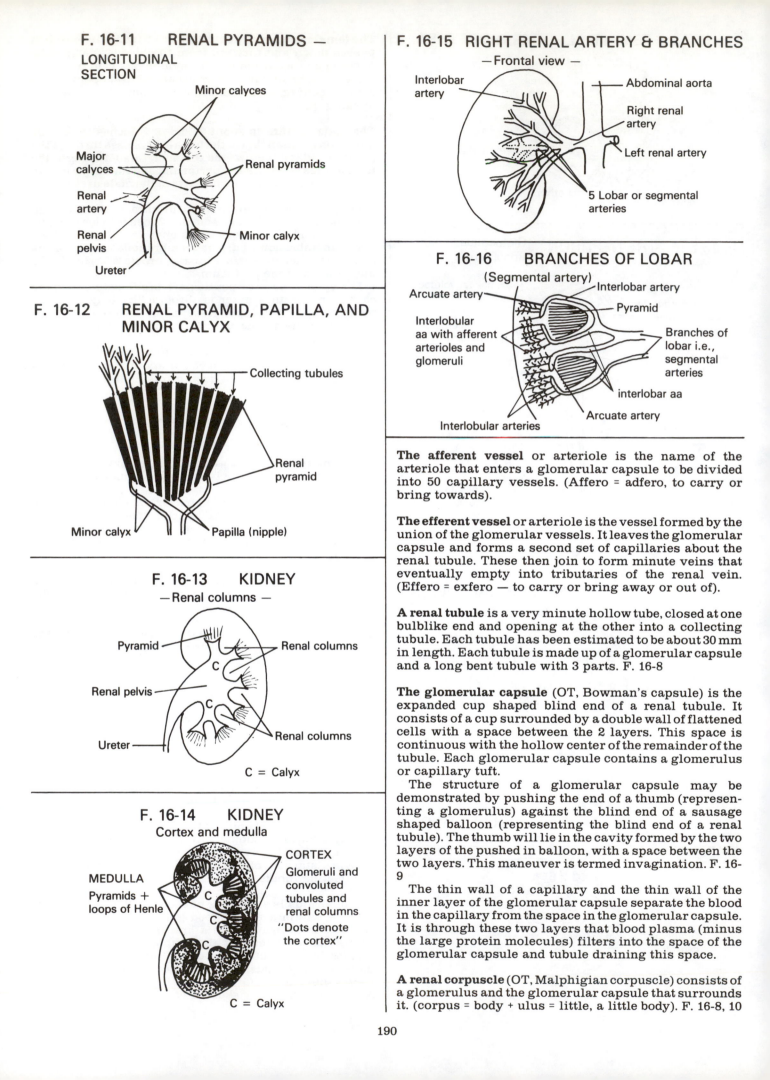

F. 16-11 RENAL PYRAMIDS —
LONGITUDINAL SECTION

Minor calyces

Major calyces

Renal pyramids

Renal artery

Renal pelvis

Minor calyx

Ureter

F. 16-12 RENAL PYRAMID, PAPILLA, AND MINOR CALYX

Collecting tubules

Renal pyramid

Minor calyx

Papilla (nipple)

F. 16-13 KIDNEY
— Renal columns —

Pyramid

Renal columns

Renal pelvis

Ureter

Renal columns

C = Calyx

F. 16-14 KIDNEY
Cortex and medulla

CORTEX
Glomeruli and convoluted tubules and renal columns

"Dots denote the cortex"

MEDULLA
Pyramids + loops of Henle

C = Calyx

F. 16-15 RIGHT RENAL ARTERY & BRANCHES
— Frontal view —

Interlobar artery

Abdominal aorta

Right renal artery

Left renal artery

5 Lobar or segmental arteries

F. 16-16 BRANCHES OF LOBAR
(Segmental artery)

Arcuate artery

Interlobar artery

Pyramid

Interlobular aa with afferent arterioles and glomeruli

Branches of lobar i.e., segmental arteries

interlobar aa

Arcuate artery

Interlobular arteries

The afferent vessel or arteriole is the name of the arteriole that enters a glomerular capsule to be divided into 50 capillary vessels. (Affero = adfero, to carry or bring towards).

The efferent vessel or arteriole is the vessel formed by the union of the glomerular vessels. It leaves the glomerular capsule and forms a second set of capillaries about the renal tubule. These then join to form minute veins that eventually empty into tributaries of the renal vein. (Effero = exfero — to carry or bring away or out of).

A renal tubule is a very minute hollow tube, closed at one bulblike end and opening at the other into a collecting tubule. Each tubule has been estimated to be about 30 mm in length. Each tubule is made up of a glomerular capsule and a long bent tubule with 3 parts. F. 16-8

The glomerular capsule (OT, Bowman's capsule) is the expanded cup shaped blind end of a renal tubule. It consists of a cup surrounded by a double wall of flattened cells with a space between the 2 layers. This space is continuous with the hollow center of the remainder of the tubule. Each glomerular capsule contains a glomerulus or capillary tuft.

The structure of a glomerular capsule may be demonstrated by pushing the end of a thumb (representing a glomerulus) against the blind end of a sausage shaped balloon (representing the blind end of a renal tubule). The thumb will lie in the cavity formed by the two layers of the pushed in balloon, with a space between the two layers. This maneuver is termed invagination. F. 16-9

The thin wall of a capillary and the thin wall of the inner layer of the glomerular capsule separate the blood in the capillary from the space in the glomerular capsule. It is through these two layers that blood plasma (minus the large protein molecules) filters into the space of the glomerular capsule and tubule draining this space.

A renal corpuscle (OT, Malphigian corpuscle) consists of a glomerulus and the glomerular capsule that surrounds it. (corpus = body + ulus = little, a little body). F. 16-8, 10

The remainder of a renal tubule consists of a hollow tubule, with many bends in its proximal and distal ends with a straight part and loop between.

The proximal convoluted part (tubule) opens from the space in a glomerular capsule, is tortuos and has many bends. It lies close to the capsule. F. 16, 8

The straight part (tubule) and loop of Henle is a "U" shaped loop that extends from a proximal convoluted tubule to a distal convoluted tubule. F. 16,8

The distal convoluted part (tubule) is continuous with the straight part. It is tortuous and bent and beacuse of the U-shaped loop lies close to the proximal convoluted part and glomerular capsule. Each distal convoluted tubule unites with many others to form a collecting tubule. (convoluted = bent). F. 16-8

A collecting tubule is formed by the union of many renal tubules. Each of these collecting tubules has its own opening into a minor calyx. F. 16-8

A renal pyramid is made up of many collecting tubules lying side by side. Renal pyramids are shaped like a blunt cone or pyramid, with the broad base directed towards the outer kidney surface. The bluntly rounded medial end fits into the cup of a minor calyx. Each collecting tubule of the group has its opening into the minor calyx. (F. 16-11, 12)

A renal papilla is the nipplelike rounded medial end of a renal pyramid, and projects into the cup shaped end of a minor calyx. A minor calyx may have one or more papillae in contact with it. Each of the collecting tubules in the pyramid will have its own opening in the papilla to the minor calyx. As the cups of the minor calyces are directed slightly anteriorly or posteriorly a longitudinal section of a kidney may not show papillae completely filling the cup of the calyx. F. 16-12

RENAL CORTEX AND MEDULLA   F. 16

Each kidney has a cortex and medulla, but before defining these some further structural details must be noted. Between each two adjacent renal pyramids there is an area separating them. This area is **called a renal column** (OT, Column of Bertini).

Most nephrons have their renal corpuscles, and their proximal and distal convoluted parts in the outer part of the kidney. The straight part and the loop of Henle are directed towards the calyces and enter a pyramid at its base between adjacent collecting tubules. Some other nephrons however lie in the renal columns between pyramids. In diagrams of the kidney with red dots representing renal corpuscles some of these will be shown between pyramids.

The renal cortex includes the outer part of the kidney and the renal columns between pyramids. The cortex therefore includes the renal corpuscles and convoluted parts of the renal tubules.

The renal medulla includes the renal pyramids, but not the renal columns between them. The medulla therefore contains the collecting tubules of the pyramids and the straight parts and the loop of Henle since they lie within the pyramid.

DETAILED STUDY OF THE RENAL BLOOD VESSELS
F. 16-15, 16

The information in the paragraphs below has been included for those students who are required to have a detailed knowledge of these vessels.

The renal artery, right or left, a branch of the abdominal aorta, supplies blood to the kidney. It divides into anterior and posterior divisions.

Lobar (segmental) arteries, branches of the renal artery supply each renal segment.

Interlobar arteries, branches of lobar arteries, enter the kidney between the renal pyramids and pass laterally.

Arcuate arteries, branches of interlobar arteries, arch over the bases of the pyramids.

Interlobular arteries, branches of the arcuate, pass out towards the kidney surface.

Afferent arterioles, branches of the interlobular arteries, enter glomerular capsules to form the glomeruli.

Glomeruli, or glomerular capillaries, are the fifty hairlike vessels which are the branches of an afferent anteriole.

Efferent arterioles are formed by the union of the glomerular capillaries at their opposite ends.

Peritubular capillaries, branches of the efferent arterioles form networks about the renal tubules. Some of them lying close to the bases of the pyramids send branches into these structures. The peritubular capillaries unite to form the smallest veins.

The two renal veins have tributaries similar to the arteries, in a reverse order.

S. 217   PHYSIOLOGY OF THE NEPHRON

These three processes involved in the secretion of urine are:
1. Filtration of blood plasma into renal tubules.
2. Reabsorption of water and some dissolved substances from tubules back into blood capillaries.
3. Selective secretion of some solutes from the blood into the renal tubules.

Blood from the abdominal aorta is conveyed to the kidneys by the renal arteries, then through their branches to the afferent arterioles and glomeruli.

Filtration. Blood plasma with many substances dissolved in it is filtered from the glomerulus, through the capillary wall and inner layer of a glomerular capsule into the capsular space and hence to the renal tubule. It is estimated that from 180 to 200 liters of fluid are filtered in 24 hours, (about 190 American or 157 British quarts). The salts dissolved in the blood plasma filter out as well, but blood protein molecules being larger are not filtered.

Reabsorption. All but about 1 to 1.5 liters (1000 to 1500 ml of water) are reabsorbed from the tubules back into blood capillaries (peritubular capillaries). Some of the solutes, for instance all of the glucose and enough sodium chloride etc., to fulfill body requirements, are reabsorbed by a process of selective reabsorption.

Selective secretion. Some solutes that are not required by the body are secreted into the tubules. Urea, creatinine and some other waste products are handled in this manner.

As stated above from 1000 to 1500 ml of fluid are not reabsorbed but are excreted as urine. The amount of urine excreted will vary with the quantity of fluid ingested, the body and external temperatures, etc. The volume will be increased with excessive intake of coffee, beer, etc. It will decrease if there is sweating, vomiting, diarrhoea or other conditions wherein fluid is lost. The percentage of salts excreted will also vary, since their percentage in the blood plasma must be kept constant. An excessive intake of sodium chloride, for instance, will result in an increased secretion. Urine does not normally contain blood proteins or glucose. When there is an excessive intake of glucose it is stored in the liver and muscles of the body.

Vasopressin the antidiuretic hormone secreted by the posterior lobe of the pituitary gland stimulates the re-absorption of water. Other factors, such as aldosterone, a corticoid from the suprarenal gland has a definite effect upon the fluid and salts passed.

Urine, the excretion of the kidneys is a straw colored fluid containing variable amounts of dissolved sub-stances. Its specific gravity varies from 1008 to 1030, depending upon the dissolved salts. Urine is acid in reaction when passed but is converted upon standing to alkaline. Beside the inorganic salts of sodium, potassium, and calcium, urine contains the products of protein metabolism such as urea, creatinine, uric acid, etc.

S. 218 FUNCTIONS OF THE URINARY SYSTEM SUMMARIZED

The kidneys have several important functions, and renal failure would cause death.
*The kidneys secrete urine that contains waste products of body activity dissolved in water.
*The kidneys regulate the fluid content of the blood, and indirectly the quantity of intracellular fluid in body tissues. This is accomplished by increasing or decreas-ing the water filtered out.
*The kidneys also regulate the concentration of the various electrolytes, etc. (solutes) dissolved in the blood; sodium, potassium, glucose, etc.
*Renin, under certain circumstances, is secreted by the kidneys. This causes an elevation of the blood pressure. Renin is secreted when the blood supply to the kidneys is decreased by injury or disease. It is converted into a vasoconstrictor that results in a decrease in the size of capillaries of the body thus raising the blood pressure. Do not confuse this term renin with rennin or rennet secreted by the stomach to curdle milk.

The nephrons — functional units of the kidneys.
*The glomeruli filter blood plasma (minus blood proteins) from blood capillaries into the glomerular capsules, and then to the renal tubules.
*The renal tubules reabsorb water and some solutes back into blood capillaries, e.g. glucose.
*The renal tubules secrete selectively some waste products such as urea, uric acid, etc. to form part of the urine.

***The renal calyces and pelves** are passages from the kidneys to the ureters to convey urine.

***The ureters** are tubes that convey urine from the kidneys to the urinary bladder by peristalsis.
*The urinary bladder is a reservoir for urine.
*It is a muscular organ that contracts when directed to do so to convey urine into the urethra. The bladder and urethral sphincters relax during this elimination.

***The urethra** is a passage to discharge urine from the bladder and convey it out of the body.
*In the male, the urethra is also a passage to carry spermatozoa into the female vagina during intercourse.

S. 219 CONGENITAL ANOMALIES OF URINARY SYSTEM

Agenesis — absence of a kidney
Microkidney — a small kidney.
Horseshoe kidney — the two kidneys are joined across the median line, usually at their lower ends, and the pelves may be on their lateral borders.
Polycystic kidneys — large cysts within the kidneys.

Double kidney, double renal pelvis, double ureter. The superior major calyx may have a separate pelvis, and a partial or completely separate ureter.
Ectopic kidney — (G. ektopos = out of place) not in the normal position, often in the pelvis.
Exstrophy — the bladder may lie outside of the abdomen, and there may be no anterior abdominal wall.
Hypospadias — the urethra opens on the inferior surface of the penis instead of at its end.
Epispadias — the urethra opens upon the upper surface of the penis.

S. 220 SOME PATHOLOGICAL CONDITIONS DEFINED

Anuria (an = without) — without urine, the kidneys are not secreting urine.

Urinary retention — the kidneys are secreting urine but it is retained in the bladder.

Uremia — an accumulation of poisonous waste products in the blood and body tissues because the kidneys are not secreting them.

Calculus — one or more stones may form in the kidney or bladder. Many urinary calculi are opaque since they contain calcium. A calculus may obstruct a calyx, renal pelvis, ureter, or the flow of urine from the bladder.

Staghorn calculus — a calculus conforming in shape to a renal pelvis and calyces, like the horn of a deer in shape.

Cystitis — inflammation of the bladder.

Pyelitis — inflammation of a renal pelvis.

Perinephritic abscess — an abscess about a kidney.

Ptosis — a kidney located below its normal level.

Polyuria — an excess of urine is secreted.

Extravasation of urine — an escape of urine from a kidney, ureter, bladder, or urethra following an injury, etc.

Hydronephrosis or pyelonephrosis — dilatation of a renal pelvis and its calyces as a result of some obstruc-tion in the renal pelvis or ureter, e.g. a calculus blocking a ureter.

S. 221 APPLICATION TO RADIOGRAPHY

The facts listed below have a direct bearing upon radiography of the urinary system:

1. Since the kidneys lie in the posterior abdomen there will be less distortion of their images if the anteroposterior position is used for radiography.

2. Calculi within the kidney or renal pelvis will lie in the posterior abdomen, while calculi in the gall bladder will lie in the anterior abdomen. Lateral views may therefore help to determine the location in a puzzling case.

3. The kidneys are solid organs except for the calycs and pelves. In radiographs of the abdomen the calyces and pelves will not be apparent as separate images, but the mass of the kidney will show.

4. The calyces, pelves, ureters, and bladder will not have separate images unless filled with some opaque material.

Because radiographs of the urinary system are used to demonstrate the various structures brief descriptions of the procedures are given here.

Urography — a procedure used to outline the calyces, pelves, ureters and bladder by filling them with an opaque medium and taking radiographs.
Two methods are used:
Intravenous urography or excretory urography consists of an intravenous injection of an opaque medium into a vein, to be excreted by the kidneys.
Retrograde urography consists of an instillation of an opaque medium by a urologist into the ureters through small catheters introduced through the urethra and bladder into the ureters. (retrograde = backwards)
Scout film — a film of the abdomen before the procedure is begun.
Pyelography, intravenous — excretory, or retrograde are used as synonymous with urography.
Cystography is the instillation of an opaque medium or of air into the urinary bladder.
Urethrography consists of the instillation of an opaque medium into the urethra, or the expelling of a medium from the bladder into the urethra and radiography during the procedure.
In **minute interval** or minute sequence urography films are exposed at minute intervals following an injection of the medium.
The psoas muscle shadows are utilized in examinations of the urinary system to asses the quality of the radiographs. In satisfactory films the lateral margins of these muscles should be clearly outlined.

S. 222 ANATOMICAL TERMS — THE URINARY SYSTEM

Kidney; Gross anatomy
ren = kidney
renal = adjective
nephros = kidney
nephritic = adjective
renal capsule
hilum (hilus)
hilar = adjective
renal sinus
renal pedicle
renal segments
superior (upper) extremity
inferior (lower) extremity

Kidney, microscopic
cortex & renal columns
medulla
nephron
glomerulus
afferent vessel or arteriole
efferent vessel or arteriole
golmerular capsule or Bowmen's capsule
renal corpuscle
renal tubule
proximal convoluted tubule (part)
straight part & loop of Henle
distal convoluted part (tubule)
collecting tubule
renal pyramid
renal papilla
renal columns

Ureter
abdominal part
pelvic part
calyx (NA, calix)
calyces (calices)
calyceal — adj.
caliceal — (NA)
major calyx
minor calyx

Urinary bladder
cyst = kystis = bladder
vesica = bladder
vesical = adj.
ureteral orifice (ostium)
internal urethral orifice (ostium)
bladder sphincter
trigon

Urethra
internal urethral orifice (ostium)
external urethral orifice (ostium)
female urethra
male urethra
prostatic part or prostatic urethra
membranous part or membranous urethra
spongy part, spongy urethra, penile part
urethral sphincter
meatus = a canal, not an opening

Blood vessels
renal artery
lobar (segmental) artery
interlobar arteries
arcuate arteries
interlobular arteries
afferent arterioles
glomeruli
efferent arterioles
peritubular capillaries
renal veins
tributaries as arteries

Nephron physiology
filtration
selective reabsorption
selective secretion
urine

NOTES

NOTES

THE REPRODUCTIVE SYSTEM

— FEMALE REPRODUCTIVE ORGANS —

Basic terms
- broad ligament | zygote ovum
- genital | puberty
- oogenesis | menstruation
- female gamete | menopause
- ovulation | sterility
- fertilization | castration

Ovary
- synonyms — oophoron - (oon = egg) / female gonad
- number
- size
- location
- structure — primary follicle / vesicular follicle / or Graafian follicle / - ova, estrogens / corpus luteum / = progesterone

Uterine tube
- synonyms — salpinx / fallopian tube
- number
- length
- origin
- termination
- openings — uterine opening; / abdominal or / pelvic opening;
- fimbriae

Uterus
- synonyms — metra / hystera
- size
- location
- parts — body & fundus; / cervix;
- cavities — uterine / cervical canal
- openings — isthmus or / internal os / uterine opening / or external os
- structure — peritoneum (serous) / or perimetrium / myometrium, muscular / endometrium, mucosa

Vagina
- synonym — colpos
- length
- location
- structure
- openings — uterine; / in hymen;

External genitals
- perineum
- labia — minora, / majora,
- mons pubis
- clitoris
- vestibule

Blood vessels
- uterine artery R or L
- ovarian artery R or L
- uterine vein, R or L
- ovarian vein, R or L

Female breast
- synonyms — mastos / mamma
- structure — nipple, areola / ducts, -lactiferous / lobes, lobules

— MALE REPRODUCTIVE ORGANS —

Basic terms
- spermatozoon | spermatogenesis
- genital | semen
- genitourinary | male gamete
- urogenital | sterilization
- puberty | castration
- zygote | fertility

Testis
- synonyms — testicle / male gonad / orchis
- number
- size
- location — scrotum
- structure — seminiferous / tubule / = spermatozoa / interstitial / cells / = androgens / efferent ducts

Glands

Seminal vesicle
- number
- size
- location
- structure — single much / coiled tubule;
- duct — duct of / seminal vesicle
- secretion

Prostate gland
- number | structure
- size | ducts
- location | secretion

Bulbourethral gland
- number | duct
- location | secretion

Ducts

Epididymis
- number | length
- location | structure
- origin | function
- termination

deferent duct or vas deferens
- number | termination
- origin | length
- course | function

Ejaculatory duct
- number | origin
- location | termination

Duct of seminal vesicle
- origin
- termination
- function

Spermatic cord
- number | termination
- structure — vas, B V. Nn.
- origin

Penis
- erectile tissue
- structure — skin — foreskin / corpora cavernosa / corpus spongiosum / urethra
- functions — excretion of urine / reproduction

17. THE FEMALE REPRODUCTIVE SYSTEM

The organs of the female reproductive system are concerned with the propagation of the human race by:
1. The production of the female reproductive cell, the ovum.
2. The provision of passages by which the sperm and ovum may meet.
3. The provision of a cavity within which the fertilized ovum may develop and mature.
4. The production and secretion of female sex hormones.

S. 223 A LIST OF THE FEMALE REPRODUCTIVE ORGANS F. 17-1, 2

1. The right and left ovaries, a pair
2. Tbe right and left uterine tubes, a pair
3. The uterus
4. The vagina
5. The external genitals:
   (1) the labia
   (2) the mons pubis
   (3) the clitoris
   (4) the vestibule and glands

S. 224 DEFINITIONS OF SOME BASIC TERMS

1. **Genital** = reproductive, e.g. genital organs.

2. **The ovary** = the organ that produces ova (eggs).

3. **Female gonad** = an alternate name for the ovary. Note — male gonad = the testis.

4. **The ovum** = the female reproductive cell, formed in the ovary — female sex cell, egg, seed; pl. ova.

   Note — **spermatozoon** = male reproductive cell, formed in the testis — male sex cell, sperm, seed; pl. spermatozoa.

5. **Gamete** = a reproductive cell, male or female, so — female gamete, male gamete.

6. **Oogenesis** = the production of ova, by the ovary, and the several miotic divisions necessary to produce a mature ovum.

7. **Ovulation** = the expulsion of an ovum from the ovary into the pelvic cavity. This occurs about once every 28 days, in the human female, during the productive period of life.

8. **Fertilization** = the union of a female ovum and a male spermatozoon, the beginning of an embryo.

9. **Zygote** = a fertilized ovum.

10. **Menstruation** = the discharge of blood and the lining membrane of the uterus (womb), occurring about every 28 days, and lasting about 4 days.

11. **The menopause** = the cessation of the periodic menstruation and production of mature ova by the female, occurring at about 45 to 50 years of age.

12. **Puberty** = the beginning of ovulation and of menstruation in the female, with the development of secondary sexual characteristics. The age of onset of puberty varies from 12 to 16.

13. **The broad ligament,** in the female is a fold of peritoneum that covers the floor of the pelvis. It crosses the pelvis from one side to the other. The uterus, uterine tubes, and their vessels lie between the two layers as if they were pushed up against the peritoneum covering the floor of the pelvis from below. These structures are covered by this peritoneal fold, a serous covering. The ovaries lie posterior to this ligament, and are covered by the posterior layer. The broad ligament divides the pelvis into an anterior part containing the bladder, and a posterior part containing the rectum, with the uterus within the fold. F. 17-1.

S. 225 DETAILED STUDY OF FEMALE REPRODUCTIVE ORGANS

1. THE OVARIES F. 17-1, 2, 8

s. ovary, pl. ovaries, adj. ovarian; the female gonads; G. oophoron — egg bearer — forms eggs (ova).

The ovaries, a pair, right and left are two flattened oval organs that lie one in each side of the pelvis, lateral to the uterus. Each ovary measures about 4 cm or 1.6 inches in length. Each lies in a pouchlike sac of peritoneum formed from the posterior layer of the broad ligament.

The germinal epithelium of the ovary is this covering of modified peritoneum, derived from the broad ligament as stated above.

The mesovarium is the mesentery of the ovary and consists of the two layers of peritoneum that pass from the broad ligament to encircle the ovary. The ovarian blood vessels and nerves pass between these two layers to reach the ovary.

Within each ovary there is a connective tissue stroma that contains many nest of cells — ovarian follicles. The ovary produces ova, the female reproductive cells, as well as gonadal hormones.

The ovum is a globular cell about 0.2 mm or 1|125 inch in diameter with a cell membrane, nucleus and at maturity containing 23 single chromosomes.

A primary ovarian follicle is a nest of cells consisting of an ovum, the sex cell, surrounded by a layer of other cells. The ovaries at birth have from 400,000 to 500,000 of these cell nests. These follicles lie dormant until puberty.

A vesicular ovarian follicle or Graafian follicle is a primary follicle that has undergone development. This development begins in some of the follicles with the approach of puberty. It is stimulated by hormones from the pituitary gland. Only one follicle reaches full development each month, the others disintegrate. During the productive period of life from 400 to 500 of the primary follicles form vesicular follicles with their ova.

During development the layer of cells surrounding an ovum in a primary follicle forms many layers. A cavity containing fluid develops in the cell mass. The ovum with a few layers of cells surrounding it remains attached to the wall at one point. As the follicle develops it approaches the covering of the ovary. At ovulation the wall of the follicle and wall of the ovary rupture and the ovum is expelled into the pelvic cavity. The ovum finds its way into the unterine tube, and descends through this tube into the uterus. Fertilization if it takes place usually occurs in the uterine tube.
L. vesicula = a vesicle — a small sac with fluid.

Ovulation is the expulsion of a ripe ovum into the pelvic cavity.

A corpus luteum is a Graafian follicle after the ovum has been expelled. Initially a small hemorrhage occurs and a new type of cell develops within the follicle. If fertilization does not occur the corpus luteum degenerates. If fertilization does occur it continues to develop. It produces a hormone called **progesterone.**

The vesicular follicles manufacture hormones called estrogens. These along with progesterone cause some changes in the lining membrane (endometrium) of the uterus in preparation for implantation of the ovum if it is fertilized. If conception occurs these hormones also cause enlargement of the breasts and other changes incidental to pregnancy. See chapter on Endocrine Glands.

With the approach of puberty there is a marked increase in the secretions from the ovary. Over a period of time these result in changes in the female. The reproductive organs enlarge, hair appears in the pubic region, the breasts enlarge, the figure matures with deposits of fat about the hips. These are the secondary sexual characteristics. At the menopause ovulation ceases, menstruation stops, hormone production decreases, and the sex organs become smaller.

2. THE UTERINE TUBES   F. 17-1, 2, 4

G. salpinx = a trumpet — uterine tube, pl. salpinges, adj. salpingeal; also called fallopian tube or oviduct.

The right and left uterine tubes are two hollow tubes that pass, one from each upper angle of the uterus, to end close to the corresponding ovary. Each tube is about 10 cm (4 inches) long, and is open at both ends. The lateral end of the tube is enlarged and surrounding its opening there are fingerlike processes called fimbriae — hence fimbriated end. One of these fimbriae is attached to the ovary.

Openings of uterine tube   F. 17-4

The abdominal opening (ostium) of the uterine tube is at its lateral or fimbriated end. It is the only opening into the abdominopelvic cavity from the exterior of the body.

The opening of the uterine tube into the uterus is at its medial end, at the upper lateral border of the uterus.
   Each tube is covered by the peritoneum of the broad ligament, similar to the peritoneum covering the small intestine. Besides this outer serous layer there is a layer of visceral muscle, and a lining of epithelium.
   The function of the uterine tube is to convey the ovum that has been expelled from the ovary to the cavity of the uterus. A spermatozoon may meet an ovum in the uterine tube and fertilize it. If fertilized the ovum passes along the tube to the uterus to become implanted in the uterine wall and develop. If fertilization does not occur the ovum passes through the uterus into the vagina and out — the female menstruates.

3. THE UTERUS   F. 17-1, etc.

L. uterus = womb, pl. uteri, adj. uterine;
G.   metra = uterus, hence endometrium, myometrium;
G.   hystera = uterus, hence hysterosalpingogram.
   The uterus is a flattened pear shaped hollow muscular organ measuring about 7.5 x 5 x 2.7 cm or 3 x 2 x 1 inches in size. It lies within the fold of the broad ligament as if it had been pushed into this ligament from below. Peritoneum therefore covers its anterior and posterior surfaces and upper margin, and forms an outer serous covering. The uterus lies posterior to and above the urinary bladder and in front of the rectum.

The **covering** walls (F. 17-6) of the uterus include:
(1) The **endometrium,** the lining membrane; endo = within + metra = uterus i.e. within the uterus.
(2) The **myometrium,** the muscular layer of visceral muscle; mys — myo = muscle + metra.
(3) The **peritoneum** or outside serous covering, the peritoneum of the broad ligament, covering the anterior and posterior surfaces and upper margin. This layer is also called the **perimetrium.**

Parts of the uterus: (F. 17-2) a body and a cervix;

The body of the uterus is its upper larger part, above the smaller constricted cervix.
The fundus is that part of the body above the openings of the uterine tubes
The cervix is the long narrow constricted lower 2.5 cm or 1 inch of the uterus, the neck. (Cervix, L. = neck)
The isthmus of the uterus is the narrow part at the junction of the body and cervix. The isthmus was formerly called the **internal os.**

Cavities: F. 17-2

The **uterine cavity** or cavity of the uterus is the space within the body. It is triangular in shape when viewed from the front — see uterosalpingogram.

The **cervical canal** is the narrow passage within the cervix that opens at its upper end into the body of the uterus, and at its lower end into the vagina.

The inferior part of the cervix protrudes into the upper part of the vagina. When viewed from the vagina it looks like the human lips when puckered. This part of the cervix may be palpated by a finger introduced into the vagina.

Openings into the uterus: (F. 17-6)

The uterine opening (NA, ostium uteri) is the opening between the cervical canal and vagina. It was formerly called the **external os.**

The openings of the uterine tubes have been defined as the two openings between the uterus and the medial ends of the uterine tubes.

The **function of the uterus** is to provide an organ in which the fertilized ovum may develop. In preparation for the reception of a fertilized ovum the endometrium or lining membrane of the uterus becomes swollen and congested. The fertilized ovum becomes implanted in this membrane and develops. If fertilization does not occur this membrane is sloughed off and menstruation occurs. The lining membrane then regenerates and after a brief interval, when a further follicle is ripening, again becomes congested. The menstrual cycle has been divided into phases: the premenstrual, menstrual, postmenstrual and resting phases.

4. THE VAGINA   F. 17-1, 3

L. vagina = a sheath, pl, vaginae, adj. vaginal.
G. kolpos — colpos = vagina.
   The vagina is a passage from the cervix of the uterus to the outside of the body, at the perineum. This canal is about 9 cm or 3.6 inches in length. It extends down and anteriorly from the uterus to the perineum. It lies posterior to the bladder and urethra and is in front of the anal canal. It is lined by squamous epithelium and has visceral muscle fibers surrounding it. The walls are normally collapsed but are capable of marked distension.
   The cervix protrudes into the upper part of the vagina, with the uterine opening in the center of the ringlike

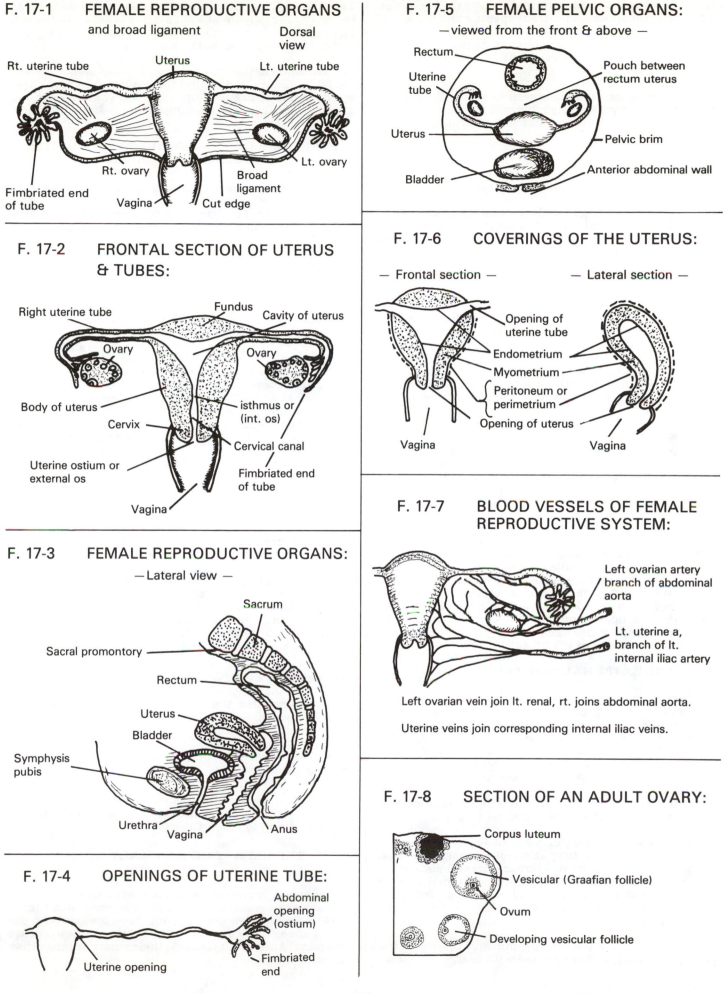

F. 17-1   FEMALE REPRODUCTIVE ORGANS
and broad ligament

Dorsal view

Rt. uterine tube
Uterus
Lt. uterine tube
Rt. ovary
Broad ligament
Lt. ovary
Fimbriated end of tube
Vagina
Cut edge

F. 17-2   FRONTAL SECTION OF UTERUS & TUBES:

Right uterine tube
Fundus
Cavity of uterus
Ovary
Ovary
Body of uterus
isthmus or (int. os)
Cervix
Cervical canal
Uterine ostium or external os
Fimbriated end of tube
Vagina

F. 17-3   FEMALE REPRODUCTIVE ORGANS:
— Lateral view —

Sacrum
Sacral promontory
Rectum
Uterus
Bladder
Symphysis pubis
Urethra
Vagina
Anus

F. 17-4   OPENINGS OF UTERINE TUBE:

Abdominal opening (ostium)
Fimbriated end
Uterine opening

F. 17-5   FEMALE PELVIC ORGANS:
— viewed from the front & above —

Rectum
Uterine tube
Uterus
Bladder
Pouch between rectum uterus
Pelvic brim
Anterior abdominal wall

F. 17-6   COVERINGS OF THE UTERUS:

— Frontal section —   — Lateral section —

Opening of uterine tube
Endometrium
Myometrium
Peritoneum or perimetrium
Opening of uterus
Vagina
Vagina

F. 17-7   BLOOD VESSELS OF FEMALE REPRODUCTIVE SYSTEM:

Left ovarian artery branch of abdominal aorta
Lt. uterine a, branch of lt. internal iliac artery

Left ovarian vein join lt. renal, rt. joins abdominal aorta.

Uterine veins join corresponding internal iliac veins.

F. 17-8   SECTION OF AN ADULT OVARY:

Corpus luteum
Vesicular (Graafian follicle)
Ovum
Developing vesicular follicle

cervix. The vagina opens externally upon the perineum, posterior to the external urethral opening and in front of the anus.

The hymen is a membrane that partly closes the opening in the virgin, usually with a small hole in it, which enlarges with intercourse.

Imperforate hymen is a condition in which there is no opening in the hymen. In such cases the young female menstruates into the vagina, and no menstrual discharge occurs.

The functions of the vagina are:
— to accommodate the penis during intercourse, and to receive spermatozoa as they are discharged from the male urethra;
— to serve as a passage for the menstrual discharge;
— to form a birth canal through which the mature fetus is expelled during childbirth, at which time it becomes grossly stretched.

S. 226 BLOOD VESSELS OF THE UTERUS AND OVARIES F. 17-7

The right and left uterine arteries are branches of the internal iliac arteries reaching the lateral margins of the uterus by passing between the two layers of the broad ligament.

The right and left uterine veins drain the uterus, and empty into the internal iliac veins.

The **right and left ovarian arteries**, branches of the abdominal aorta, below the origins of the renal arteries extend down along the posterior abdominal wall to the pelvis. They pass between the two layers of the broad ligament to reach the ovaries, tubes, and uterus.

Right and left ovarian veins drain blood from the ovaries. They pass up in a reverse direction to the ovarian arteries. The right joins the inferior vena cava while the left joins the left renal vein.

In the female a passage exists from the exterior through the cavities of the vagina, uterus, and uterine tubes into the pelvic cavity. Microorganisms sometimes enter the body by this route, and may cause peritonitis.

The uterine tubes may be obstructed, so that neither ova nor spermatozoa can pass through them. Pregnancy cannot occur. A blockage may be demonstrated by instilling some opaque medium through the vagina into the uterus and tubes and taking radiographs to demonstrate patency of the tubes.

S. 227 THE EXTERNAL GENITALS

The perineum is the area lying between the medial surfaces of the upper parts of the thighs, and extending from the pubic arch anteriorly to the tip of the coccyx posteriorly. When the thighs are abducted a four sided area may be mapped out. It conforms to the pelvic outlet. In the female the external genitals occupy this area. The openings of the urethra, vagina, and anus are located in the median line here.

The external genitals include the labia majora and minora, the clitoris, vestibule, and mons pubis.

The labia are two pairs of folds of the skin that run anteroposteriorly from the pubic arch almost to the anus. The labia majora are the two larger lateral folds and are covered with hair. The labia minora are two smaller folds that lie medial to the larger labia majora, and meet anteriorly at the clitoris, and posteriorly behind the vaginal opening.

The mons pubis is a rounded prominence in front of the symphysis pubis, and is covered with hair.

The clitoris is a small rounded prominence in the midline below the symphysis at the anterior junction of the two labia minora. It has erectile tissue similar to the penis and enlarges with sexual excitement.

The vestibule is the area between the medial margins of the labia minora and behind the clitoris. It is perforated by the external urethral opening, the vaginal opening, and the anus. It has the openings of the vestibular glands.

FUNCTIONS OF THE FEMALE REPRODUCTIVE SYSTEM SUMMARIZED

The ovaries produce ova, and secrete gonadal (sex) hormones.
*The vesicular, i.e. Graafian follicles, form ova, the female reproductive cells, and estrogens, female gonadal or sex hormones.
*The ripe ovum surrounded by a wall of cells is expelled from the follicle into the pelvic cavity — ovulation.
*The ruptured Graafian follicle becomes a corpus luteum, and secretes progesterone, a hormone, as well as some estrogens.
***Estrogens** stimulate growth of the young female, and development of secondary sexual characteristics as puberty approaches. They cause enlargement and maturity of the sexual organs, and help to maintain them as functioning organs.
*They are necessary for the development of the Graafian follicles and ova. They are also responsible for sexual behavior, and the sex urge.
*They cause thickening of the lining membrane (endometrium) of the uterus with an increase in blood capillaries in preparation for implantation of a fertilized ovum. They initiate some changes in the female breast.
***Progesterone** also causes further changes in the uterus and in breast tissue, in anticipation of an ovum being fertilized.
***The uterine tube** is a passage through which ova are conveyed from the pelvic cavity to the uterus.
***The uterus** is an organ in which the ovum, if fertilized, may develop into a fetus.
*Spermatozoa may pass up through the uterus to enter a uterine tube and fertilize an ovum.
*If fertilization does not occur the unfertilized ovum passes through the uterus into the vagina.
*Menstruation follows; the lining membrane of the uterus is sloughed off, and some bleeding takes place.

***The vagina** is a passage from the uterus to the perineum outside of the body.
*It accommodates the penis during intercourse, and receives semen containing spermatozoa at ejaculation.
*It is a passage through which menstrual discharge is expelled from the body.
*It is a birth canal through which the fetus must pass during childbirth.

The ovum, if not fertilized, must travel from the pelvis following ovulation through the uterine tube, the uterus, and the vagina to reach the exterior.

S. 228 THE BREAST OR MAMMARY GLAND. F. 17-9

L. mamma — breast, pl, mammae, adj. mammary.
G. mastos — breast, hence mastectomy and mastitis.
While both males and females have breasts only the female breasts are capable of producing milk to suckle the young. Animals that suckle their young are classified as mammals.

F. 17-9 THE FEMALE BREAST — THE MAMMARY GLAND

(1) Frontal view   (2) Secreting glands

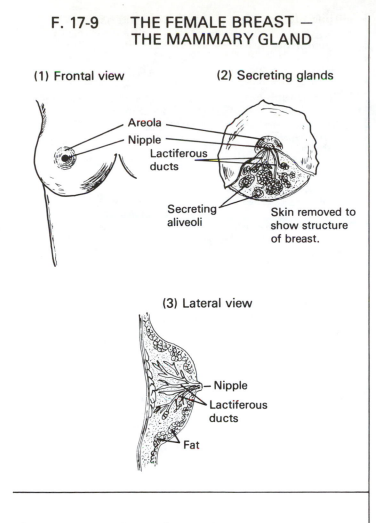

(3) Lateral view

The breast is made up of several structures:

(1) the nipple
(2) the areola
(3) lactiferous ducts
(4) lobes
(5) lobules, with secreting glands.

The nipple is a small rounded prominence that extends out from the breast at its center. L. papilla = nipple.

The areola is a circular area of skin surrounding the base of the nipple, colored pink in the young, but brown after a pregnancy has begun.

The lactiferous ducts are from 15 to 20 small tubes that pass from the lobes to open upon the surface of the nipple by minute openings. They store milk and convey it to the nipple as the infant suckles.

The lobes are the larger divisions of the breast, and are made up of many secreting lobules.

The lobules are the small secreting units, and are composed of compound exocrine glands that secrete milk that is carried by minute ducts to the lactiferous ducts. The milk consists of protein, fat, and a sugar lactose.

The breast contains many fat cells. It enlarges with the onset of puberty. During pregnancy the glandular tissue enlarges. When the infant is weaned the breast becomes smaller.
Breasts are frequently mapped out similar to the face of a clock. Facing the subject 12 o'clock would be a point directly above the nipple, three o'clock would be to the left of the nipple, etc.

BLOOD AND LYMPHATIC VESSELS OF THE BREAST

The internal thoracic (OT, internal mammary), intercostal and axillary arteries supply blood.
Internal thoracic, intercostal, and axillary veins drain blood from the breasts.

Lymphatic vessels pass from the breast in several directions. Some pass to the axilla, some pass up to nodes about the clavicle, some pass deep through the pectoral muscles and intercostal lymphatics, others cross the midline to the opposite breast, while others pass down towards the diaphragm. Infection of the breast, or breast cancer, may spread in any of these channels.

S. 229 SOME ANOMALIES AND PATHOLOGICAL TERMS

Imperforate hymen — no opening from the vagina to the exterior.

Bicornuate uterus — the uterus is partly or completely divided into two lateral parts. Each half has an opening into a uterine tube.

Endometritis — an inflammation of the lining membrane (endometrium) of the uterus.

Endometriosis — islands or implants of endometrium scattered on the surfaces of the abdominal and pelvic organs.

Cervicitis — an inflammation of the cervix of the uterus.

Ovarian cyst — a cyst (fluid filled sac) within the ovary. It may cause gross ovarian enlargement.

Ectopic pregnancy — a pregnancy developing outside of the uterus; may be abdominal or tubal. (G. ektopos = out of place)

Abdominal pregnancy — the ovum is fertilized before it enters the uterine tube and the embryo develops in the abdomen or pelvis. It does not descend into the uterus.

Tubal pregnancy — the ovum becomes fertilized and develops within the uterine tube instead of passing into the uterus.

S. 230 SOME OPERATIVE PROCEDURES

D & C = dilatation and curettage — the cervix of the uterus is dilated, a curette (spoon shaped instrument) is introduced through the cervix into the uterus, and the lining is scraped to get specimens for microscopic examination.

Hysterectomy — removal of the uterus at operation.

Oophorectomy — operative removal of an ovary.

Salpingectomy — operative removal of a uterine tube.

Mammectomy or mastectomy — operative removal of a breast.

S. 231 APPLICATION TO RADIOGRAPHY

The uterus, uterine tubes and ovaries do not have separate images visible by radiography of the pelvis. The cavities within these are also not demonstrable in routine radiographs. In order to outline these cavities they must be filled with some opaque medium first. In the case of a female who has been unable to become pregnant it may be necessary to determine whether the uterine tubes are blocked. If they are obstructed spermatozoa cannot ascend, nor can ova descend, so fertilization cannot occur.

Hysterosalpingography or uterosalpingography is a procedure whereby some opaque medium is instilled through a tube into the uterus. If the tubes are patent the medium will outline them and spill out into the pelvic cavity. The cavity of the uterus will also be made visible when radiographs are made.

Hysterography and **salpingography** are terms used to describe radiography of the parts separately. It is not possible however to outline the one without the other if the structures are normal.

Mammography is a procedure whereby the breast is outlined on a radiograph, using a special film and technique.

S. 232 ANATOMICAL TERMS — FEMALE REPRODUCTIVE

GENERAL TERMS — FEMALE

ovary
ovum
oogenesis
female gonad
ovulation
menstruation
puberty
menopause
fertilization
genital
broad ligament

OVARY

ovaries
ovarian
female gonad
oophoron or
egg bearer
ovum
ova
primary ovarian
  follicle
vesicular ovarian
  follicle or
Graafian follicle

corpus luteum
estrogens
progesterone

UTERINE TUBE

fallopian tube or
oviduct
salpinx
salpinges
salpingeal
fimbriae
fimbriated
opening of uterine
  tubes

UTERUS — womb

metra or
hystera
uteri
uterine
endometrium
myometrium
peritoneum or perimetrium
serous membrane

parts;
  body
  fundus
  cervix
  cervical

uterine cavity
cervical canal
isthmus or

internal os
uterine opening or
external os

VAGINA:

vaginae
vaginal
colpos (kolpos)
hymen

BLOOD VESSELS:

uterine aa
uterine vv
ovarian aa
ovarian vv

EXTERNAL GENITALS:

perineum
labia, majora, minora
clitoris
vestibule

FEMALE BREAST:

mamma, mammae
mastos
mammary
papilla or nipple
areola
lactiferous ducts
lobes
lobules

18. THE MALE REPRODUCTIVE SYSTEM

The student may recall that the terms genitourinary and urogenital are employed when the reproductive and urinary systems are described together. In the study detailed below the male reproductive system only is included.

S. 233 PARTS OF THE MALE REPRODUCTIVE SYSTEM

To simplify the study the male reproductive organs have been divided into 3 groups: the glands, the ducts, and the penis.

1. **Male reproductive glands**
   - (1) two testes or testicles, right & left
   - (2) the prostate gland
   - (3) two seminal vesicles, right & left
   - (4) two bulbourethral glands, right & left

2. **Ducts**
   - (1) right and left epididymides, epididymis
   - (2) right and left deferent ducts or vasa
   - (3) right and left ejaculatory ducts

3. **The penis**

S. 234 BASIC DEFINITIONS

The testis or testicle, also called the male gonad, is the organ that produces spermatozoa.

The spermatozoon is the male reproductive cell, the sperm, male sex cell, or seed, and is formed in the testis; pl. spermatozoa. (F. 18-5)

Spermatogenesis is the production of spermatozoa, and includes the several divisions necessary to produce mature spermatozoa, each with 23 chromosomes.

Semen is a thick white fluid containing spermatozoa and the secretions of the other male reproductive glands. It is introduced into the vagina at intercourse.

Puberty, in the male is the age at which the testes begin to produce mature spermatozoa, with an increase in the secretion of the male sex hormones, the androgens. These stimulate the development of the secondary sexual characteristics: the enlargement of the penis, and testes, the appearance of the whiskers, the change in voice, and a change in the attitude towards the female.

Genital = reproductive, hence genital organs.

Male gonad = the testis

Sterilization = the prevention of mature spermatozoa from being disharged from the male organs.
Castration = any procedure by which spermatozoa are prevented from forming. The testes may be removed, or radiation may be applied to them.

Fertility is the ability to produce offspring by a spermatozoon uniting with an ovum.

male gamete = male reproductive cell, spermatozoon

S. 235 DETAILED STUDY OF MALE REPRODUCTIVE GLANDS

(1) THE TESTES: F. 18-1, 2, 3
L. testis, pl. testes, adj. testicular
G. orchis = testis; hence orchidectomy, orchitis

The testes, right and left are the two glands in the male that produce spermatozoa, and the male sex or gonadal hormones — the androgens. They are also called the male gonads. They lie in a sac, the scrotum.
The scrotum is a sac that is suspended below the pubic arch, anterior to the upper thighs. It is covered by skin, and has a midline partition. One testis lies in each compartment.

Each testis is from 4 to 5 cm or 2 inches in length, and lies vertically, with its epididymis posterior to it. The testis has a large number of lobules or compartments. From one to three minute coiled tubules (seminiferous tubules) are located in each lobule. These tubules converge towards the upper end of the testis posteriorly, and several hundreds of them unite to form 15 to 20 larger tubules (efferent duct). These join together to form a single tube, the epididymis. tubule = a little tube

Each seminiferous tubule is lined with a membrane, the cells of which by division and redivision produce the mature spermatozoa. These are set free in the tubules and migrate to the epididymis.

Each spermatozoon consists of an expanded end (the head), a constricted neck, a short body and a long tail to propel the sperm. During the process of maturing the number of chromosomes is reduced to 23, one-half the original number. The parent of the mature sperm has 46 chromosomes, or 23 pairs, with a pair of sex chromosomes. One of these sex chromosomes is an X chromosome, and one a Y chromosome.

Upon division of this cell one of the spermatozoa produced will have an X sex chromosome, the other a Y sex chromosome. Millions of these spermatozoa are formed. (F. 18-5)

Androgens or male sex hormones including testosterone are secreted by interstitial cells within the testes and are absorbed by blood capillaries for distribution.

The testes develop in the posterior abdomen, behind the peritoneum. Each testis later descends into the scrotum, carrying with it the peritoneum that covers its anterior surface. This prolongation of peritoneum becomes closed, and pinched off. The peritoneum attached to the anterior surface of the testis forms a closed sac in contact with the testis. Fluid may accumulate within the sac and result in enlargement of the scrotum. Occasionally the prolongation of peritoneum pulled down with the testis does not close off, and forms a canal by which bowel, etc., may pass down into the scrotum producing an inguinal hernia. Inguen = groin; (see F. 18-6)

BLOOD VESSELS OF THE TESTES

The testicular arteries, right and left, branches of the abdominal aorta, given off below the origins of the renal arteries supply the testes. The testicular veins drain the testes, the right emptying directly into the inferior vena cava, the left into the left renal vein. Lymphatic vessels from the testes drain into abdominal lymph nodes.

(2) THE PROSTATE GLAND F. 18-1, 4

The prostate gland is a single globular gland that is located in the male pelvis below the urinary bladder. It lies below the internal urethral orifice, and the urethra passes through it, hence the name "prostatic part" or prostatic urethra. The prostate is composed of many secreting tubules that join together to form about 20 small ducts. These empty into the prostatic part of the urethra. They secrete a watery fluid that forms part of the semen. If a finger is inserted into the rectum the size, shape, hardness and nodularity of the prostate gland may be determined. In elderly males the prostate frequently

F. 18-1   PARTS OF THE MALE REPRODUCTIVE SYSTEM :   Lateral views—

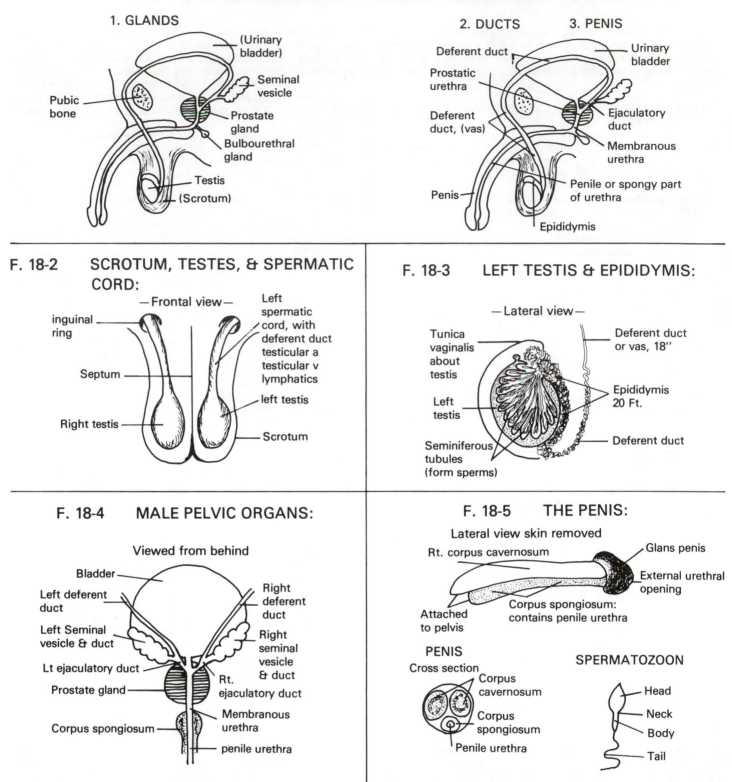

1. GLANDS

(Urinary bladder)
Seminal vesicle
Pubic bone
Prostate gland
Bulbourethral gland
Testis
(Scrotum)

2. DUCTS   3. PENIS

Deferent duct
Prostatic urethra
Deferent duct, (vas)
Penis
Urinary bladder
Ejaculatory duct
Membranous urethra
Penile or spongy part of urethra
Epididymis

F. 18-2   SCROTUM, TESTES, & SPERMATIC CORD:

—Frontal view—

inguinal ring
Septum
Right testis
Left spermatic cord, with deferent duct
testicular a
testicular v
lymphatics
left testis
Scrotum

F. 18-3   LEFT TESTIS & EPIDIDYMIS:

—Lateral view—

Tunica vaginalis about testis
Left testis
Seminiferous tubules (form sperms)
Deferent duct or vas, 18"
Epididymis 20 Ft.
Deferent duct

F. 18-4   MALE PELVIC ORGANS:

Viewed from behind

Bladder
Left deferent duct
Left Seminal vesicle & duct
Lt ejaculatory duct
Prostate gland
Corpus spongiosum
Right deferent duct
Right seminal vesicle & duct
Rt. ejaculatory duct
Membranous urethra
penile urethra

F. 18-5   THE PENIS:

Lateral view skin removed

Rt. corpus cavernosum
Attached to pelvis
Corpus spongiosum: contains penile urethra
Glans penis
External urethral opening

PENIS
Cross section

Corpus cavernosum
Corpus spongiosum
Penile urethra

SPERMATOZOON

Head
Neck
Body
Tail

---

enlarges, and by pressing upon the base of the bladder or the urethra may obstruct the free flow of urine at urination.

(3) THE SEMINAL VESICLES   F. 18-1, 4

The right and left seminal vesicles are two elongated noduar glands that lie obliquely in the pelvis posterior to the bladder. They are located above the prostate gland, one on each side of the midline, and in front of the rectum. Each gland is about 7.5 cm or 3 inches in length and consists of a single tubule that is coiled upon itself. The tubule secretes a part of the semen. Each seminal vesicle tubule ends in an excretory duct that joins the corresponding deferent duct to form the ejaculatory duct. This gland may also be palpated by a finger introduced into the rectum.

(4) THE BULBOURETHRAL GLANDS   F. 18-1

These small pea shaped glands lie one on each side of the membranous part of the urethra. Each ends in a duct that opens into part of the urethra, and their secretion forms part of the semen.

(1) THE EPIDIDYMIS — EPIDIDYMIDES F. 18-3

The epididymis lies posterior to the testis. It is composed of a single tube that is repeatedly coiled upon itself. If stretched out it would be about 20 feet long. It begins at the upper end of the testis by the union of the 15 or 20 efferent tubules formed by the minute tubules — the seminiferous tubules. The epididymis extends from the upper to the lower end of the testis. From here it is continued on as the deferent duct. There are two epididymides, a right and a left.

(2) THE DEFERENT DUCT F. 18-1 (2)

L. ductus deferens; vas deferens; often simply vas; (vas = vessel); also called seminal duct.

Each deferent duct is a continuation of the epididymis. It terminates by joining the excretory duct of the same side to form the ejaculatory duct. This tube pursues a roundabout course to get from the scrotum to the prostatic urethra. It first ascends behind the testis within the scrotum. It passes over the upper surface of the pubic bone to reach the lower abdomen. It enters the inguinal canal above the inguinal ligament through openings (rings) in the anterior abdominal muscles. It then descends along the lateral pelvic wall into the pelvis. Posterior to the bladder it lies beside the seminal vesicle, and joins the duct from this gland. There are right and left deferent ducts. (defero = to carry down).

(3) THE EJACULATORY DUCTS F. 18-1(2), 4

The ejaculatory duct, right or left is formed by the union of a deferent duct with the corresponding excretory duct from the seminal vesicle. This junction occurs posterior to the inferior part of the bladder, and above the prostate gland. The duct then passes into the prostate gland to open into the prostatic part of the urethra. The spermatozoa and secretion of the seminal vesicles and from the prostate gland and bulbourethral glands is expelled into the male urethra during intercourse.

Students should realize that from the testis spermatozoa must travel through the epididymis, deferent duct, and ejaculatory duct to reach the urethra.

THE PENIS F. 18-1, 5

L. penis; G. phallos.

The penis is the male organ of copulation. It is cylindrical and when flaccid hangs down below the pubes, and anterior to the upper thighs. It may be 15 cm or six inches in length. The penis consists of three cylinders of erectile tissue that extend through the length of the organ. These are bound together by fibrous tissue and the whole covered by skin. The urethra passes through the penis.

The corpora cavernosa are the two cylinders of erectile tissue that occupy the lateral and upper parts of the organ. Their posterior ends are attached to the pelvis while their anterior ends butt against the glans penis.

The corpus spongiosum is a single smaller cylinder of erectile tissue that extends from the membranous urethra to the end of the penis. It is centrally located. Its distal end is expanded to form a knoblike enlargement of the glans. The penile or spongy part of the urethra (third part) runs through the center of the corpus spongiosum, and opens at the gland by a slitlike orifice, the external urethral opening (orifice).

Erectile tissue is a tissue that when filled with blood enlarges and becomes firm. It is found in the penis,

female clitoris, and the nipple of the female breast. It is composed of blood spaces or sinuses with incomplete partitions between. It resembles a sponge. During sexual excitement the arteries feeding the spaces dilate, and fill the spaces with blood. The veins draining the organ constrict. The penis therefore becomes enlarged, firm and erect. During intercourse at the climax semen is expelled through the male urethra into the vagina. The penis is covered by skin, and the foreskin (a double layer that forms a fold) covers the glans.

The spermatic cord (F. 18-2) right or left, is a small ropelike structure that extends from the scrotum on each side up over the pubic bone to the anterior abdominal wall. It contains the deferent duct, the testicular artery and vein, and the lymphatics and nerves of the testis. It may be rolled between the examiner's thumb and finger. It is a minor surgical procedure to cut the deferent duct in the spermatic cord in order to sterilize a male subject.

FUNCTIONS OF THE MALE REPRODUCTIVE SYSTEM

SUMMARIZED

The testis produces spermatozoa, and secretes androgens, the male gonadal (sex) hormones.
*Seminiferous tubules in the testis form spermatozoa, the male reproductive cells.
*Androgens, i.e. testosterone, and androsterone, male gonadal hormones are secreted by interstitial cells in the testis.
*Androgens stimulate growth in the young male, and development of the secondary sex characteristics at puberty. They cause maturing of the sex organs, their maintenance in a functioning state, and the production of spermatozoa. They are responsible for sexual behavior and sexual drive.

***The epididymis** is a very tortuous passage that conveys spermatozoa from the upper posterior end of the testis to the commencement of the deferent duct at its lower end.

***The deferent duct** or vas deferens is a passage that conveys spermatozoa from the epididymis to the ejaculatory duct in the pelvis.

***The seminal vesicle** is a small gland that secretes part of the semen. It lies posterior to the lower part of the urinary bladder.
***The duct** of the seminal vesicle conveys the secretion of the seminal vesicle to the deferent duct to mix with spermatozoa.

***The ejaculatory duct** conveys spermatozoa and the secretion of the seminal vesicle to the prostatic urethra. It is formed by the union of the duct of the seminal vesicle and the deferent duct.

***The prostate gland,** below the urinary bladder secretes part of the semen.
Small prostatic ducts open into the prostatic urethra and convey prostatic secretion into the urethra to mix with spermatozoa.

***Bulbourethral glands** secrete part of the semen and have ducts opening into the urethra.

***The urethra** conveys semen, containing spermatozoa through the penis into the vagina at intercourse.

Semen contains spermatozoa as well as the secretions of the seminal vesicles, prostate gland, and bulbourethral glands. (L. semen = seed).

Spermatozoa must travel from the testis through the epididymis, deferent duct, ejaculatory duct and male urethra to reach the exterior.

S. 237 CONGENITAL ANOMALIES

Epispadias — external urethral opening on the upper i.e. dorsal surface of the penis, not at its end.

Hypospadias — the external urethral opening is on the under surface (ventral surface) of the penis.

Cryptorchidism — non-descent of the testes into the scrotum, they remain in the abdomen or in the anterior abdominal wall. Mature spermatozoa are not formed if this anomaly occurs.

S. 238 PATHOLOGICAL CONDITIONS & OPERATIONS

Epididymitis = an inflammation of the epididymis
Epididymectomy = removal of the epididymis
Orchitis = an inflammation of the testis
Orchidectomy = removal of a testis
Prostatism = an enlargement of the prostate gland that frequently occurs in older males. It is not a cancer. In may obstruct the urethra, resulting in retention of urine or incomplete emptying of the bladder.
Prostatectomy = removal of the prostate gland
Transurethral prostatectomy = pieces of prostatic tissue are removed through the urethra by a cystoscope with a special cutting blade.
Seminoma = a cancer of the tubules of the testis.
Vasectomy = sterilization of the male by the removal of a segment of the vas deferens (the deferent duct) where it lies in the spermatic cord close to the skin.
Circumcision = the removal of the fold of skin covering the glans penis (the foreskin or prepuce) in male children. This is done routinely on Hebrew male babies, and frequently in other males to promote cleanliness.

S. 239 ANATOMICAL TERMS
— MALE REPRODUCTIVE

male puberty	orchis
genitals — male	scrotum
urogenital	seminiferous tubules
genitourinary	efferent ducts
testis	interstitial cells
testes	spermatozoon
testicle	spermatozoa
testicular	spermatogenesis
male gonad	androgens

testosterone	seminal duct
androsterone	ejaculatory duct =
sterilization	deferent duct +
castration	excretory duct of
prostate gland	seminal vesicle
tubules	penis
ducts	erectile tissue
prostatic urethra	corpus cavernosum
seminal vesicles	corpora cavernosa
tubules	corpus spongiosum
excretory duct	glans
bulbourethral glands	foreskin
epididymis	spermatic cord
epididymides	inguinal ligament
deferent duct	inguinal ring
vas — vasa	inguinal canal
vas deferens	inguinal hernia

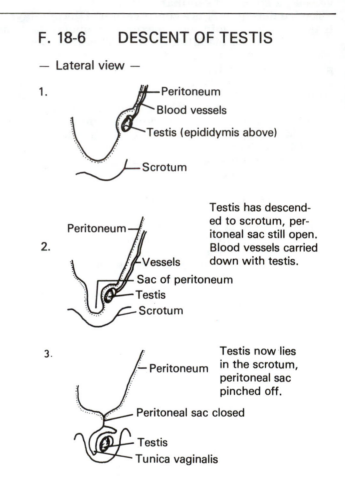

F. 18-6   DESCENT OF TESTIS

— Lateral view —

1.
— Peritoneum
— Blood vessels
— Testis (epididymis above)
— Scrotum

2.
Peritoneum —
— Vessels
— Sac of peritoneum
— Testis
— Scrotum

Testis has descended to scrotum, peritoneal sac still open. Blood vessels carried down with testis.

3.
— Peritoneum
— Peritoneal sac closed
— Testis
— Tunica vaginalis

Testis now lies in the scrotum, peritoneal sac pinched off.

19. INTRODUCTION TO EMBRYOLOGY

The story of the development of the human body is fascinating but somewhat complicated for the junior student. Only a few facts will be included in the present outline.

S. 240 DEFINITIONS FOR SOME BASIC TERMS

Embryology is a study of the development of the fertilized ovum from fertilization to birth.

Embryo is the term used to define the developing fertilized ovum to the end of the second month of intrauterine life.

Fetus, in the human refers to the developing fertilized ovum from the beginning of the third month of intra-uterine life until birth.

Ovulation has been defined as the expulsion of an ovum from a vesicular (Graafian) follicle and ovary into the pelvic cavity. This ovum then enters one of the uterine tubes and descends through it to the uterus. This takes about four or five days.

Fertilization is the union of an ovum and spermatozoon, resulting in a cell with 46 chromosomes, or 23 pairs, including 23 chromosomes from the ovum and 23 from the spermatozoon.

A gamete is the female or male mature reproductive cell, an ovum or spermatozoon, i.e. a female gamete, or a male gamete.

A zygote is the cell produced by the union of an ovum (female gamete) and a spermatozoon (male gamete) i.e. a fertilized ovum.

Oogenesis is the process whereby mature ova are produced. This is termed maturation. In the original germ cell there are 46 chromosomes, or 23 pairs — diploid, including a pair of "X" sex chromosomes. By repeated miotic cell division the number of chromosomes becomes reduced to 23 single chromosomes — haploid, including one "X" sex chromosome. The other cell from each division is called a polar body. It does not mature and is destroyed. This process begins within the ovary. The final division may occur in the uterine tube during descent of the ovum. The mature ovum is now ready for fertilization. (Oon = egg + genesis = the production).

Spermatogenesis is the process by which mature spermatozoa are produced. The germ cells in the seminiferous tubules of the testis contain 46 chromosomes or 23 pairs. One pair has two sex chromosomes: one an "X" sex chromosome, the other a "Y" chromosome. Maturation is accomplished by repeated miotic cell division. At the final division two mature spermatozoa are produced, each containing 23 single chromosomes — haploid. One of these will contain a single "X" sex chromosome, the other a single "Y" sex chromosome. Each one of these is capable of fertilizing an ovum. During intercourse 200 000 000 spermatozoa may be introduced into the vagina.

In an earlier chapter it was stated that fertilization of the ovum usually occurs in one of the uterine tubes. This ovum then descends through the tube into the uterus. it burrows its way into the endometrium, the lining membrane of the uterus, and becomes **implanted** in this tissue. For fertilization to occur the spermatozoon must pass from the vagina into the uterus, and out into one of the uterine tubes.

The intrauterine or neonatal period includes the time the embryo and fetus spend within the uterus, ten lunar months — (10 X 28) or 280 days. (a lunar month = 28 days). Within the uterus the ovum develops into an embryo and fetus and grows.

Segmentation or cleavage. The fertilized ovum divides and redivides repeatedly by mitotic cell division. It forms two, four, eight, sixteen, etc., cells and eventually a ball-like structure resembling a berry. This process is termed segmentation or cleavage, with formation of a ball. (F. 19-2)

Hollow ball of cells. The center of the ball becomes hollowed out and fills with fluid. The cells forming this ball become arranged into two groups. The outer layer forms a covering for the ball, and finally some of the fetal membranes and placenta. The inner layer, named the germinal disc, embryoblast, or inner cell mass lies at one end, forming a clump of cells. This inner cell mass forms the embryo and some fetal membranes. It divides into three primary or basic layers of cells, the ectoderm, mesoderm and entoderm (endoderm). F. 19-3

The ectoderm, the outer layer, by a process of repeated cell division forms a covering membrane that surrounds the other layers, to become the outer layer of skin. F. 19-4

In the dorsal area of this structure a groove appears running lengthwise along the length of the embryo. This groove deepens, and finally the sides of the groove meet posteriorly to form a tube. It becomes pinched off from the skin layer. This tube forms the brain and spinal cord. (Ecto = outside)

The entoderm, endoderm or inner layer, following a period of development including the formation of a yolk sac, forms a tube from the head end to the lower end of the embryo. This becomes the lining of the digestive tract. At first this tube is closed at the pharyngeal and anal ends. During further development the ends open. Very occasionally the anal end does not open, resulting in an imperforate anus. The larynxs, trachea, lungs, liver, pancreas, gall bladder and bile ducts bud off from the primitive digestive tract. (Ento = in, within).

The mesoderm, the middle layer develops between the two other layers, and separates into several parts. Two of these, one on either side of the brain and spinal cord form the skull and vertebral column, surrounding these structures. The skeletal muscles, and bony parts form from mesoderm. One layer grows around the entoderm of the digestive tract to form visceral muscle and connective tissue here. Mesoderm also forms the parts of the cardiovascular system.

To summarize, in simplest terms, the ectoderm grows to the head and tail ends, and around the embryo forming the epidermis and nervous system. Meanwhile the entoderm grows to form the lining of the digestive tract from head to tail inside the envelope of ectoderm. The mesoderm, sandwiched between entoderm and ectoderm grows to form muscle layers, connective tissue and bone.

S. 241 STRUCTURES FORMED FROM THE 3 BASIC LAYERS

1. **The Ectoderm**
   the skin, hair, nails
   the central nervous system, brain and cord
   enamel of the teeth
   lining of mouth, nasal cavities, etc.

2. **The Mesoderm**
   skeletal and visceral muscle
   connective tissue & derivatives;
       fat, cartilage, bone, fibrous and
       elastic tissues
   blood vessels and the heart
   parts of the urinary system

3. **The Entoderm**
   lining membranes of
   digestive tract,
   trachea and lungs
   liver, pancreas, gall bladder, bile ducts
   urinary bladder, etc.

The amniotic cavity is a sac filled with fluid that surrounds the developing embryo and fetus, and in which the embryo floats. The fluid acts as a cushion. This cavity is lined by a membrane called the amnion. During childbirth the sac ruptures (the membranes rupture) and fluid is expelled through the vagina. The amnion is expelled along with the placenta, (the afterbirth) of the fetus. The fetus is in contact with the uterus surrounding the sac by the umbilical cord and placenta.

THE SEX OF THE EMBRYO

A female embryo forms when any ovum with its "X" sex chromosome is fertilized by a spermatozoon that contains an "X" sex chromosome. (X + X)

A male embryo forms when any ovum with its "X" chromosome is fertilized by a spermatozoon that contains a "Y" sex chromosome. (X + Y)

THE FORMATION OF TWINS

Fraternal twins form if two ova mature at the same time and are both fertilized. Two separate sacs form. The twins may be of the same or of opposite sexes. They wiil have a familial resemblance only.

Identical twins form if a single ovum is fertilized by a single spermatozoon, and if, during segmentation the ball of cells divides into two identical parts. Each part develops into an embryo. A single sac forms, the embryos are of the same sex, and are identical in appearance. Siamese twins occur when the separation is incomplete. Triplets form by a division into three identical embryos.

Haploid is a term that is sometimes used in Genetics. It means simple or single, e.g. the mature ovum or spermatozoon has 23 **single** chromosomes — haploid.

Diploid refers to a doubling up, such as a fertilized ovum (zygote) with its 23 **pairs** of chromosomes — diploid. Similarly a somatic cell is diploid as it has 23 pairs, or 46 chromosomes.

S. 242   ANATOMICAL TERMS — EMBRYOLOGY

oogenesis	segmentation
spermatogenesis	cleavage
ovulation	hollow balls of cells
fertilization	germinal disc, or
haploid	embryoblast, or
diploid	inner cell mass
implantation	ectoderm
maturation	mesoderm
embryology	entoderm
embryo	amnion
fetus	amniotic cavity

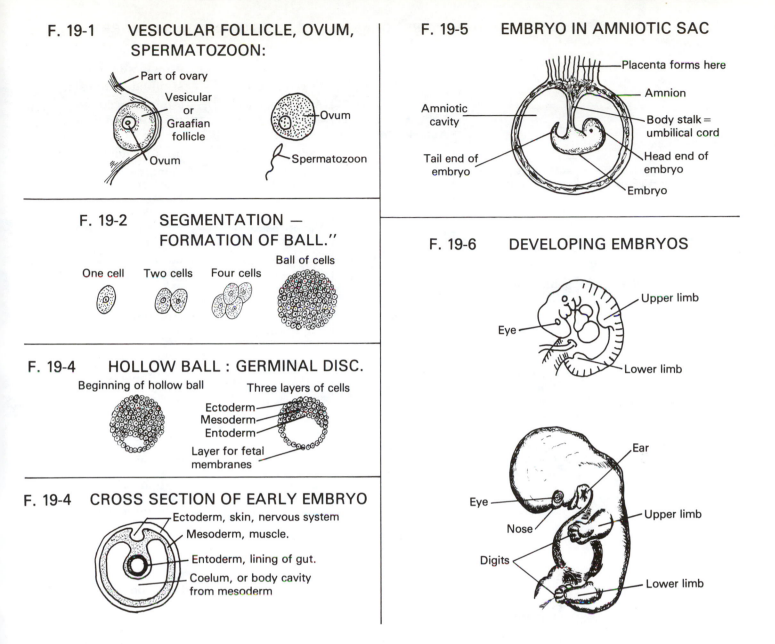

F. 19-1   VESICULAR FOLLICLE, OVUM, SPERMATOZOON:

Part of ovary
Vesicular or Graafian follicle
Ovum
Ovum
Spermatozoon

F. 19-2   SEGMENTATION — FORMATION OF BALL."

One cell   Two cells   Four cells   Ball of cells

F. 19-4   HOLLOW BALL : GERMINAL DISC.

Beginning of hollow ball   Three layers of cells
Ectoderm
Mesoderm
Entoderm
Layer for fetal membranes

F. 19-4   CROSS SECTION OF EARLY EMBRYO

Ectoderm, skin, nervous system
Mesoderm, muscle.
Entoderm, lining of gut.
Coelum, or body cavity from mesoderm

F. 19-5   EMBRYO IN AMNIOTIC SAC

Placenta forms here
Amnion
Body stalk = umbilical cord
Head end of embryo
Embryo
Amniotic cavity
Tail end of embryo

F. 19-6   DEVELOPING EMBRYOS

Upper limb
Eye
Lower limb

Ear
Eye
Nose
Digits
Upper limb
Lower limb

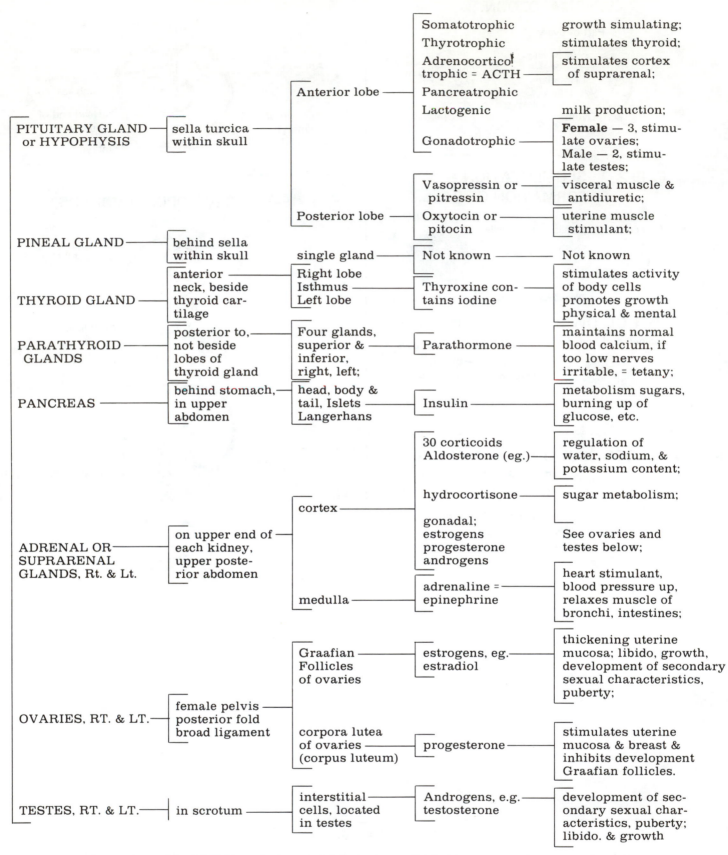

Gland	Location	Structure	Hormone	Function
PITUITARY GLAND or HYPOPHYSIS	sella turcica within skull	Anterior lobe	Somatotrophic	growth simulating;
			Thyrotrophic	stimulates thyroid;
			Adrenocorticotrophic = ACTH	stimulates cortex of suprarenal;
			Pancreatrophic	
			Lactogenic	milk production;
			Gonadotrophic	**Female** — 3, stimulate ovaries; Male — 2, stimulate testes;
		Posterior lobe	Vasopressin or pitressin	visceral muscle & antidiuretic;
			Oxytocin or pitocin	uterine muscle stimulant;
PINEAL GLAND	behind sella within skull	single gland	Not known	Not known
THYROID GLAND	anterior neck, beside thyroid cartilage	Right lobe Isthmus Left lobe	Thyroxine contains iodine	stimulates activity of body cells promotes growth physical & mental
PARATHYROID GLANDS	posterior to, not beside lobes of thyroid gland	Four glands, superior & inferior, right, left;	Parathormone	maintains normal blood calcium, if too low nerves irritable, = tetany;
PANCREAS	behind stomach, in upper abdomen	head, body & tail, Islets Langerhans	Insulin	metabolism sugars, burning up of glucose, etc.
ADRENAL OR SUPRARENAL GLANDS, Rt. & Lt.	on upper end of each kidney, upper posterior abdomen	cortex	30 corticoids Aldosterone (eg.)	regulation of water, sodium, & potassium content;
			hydrocortisone	sugar metabolism;
			gonadal; estrogens progesterone androgens	See ovaries and testes below;
		medulla	adrenaline = epinephrine	heart stimulant, blood pressure up, relaxes muscle of bronchi, intestines;
OVARIES, RT. & LT.	female pelvis posterior fold broad ligament	Graafian Follicles of ovaries	estrogens, eg. estradiol	thickening uterine mucosa; libido, growth, development of secondary sexual characteristics, puberty;
		corpora lutea of ovaries (corpus luteum)	progesterone	stimulates uterine mucosa & breast & inhibits development Graafian follicles.
TESTES, RT. & LT.	in scrotum	interstitial cells, located in testes	Androgens, e.g. testosterone	development of secondary sexual characteristics, puberty; libido. & growth

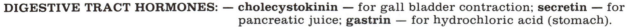

DIGESTIVE TRACT HORMONES: — **cholecystokinin** — for gall bladder contraction; **secretin** — for pancreatic juice; **gastrin** — for hydrochloric acid (stomach).

20. THE ENDOCRINE GLANDS

S. 243   GENERAL INFORMATION

Secretory glands were defined in section 45, following the study of the skin. Sebaceous and Sudoriferous glands had been described and it seemed logical to follow with definitions for secretory glands in general. These were classified as:
1. Exocrine glands
2. Endocrine glands

Exocrine glands were defined as having ducts or tubes by which the secretions manufactured by them are discharged upon the skin surface or into the cavities of hollow organs. Some of them form enzymes. Exocrine glands include sweat, sudoriferous, salivary glands, those in the lining membranes of the parts of the digestive system, the female breast, and many others.

Endocrine glands, ductless glands or glands of internal secretion, are nests or clumps of cells having no ducts. They are called ductless glands. They take up raw materials delivered to them by blood capillaries, and from these they manufacture new substances called **hormones**. These products are absorbed directly into blood capillaries to be conveyed to all parts of the body. Because of the absence of secreting ducts they are also named glands of internal secretion. (endo = within + krine = to separate).

A hormone is a secretion manufactured by an endocrine gland and secreted directly into blood capillaries. No ducts are involved. Each hormone is delivered to all body tissues. A hormone may activate the cells of body tissues to perform their functions, or may stimulate body cells or other glands to increase their activity. While there are several hormones secreted by recognized endocrine glands, others are secreted by special cells in the lining membranes of the digestive tract. These will be briefly described in Section 247.

Double glands — these have two distinct parts, each secreting its own hormones. For example, the cortex and medulla of the suprarenal glands, or the anterior and posterior lobes of the pituitary gland.

Mixed glands — some glands have an endocrine part to secrete one or more hormones, and an exocrine part to secrete enzymes, e.g. the pancreas.

Glands with multiple functions — some glands secrete more than one hormone, for example, the anterior lobe of the pituitary gland secretes several hormones.

Interrelationship of glands — a hormone secreted by one gland such as the anterior lobe of the pituitary may stimulate some other gland to produce, or to increase its secretion. If this gland, for example the thyroid, produces an excessive amount of its hormones, this excess will cause the anterior pituitary to suppress its secretion.

Proof of existence of endocrine glands.

By observing the behavior of subjects:
1. With an absence or underdevelopment of a particular gland.

2. In whom disease has destroyed an endocrine gland.

3. With a tumor of an endocrine gland, producing an excess of its hormone.

4. From whom an endocrine gland has been removed, followed by an implantation of a similar gland, or by administration of the hormone secreted by that gland.

Some hormones have been isolated in pure form from animal glands, and some have been produced chemically (synthetically).

Some hormones have been found to be effective when taken by mouth, but others are digested if given orally, so must be injected.

S. 244   METABOLISM

This term is discussed here because the hormones secreted by endocrine glands affect metabolism. Metabolism is a term used to include the chemical changes that occur in body cells as a result of cellular activity. It includes all the wearing down, and building up activities of each cell. On the one hand cellular constituents and food products are broken down, as during muscular activity, and in the production of glandular secretions, etc. On the other hand these substances are used to build up, repair, and replace used up constituents. Protein, fat, and carbohydrate metabolism may be involved. The rate of metabolism of a subject at rest physically and mentally is lower than that of the same person during muscular activity, digestion of food, etc. Basal metabolism is that of a subject at rest. Because all activity of body cells requires oxygen in proportion to the degree of activity the rate of metabolism may be measured by measuring the amount of oxygen a subject uses. Other methods are also utilized.

F. 20-1   LOCATION OF ENDOCRINE GLANDS:

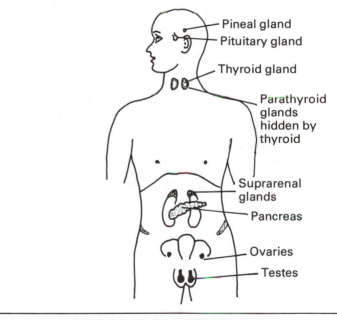

- Pineal gland
- Pituitary gland
- Thyroid gland
- Parathyroid glands hidden by thyroid
- Suprarenal glands
- Pancreas
- Ovaries
- Testes

S. 245   A LIST OF ENDOCRINE GLANDS
F. 20-1

Listed from the head end downwards.

1. Pituitary gland or hypophysis (one only)
   (1) Posterior lobe or neurohypophysis
   (2) anterior lobe or adenohypophysis
2. Pineal gland (one)
3. Thyroid gland (one)
   (1) Right lobe
   (2) Left lobe
   (3) Isthmus

4. Parathyroid glands (four)
   (1) Two right superior & inferior
   (2) Two left superior and inferior
5. Pancreas (one) Islands of Langerhans
6. Suprarenal or adrenal glands (two) Rt. & Lt.
   (1) Cortex
   (2) Medulla
7. Gonads; ovaries — right & left
         or testes — right & left
8. Gastrointestinal

S. 246 DETAILED STUDY OF THE ENDOCRINE GLANDS

1. THE PITUITARY GLAND (HYPOPHYSIS) F. 20-2

The pituitary gland or hypophysis lies within the cranial cavity, in the pituitary fossa of the sella turcica - a part of the sphenoid bone. (hypo = below + physis = growth, i.e. to grow below the brain). It is below the forebrain, and is attached by a stalk to the hypothalamus of the diencephalon. It measures about 1.2 cm or .5 inches in diameter, but in spite of its size it is a very important structure. It is a double gland and consists of two distinct parts, an anterior and a posterior lobe. Each lobe has its own hormones.

(1) **The posterior lobe** is derived from a pouch of the third ventricle of the brain, hence the name neurohypophysis. Its two hormones oxytocin and antidiuretic hormone (vasopressin) are actually formed in the hypothalamus of the brain and are stored in this posterior lobe of the pituitary gland. They are released into blood capillaries as required. (This is a more recent concept).

Functions:

a. **Oxytocin** or pitocin stimulates the muscle of the uterus to contract during childbirth.

b. **Antidiuretic hormone** (ADH) or vasopressin stimulates the reabsorption of water from the renal tubules, thus decreasing urinary output. It also stimulates contraction of the involuntary muscles of the intestine, bronchi, and blood vessels.

(2) **The anterior lobe** of the pituitary gland is derived from a pouch of the mouth of the fetus. The pouch extends upward and becomes pinched off. Adenohypophysis is an alternate name denoting a glandular structure. It has two types of cells acidophil and basophil. Its hormones have been named **trophic hormones** or tropic hormones because they stimulate the other endocrine glands to perform their functions (trophic = to nourish, i.e. to stimulate). The hormones secreted include thyrotrophic, adrenocorticotrophic, gonadotrophic (male or female), somatotrophic, luteotrophic, or prolactin, and possibly melanocyte stimulating.

Functions:

a. **Thyrotrophic hormone** (TSH) stimulates the thyroid gland to grow and produce its hormones.

b. **Adrenocorticotrophic hormone** (ACTH) stimulates the cortex of the suprarenal gland to secrete its hormones.

c. **Gonadotrophic hormones in the female:**

Follicle stimulating hormone (FSH) and **Luteinizing hormone** (LH) stimulate: the ovaries in adolescents to mature, development of Graafian follicles and ova, production of estrogens — ovarion hormones, ovulation — discharge of ova, development of the corpus luteum and secretion of progesterone.

Luteotrophic hormone (LTH) or prolactin is concerned with the enlargement of breast tissue for secretion of milk.

c. **Gonadotrophic hormones in the male:**

Follicle stimulating hormone (FSH) stimulates growth of testes, production of spermatozoa by the testis. **Interstitial cell stimulating hormone** (ICSH), the equivalent of luteinizing hormone stimulates production of androgens by the enterstitial cells of the testis.

d. **Somatotrophic hormone** (STH) the growth stimulating hormone is concerned with the growth of the skeleton and other structures during childhood.

e. **Melanocyte stimulating hormone** is said to stimulate the formation of pigment in humans. Possibly this is formed in an intermediate lobe, a small part located between the other two lobes.

The secretion of trophic hormones is governed by the amount of the hormone of the gland stimulated circulating in the blood. For example, thyrotrophic hormone stimulates the thyroid gland to secrete thyroxine. When the concentration of thyroxine in the blood reaches a certain level the thyroxine acts upon the anterior lobe of the pituitary gland causing it to decrease its secretion of thyrotrophic hormone. This in turn will decrease stimulation of the thyroid gland and reduce the quantity of thyroxine secreted. A decrease in the amount of thyroxine in the circulating blood will have the opposite effect. It will cause the pituitary gland to increase its secretion of thyrotrophic hormone, and eventually the thyroid gland will be stimulated to secrete more thyroxine.

2. THE PINEAL GLAND F. 20-1

The pineal gland or body lies within the cranial cavity, in the midline posterior to the midbrain. It is a minute structure of unknown function with a secretion that has not been isolated. In adults it may become calcified. It then becomes visible in radiographs of the skull. Normally lying in the midline a displacement may indicate some inflammatory lesion, hemorrhage, or tumor within the skull.

3. THE THYROID GLAND F. 20-3

The thyroid gland consists of **two lobes**, a right and a left lying in the anterior neck, one on either side of the thyroid cartilage. The two lobes are joined across the midline by a narrow **isthmus.** Each lobe is about 5 cm or 2 inches in vertical length, with a pointed upper end. The thyroid lobes are not easily palpated, but since they move upward with the trachea during swallowing they may be felt with a thumb and finger, one on either side of the thyroid cartilage. Occasionally some thyroid tissue lies at a lower level behind the upper sternum — a retrosternal thyroid.

Hormones: thyroxine and tetraiodothyronine. These contain iodine that is obtained from water and food consumption. Some areas in North America, e.g. the Rocky Mountains, and areas about the Great Lakes, etc. have insufficient iodine in the water to supply iodine required to produce the hormones. In some countries laws are in force requiring packagers of common table salt (sodium chloride) to add a certain percentage of potassium iodide to ensure that humans get sufficient iodine to produce the hormones.

Functions: to promote mental and physical development in the young, i.e. growth, and to stimulate metabolism in

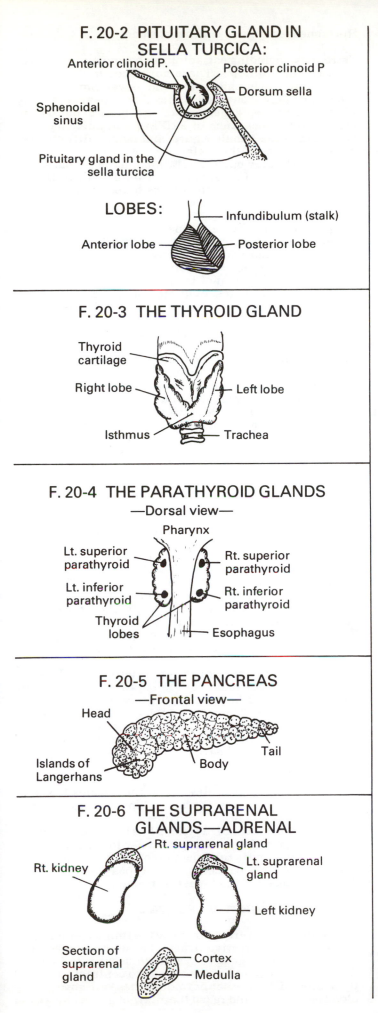

F. 20-2 PITUITARY GLAND IN SELLA TURCICA:

Anterior clinoid P.
Posterior clinoid P
Dorsum sella
Sphenoidal sinus
Pituitary gland in the sella turcica

LOBES:

Infundibulum (stalk)
Anterior lobe
Posterior lobe

F. 20-3 THE THYROID GLAND

Thyroid cartilage
Right lobe
Left lobe
Isthmus
Trachea

F. 20-4 THE PARATHYROID GLANDS
—Dorsal view—

Pharynx
Lt. superior parathyroid
Rt. superior parathyroid
Lt. inferior parathyroid
Rt. inferior parathyroid
Thyroid lobes
Esophagus

F. 20-5 THE PANCREAS
—Frontal view—

Head
Tail
Islands of Langerhans
Body

F. 20-6 THE SUPRARENAL GLANDS—ADRENAL

Rt. suprarenal gland
Rt. kidney
Lt. suprarenal gland
Left kidney
Section of suprarenal gland
Cortex
Medulla

all age groups. A basal metabolism test will indicate the performance of the thyroid gland.

Growth of the body is influenced by the somatotrophic hormone of the pituitary gland, the thyroid hormones and the hormones of the ovaries or testes.

4. THE PARATHYROID GLANDS F. 20-4

The parathyroid glands are four small bean shaped bodies that lie in the neck posterior to the thyroid gland. There are two right, and two left, superior and inferior. Each gland measures 6 mm or .25 inches in vertical diameter.

Hormone: parathormone or parathyroid hormone

Functions: regulation of calcium metabolism, and its concentration in blood and tissues, especially in bones.

The excitability of nervous and muscular tissues is influenced by the amount of calcium in blood and tissue fluid. When the calcium content is low spasms (tetany) of muscles or convulsions may occur. Since the calcium content is regulated by the parathormone secreted by the parathyroid glands a decrease in its secretion may cause these effects. If for instance, the parathyroid glands are accidentally removed at an operation upon the thyroid gland tetany may occur.

5. THE PANCREAS F.20-5

A mixed gland with exocrine and endocrine parts. The pancreas has already been described as one of the abdominal accessory organs of the digestive system. It was pointed out that its exocrine function consists of the manufacture of pancreatic juice, containing digestive enzymes. These are conveyed by the pancreatic duct to the small intestine.

The endocrine part consists of minute nests of cells, the Islands (islets) of Langerhans, that have no ducts, but secrete their hormones directly into blood capillaries.

Hormones: insulin and glucagon

Function: insulin stimulates body cells to use up (burn) glucose to produce energy, and promotes its storage in the liver, thus regulating the blood sugar concentration. **Glucagon** is said to have an opposite effect, promoting the release of glucose from the liver thus raising the blood sugar level.

6. THE SUPRARENAL OR ADRENAL GLANDS F.20-6

Supra = above + ren = kidney, i.e. above a kidney;
Ad = towards + ren; towards a kidney.

There are two suprarenal glands, a right and a left. They are flat caplike structures and lie upon the upper end of the corresponding kidney. Each measures about 3 to 5 cm (or 1.2 to 2 inches) in diameter. Each suprarenal gland is a double gland consisting of a medulla and cortex.

(1) **The medulla** is the central part, medulla = marrow.

Hormones: adrenalin(e) or epinephrine
noradrenalin(e) or norepinephrine

Functions: these hormones have a similar effect to stimulation of the sympathetic part of the autonomic nervous system, and affect the same structures. They cause an increase in the rate of the heart beat, and cause contraction of blood vessels thus raising the blood pressure. They cause dilatation of bronchi, and are used to relieve the spasm of bronchial asthma. How much of these hormones is normally released into the blood is questionable. Certainly in times of stress "fear, flight or fight" they are definitely secreted. The medulla of the suprarenal is not essential to life, and may be removed.

(2) **The cortex** of the suprarenal gland is its outer part, (Cortex = bark, i.e. outer part). This part of the gland is essential to life, and its removal or destruction will result in death.

Hormones: the hormones secreted by the suprarenal are named corticoids, steroids, or corticosteriods: corticoids because they are secreted by the cortex, steroids because they were thought to resemble fats and have the steroid formula, and corticosteriods, a combination of the other two names.

About thirty corticoids are formed in the suprarenal cortex, but only seven of these appear to be important in medicine.

The corticoids have been divided into three groups which, while of some interest should not be required knowledge for radiological technicians.

1. **Glucocorticoids** — concerned with the metabolism of glucose, amino acids and fats, e.g. **cortisone.**

2. **Mineralocorticoids** — that help to regulate the electrocyte (salt) balance of blood and tissues by promoting the reabsorption of water and sodium from the renal tubules of the kidneys, e.g. **aldosterone**

3. **Gnadal hormones** — female and male, similar to the hormones secreted by the ovaries and testes, but in lesser amounts.

Functions: as defined above.

7. THE GONADS — THE OVARIES   F.20-7

In the chapter describing the female reproductive system the ovaries were discussed, and the primary and vesicular (Graafian) follicles and corpora lutea were defined. The dual function of producing ova and secreting female sex hormones was pointed out. (S.225) As the age of puberty approaches gonadotrophic hormones of the anterior pituitary lobe are secreted in increasing amounts. These cause enlargement of the ovaries, the development of vesicular follicles, ovulation, and the secretion of female sex hormones.

Hormones: estrogens & progesterone

Estrogens, including estradiol, are hormones secreted by the vesicular follicles. These hormones in the teen age female are responsible for the development of the secondary sexual characteristics that appear at puberty. These include enlargement of the breast, uterus and vagina, the appearance of axillary and pubic hair, the maturing of the female figure, and personality changes with awareness of the male.

In addition they cause alterations in the endometrium of the uterus, and an increase in its blood capillaries. These changes prepare the uterus to receive a fertilized ovum if conception occurs.

Progesterone is the hormone secreted by the corpus luteum and, if pregnancy occurs, by the placenta or afterbirth. It causes further changes in the endometrium. It is necessary for the completion of ovulation. It promotes development of the milk secreting cells of the breasts. It suppresses ovulation and menstruation if pregnancy occurs.

7. THE GONADS — THE TESTES   F.20-8

The anatomy of the testes was described in the chapter on the male reproductive system (S.235). The production of mature spermatozoa, and the secretion of male sex hormones by the testes were outlined. As in the female, during the second decade of life there is an increase in the secretion of male gonadotrophic hormones by the anterior lobe of the pituitary gland. These stimulate the testes to produce mature spermatozoa, and the interstitial cells (of the testes) to produce male sex hormones.

Hormones: androgens = testosterone, androsterone

Androgens are the male sex hormones secreted by the interstitial cells within the testes, and include testosterone and androsterone: these are responsible for the changes that occur as puberty approaches — the secondary sexual characteristics of the male. The penis, and other reproductive organs enlarge, the beard begins to grow, and the voice deepens. The prostate and seminal vesicles become active, and personality changes occur.
REMEMBER: Andy produces androgens
         Esther produces estrogens

Notes: The sex hormones are steroids with some similarities to corticoids of the suprarenals. Minimum amounts of the sex hormones are secreted by members of the opposite sex. The suprarenal glands secrete small amounts of male and female sex hormones.

Castration before puberty in the female or male prevents the occurrence of the normal changes of puberty.

In some instances castration of the ovaries has been employed to decrease the secretion of estrogens and inhibit the growth of breast cancer.

Similarly removal of the testes has been used to inhibit the rate of growth of prostatic cancer in the male.

In some instances the injection of sex hormones of the opposite sex has caused the disappearance of secondaries in cancer of the breast or prostate.

S. 247   THE DIGESTIVE SYSTEM

Some hormones are secreted by the lining membrane of parts of the digestive tract. These hormones may stimulate other digestive glands to secrete, to increase their secretion, or cause their ducts to contract and expel the secretion. Like other hormones these are absorbed by blood capillaries and reach their target glands by way of

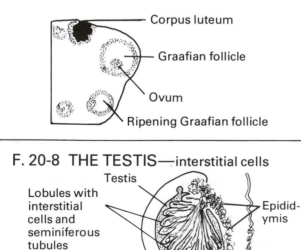

F. 20-7 THE OVARY—follicle & corpus luteum

- Corpus luteum
- Graafian follicle
- Ovum
- Ripening Graafian follicle

F. 20-8 THE TESTIS—interstitial cells

Testis

Lobules with interstitial cells and seminiferous tubules

Epididymis

Ductus Deferens

the circulatory system. These hormones include gastrin, secretin, pancreozymin, choleystokinin, and others.

(1) **Gastrin** is a hormone formed in the lining membrane of the pyloric part of the stomach. Secretion is stimulated by the presence in the stomach of such foods as meats, vegetables, or soups. Gastrin stimulates gastric glands in the fundus of the stomach to secrete hydrochloric acid and pepsin.

(2) **Secretin** is a hormone formed in the lining membrane of the duodenum when stimulated by the presence of fat, protein, or acid in the duodenum. It is carried to the pancreas and stimulates this organ to secrete a watery pancreatic juice. It also stimulates the liver to produce bile.

(3) **Pancreozymin** is another hormone secreted by the lining membrane of the small intestine. It stimulates the pancreas to produce pancreatic juice rich in digestive enzymes.

(4) **Cholecystokinin** is a hormone secreted by intestinal mucosa when fat reaches the duodenum from ingested food. It stimulates contraction of the gall bladder and bile ducts to contract and expel bile into the duodenum. This knowledge is utilized in cholecystography when fatty preparations or foods are fed to patients to empty these organs. These preparations are called cholagogues. Several of them are on the market, each with its trade name.

S. 248 SOME PATHOLOGICAL CONDITIONS AFFECTING THE ENDOCRINE GLANDS

Several pathological conditions result from an insufficient or an excessive secretion of an endocrine gland. Because the results of these changes in secretion help to explain the normal function of endocrine glands they are defined in this section. Too little secretion may be due to underdevelopment in early life. It may be due to destruction of a gland from disease, etc. at any age. Oversecretion is frequently due to a tumor of the gland. Sometimes these changes in secretion cause changes in body structure or function that are demonstrable by radiography.

1. THE PITUITARY GLAND

(1) **Giantism** is an overgrowth of a subject due to an excessive secretion of the somatotrophic (growth hormone) of the anterior lobe of the pituitary gland. The skeleton may grow to a height of 210 to 240 cm or 7 to 8 feet — the giant seen in the circus.

(2) **Acromegaly** is a similar condition occurring after growth is complete, when there is oversecretion of the somatotrophic hormone. The head, hands, and feet enlarge so that hats, gloves, or boots previously large enough are too small.

(3) **Pituitary dwarfism** is due to insufficient secretion of somatotrophic (growth stimulating) hormone during the growth period of life, the subject may be only four feet tall. Sexual and mental development is normal unless there are other deficiencies. There are several other causes of dwarfism in addition to the one defined here. (a dwarf = a midget).

2. THE THYROID GLAND

(1) **Simple goiter** (goitre) is an enlargement of the thyroid gland resulting from an inadequate supply of iodine in the diet. Insufficient thyroxine is produced. The thyroid gland enlarges in an attempt to supply sufficient hormones. Compulsory addition of potassium iodide to table salt has resulted in a definite decrease in goiter.

(2) **Cretinism** is a condition caused by an absence or underdevelopment of the thyroid gland in infancy. Both physical and mental development are affected. The victim is a mental and physical dwarf being of small stature and mentally retarded. If the condition is recognized early in life it may be successfully treated by giving thyroid extract.

(3) **Myxodema** is a similar condition that sometimes occurs at a later age. It may result from disease of the thyroid gland. It may be due to removal of too much thyroid during an operation for hyperthyroidism. (See below) The individual becomes mentally and physically slow, appears lazy, and becomes overweight. The treatment consists of taking thyroid extract on a lifetime basis.

(4) **Hyperthyroidism,** Grave's Disease, or exophthalmic goiter, is due to an oversecretion of the hormones of the thyroid gland. The thyroid enlarges and the disease is characterized by nervousness, tremors, sleeplessness, rapid heart beat, loss of weight, protruding eyeballs, and a high basal metabolic rate.

3. THE PARATHYROID GLANDS

(1) **Hypoparathyroidism** is a condition resulting from an insufficient secretion of parathyroid hormone. It may be developmental. It may occur from accidental removal of one or more parathyroids during surgery on the thyroid gland. It results in muscular spasms, twitchings, and sometimes convulsions due to the low percentage of calcium in the circulating blood.

(2) **Hyperparathyroidism** is due to an excessive secretion of the parathyroid hormone from a tumor, etc. Because of this there is an absorption of the calcium from bones resulting in a high blood calcium. Bone becomes decalcified and cystic. Sometimes calculi form in the kidneys, or calcium is deposited in the kidney substance — calcinosis.

4. THE SUPRARENAL GLANDS

(1) **Addison's Disease** is a condition resulting from a decrease in hormones secreted by the cortex of suprarenal glands. There is muscular weakness, low blood pressure, and pigmentation of the skin. There is an increase in the sodium and water secretion by the kidneys. The disease is fatal.

(2) **Cortical tumors** may occur in the young or in the adult. In the younger age group there is an early and marked development of the sex organs, and an early appearance of secondary sexual characteristics. The epiphyseal cartilages ossify early, and fuse with the bodies of bones before the normal age causing stunting of growth. These tumors may cause masculinity in the female, or feminism in the male.

S. 249 APPLICATION TO RADIOGRAPHY

Endocrine glands are composed of soft tissues. There is no marked difference in density from the other tissues surrounding them. Therefore they do not have separate images on radiographs. When they are enlarged or have some other abnormality, such as calcium deposition they are sometimes visible in radiographs.
Radiological evidence of the presence of endocrine glands may be demonstrated.
(1) if calcium has been deposited in the gland, the calcium will be visible, e.g. calcium in the pancreas;

(2) if there is pressure upon adjacent organs such as bones resulting in displacements, deformity or destruction, e.g. enlargement of the sella tircica due to a tumor of the pituitary gland;

(3) if there are changes in bone density due to insufficient or excessive secretion of a hormone by some endocrine gland, e.g. bone in hyperparathyroidism.

Pituitary gland — while this gland is not visible the sella turcica in which it lies is visible in radiographs. Enlargement of the pituitary causes enlargement of the sella with thinning or disappearance of the adjacent clinoid processes. Radiographs may show these changes and suggest a tumor of the pituitary gland.

In dwarfism, pituitary or thyroid, radiographs may demonstrate early closure at the epiphyseal cartilages.

Pineal gland — while the gland is not normally visible, with increasing age it sometimes becomes calcified. Its position may then be determined.

Thyroid gland — an enlarged thyroid gland is often visible in the neck, and may show displacement of the trachea or esophagus, or it may be calcified. Retrosternal goiter may be demonstrated in the superior mediastinum.

In cretinism bones may show a delay in the appearance of ossifying epiphyses, and a delay in the disappearance of the epiphyseal cartilages.

Parathyroid glands — these are not normally visible in radiographs, but may be demonstrated indirectly by changes in bone density due to changes in parathyroid secretion.

Pancreas — occasionally calcifications occur in this organ. They may be seen on radiographs.

Suprarenal glands — these may be visualized if enlarged or calcified. They may be outlined if air is injected behind the anus and allowed to pass up posterior to the peritoneum to surround the kidneys and suprarenals. The air then becomes a contrast medium.

S. 250 TERMS REFERRING TO ENDOCRINE GLANDS

endocrine gland
exocrine gland
ductless gland
internal secretion

pituitary gland
anterior lobe, pituitary
master gland
trophic hormone

hormone
double gland
mixed gland
multiple function
interrelationship
metabolism
basal metabolism
somatotrophic hormone
growth stimulating, STH
Posterior lobe
  of pituitary
oxytocin
pitocin
vesopressin or
antidiuretic hormone
pituitrin
pineal gland
thyroid gland
right lobe
left lobe
isthmus
thyroxine
tetraiodothyronine
potassium iodide
parathyroid glands
superior parathyroid G
inferior parathyroid G
parathormone
parathyroid hormone
calcium metabolism
pancreas
exocrine function
Islands of Langerhans
insulin
glucagon
suprarenal glands
adrenal glands
medulla (suprarenals)
adrenalin, epinephrine
noradrenalin or
norepinephrine
** TSH = thyrotrophic h
ACTH = adrenocortico-
    trophic h
LSH = luteotrophic h
LH = luteinizing h
FSH = follicle stimul-
    ating h
ICSH = interstitial
    cell stimulating
    hormone
** For reference only

trophic hormone
thyrotrophic hormone
  TSH
adrenocorticotrophic
  hormone (ACTH)
luteotrophic hormone,
  LSH
luteinizing hormone, LH
cortex (suprarenals)
corticoids
steroids
corticosteroids
glucocorticoids
mineral corticoids
gonadal steroids
Gonads
Ovaries
primary ovarian follicle
vesicular ovarian
  follicle
or Graafain follicle
corpora lutea
corpus luteum
estrogens
estradiol
progesterone
Testes
androgens
testosterone
androsterone
Gastrointestinal
gastrin
secretin
pancreozymin
cholecystokinin
cholagogue
giantism **
acromegaly **
dwarfism **
cretinism **
myzodema **
hyperthyroidism **
goiter (re) **
hyperparathyroidism **
Addison's Disease
Adrenal cortical tumor

NOTES

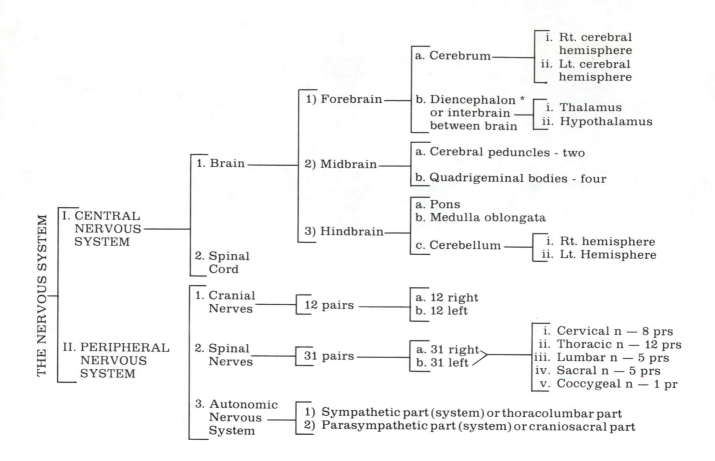

Note:- * Diencephalon has been added under forebrain in order to include the thalami.

SOME ANATOMICAL TERMS RELATING TO THE NERVOUS SYSTEM:

1. **Neuron** — a nerve cell, with a cell body, axon and dendrites (s. dendron)

2. **Neuroglia** — the supporting cells of the nervous system; (s. neuroglion)

3. **Synapse** — the point of contact of an axon with a dendron of some other neuron

4. **Sensory neuron** — a nerve cell that carries impulses into the spinal cord or brain

5. **Motor neuron** — a nerve cell that conveys nerve impulses from the brain or spinal cord to a muscle or secreting gland to cause muscular contraction or secretion

6. **Ganglion** — a group of nerve cell bodies outside of the brain or spinal cord

7. **Nucleus** — a group of nerve cell bodies within the brain or spinal cord that forms one nerve or a single tract

8. **Center** — a group of nerve cell bodies concerned with some body function

9. **Meninges** — the 3 coverings of the brain and spinal cord: — the dura mater, the arachnoid, and the pia mater; with the subarachnoid space containing cerebrospinal fluid between the arachnoid and pia mater; (CSF), meninx = a membrane

10. **Ventricles of the brain** — 4 cavities within the brain filled with cerebrospinal fluid: — right lateral ventricle, left lateral ventricle, third ventricle, and fourth ventricle, with communicating (connecting) foramina and passage (aqueduct):—
    Interventricular foramina (Munro); **cerebral aqueduct** (aqueduct of Sylvius); **lateral apertures** 4th ventricle (Luschka); **medial aperture** 4th ventricle (Magendie).

21. THE NERVOUS SYSTEM

A neuron is a nerve cell and consists of a cell body with its nucleus, two sets of processes (dendrites) and an axon. F. 21-2

Dendrites (s. dendron) are several processes that extend out from the cell body, and carry impulses towards the cell body. (G. dendron = a tree; branches).

An axon is a single process extending out from the cell body and may be two or three feet in length. It conveys impulses away from the cell body.

A sensory neuron or an afferent neuron, is a nerve cell that conducts impulses towards or into the spinal cord, or up the spinal cord to the brain. These impulses may be sensations of heat, cold, touch, pressure, pain, or of position, or from the special sense organs — the eye (sight), ear (hearing), nose (smell) or mouth (taste). (af fero = to carry to). F.21-2(3)

A motor neuron, or an efferent neuron, is a nerve cell that carries impulses from the brain down to the spinal cord, or from the cord to muscles or secreting glands, or directly from the brain to muscle or to a gland. (e, ex = out of + fero = to carry away). F.21-2(1)

F. 21-1 DIAGRAM OF THE PARTS OF THE CENTRAL NERVOUS SYSTEM

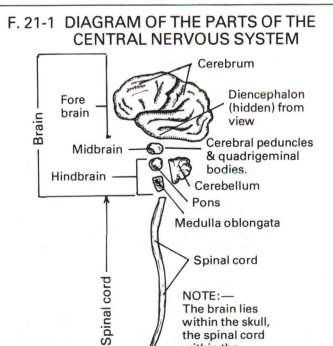

F. 21-2 EXAMPLES OF NEURONS— NERVE CELLS

(1) MOTOR NEURON; to muscle or gland

(2) NEURON FROM BRAIN OR CORD

(3) SENSORY NEURON—from skin, etc.

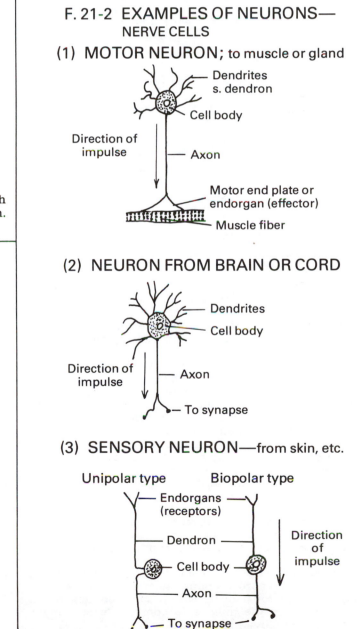

F. 21-3 A SYNAPSE—

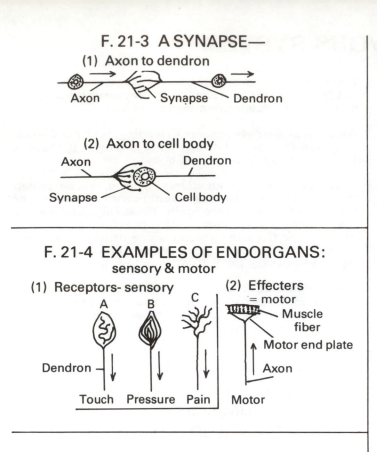

(1) Axon to dendron

Axon — Synapse — Dendron

(2) Axon to cell body

Axon — Dendron

Synapse — Cell body

F. 21-4 EXAMPLES OF ENDORGANS:
sensory & motor

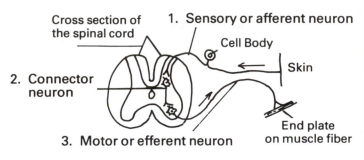

(1) Receptors- sensory

A   B   C

Dendron —

Touch   Pressure   Pain

(2) Effecters
= motor

Muscle fiber
Motor end plate
Axon

Motor

F. 21-5 REFLEX ARC WITH THREE NEURONS

Cross section of the spinal cord

1. Sensory or afferent neuron

Cell Body

Skin

2. Connector neuron

3. Motor or efferent neuron

End plate on muscle fiber

A synapse is the point of contact of the axon of one nerve with the dendron (dendrites) of another neuron.

An Endorgan is a structure at the free end of an axon or dendron designed to pass on or pick up impulses or sensations. F.21-4

a. **A receptor** is an endorgan at the outer end of a dendron of a sensory neuron that picks up sensations. These may be sensations of touch, pressure, heat, cold, pain or position, of from one of the special senses. Receptors vary in structure from minute fibrils to complex endorgans that vary with their function.

b. **An Effector** is an endorgan at the distal end of an axon of a motor neuron. It may be a plate on a muscle fiber carrying impulses for contraction, or minute delicate branching fibers about the secreting cells of a gland that stimulate secretion.

A reflex arc is a complete circuit consisting of a sensory neuron ending in the spinal cord, a connecting neuron within the cord, and a motor neuron to a muscle. F.21-5

A reflex act: sensory neuron carries a sensation of pain, etc., to the spinal cord. A connecting neuron conveys the impulse to a motor neuron. This in turn conducts it to a muscle, causing the muscle to contract, e.g. a finger touching a hot stove.

A ganglion is a group of nerve cell bodies outside of the spinal cord or brain, e.g. the dorsal root ganglion of a spinal nerve.

A nucleus is a group of cell bodies within the brain and spinal cord, e.g. the nucleus of a cranial nerve.

A center is a group of nerve cell bodies concerned with some specific function, e.g. respiratory center.

A plexus is a network of cell processes, e.g. the cervical, brachial, lumbar and sacral plexuses.

White matter (NA, white substance) consists of nerve fibers, axons and dendrites, and are white in color. They form the central parts of the cerebrum, but the outer parts of the spinal cord. F.21-7

Gray matter (NA, gray substance) is made up of the nerve cell bodies and some processes. It appears darker than the white matter. It forms the outer part of the cerebrum, (the cortex) and surrounds the white matter here. It forms the central parts of the spinal cord, and is surrounded by white matter. F.21-7

Funiculi are large columns of nerve fibers that pass up and down the spinal cord or brain. (s. funiculus, L. funis - a cord, so a little cord).

Tracts are bundles of nerve fibers, and parts of funiculi, in close contact with each other that pass up or down the spinal cord. The nerve fibers within each tract have a common destination and some specific function. They are made up of axons or dendrites. They are described as ascending (going up) and descending tracts (passing down).

Irritability is that property of nerves that renders them reponsive to stimuli such as heat, cold, etc.

Conductivity in nerves is the ability to convey impulses, generated by stimuli, along the neurons. The impulse is carried in one direction — from dendrites to cell body — to axon — to synapse. The impulse is not electrical but electrical changes result. These electrical changes are utilized in the procedure called electroencephalography (EEG) by which these changes are recorded and analysed.

S. 253   DIVISIONS OF THE NERVOUS SYSTEM
F.21-1

The student should refer to the chart printed at the beginning of this chapter in which the divisions of the nervous system were listed. It should be noted however that the parts of this system form a complete unit, and are not isolated, independent structures. The functions of the nervous system are communication and control. The separation into parts simplifies the study. There are two divisions of the system: the central nervous system, and the peripheral nervous system. These are detailed below.

The central nervous system is the part enclosed by bone. It includes the **brain,** within the cerebral cranium, and the **spinal cord** within the vertebral column.

The brain is further divided into the forebrain, midbrain and hindbrain.

The forebrain consists of the cerebrum and diencephalon or interbrain, or between brain.

The midbrain includes two cerebral peduncles, and four small bodies, the corpora quadrigemina or colliculi.

The hindbrain has a pons, medulla oblongata, and cerebellum.

The spinal cord the other part of this division, is described in S. 256.

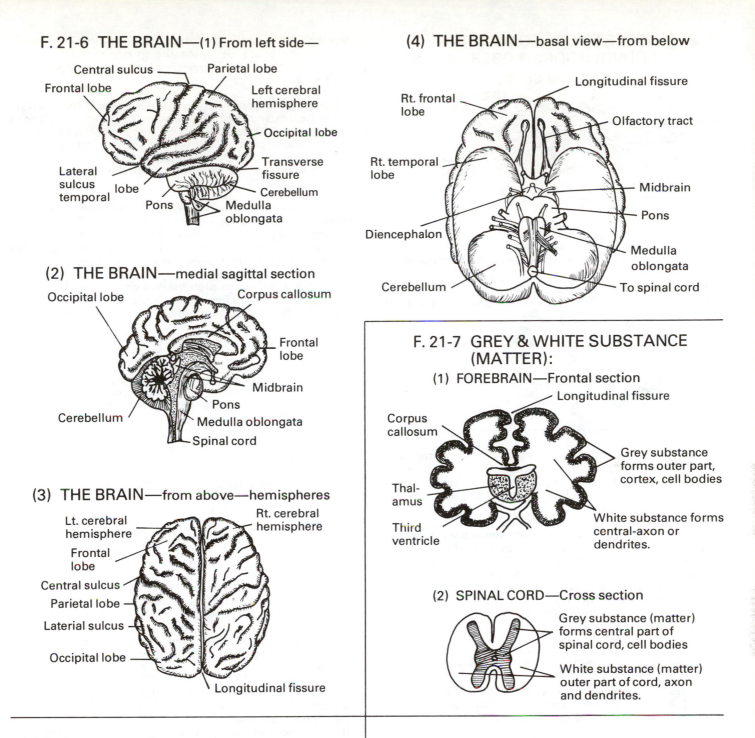

F. 21-6 THE BRAIN—(1) From left side—

Central sulcus
Frontal lobe
Parietal lobe
Left cerebral hemisphere
Occipital lobe
Transverse fissure
Lateral sulcus
temporal lobe
Cerebellum
Pons
Medulla oblongata

(2) THE BRAIN—medial sagittal section

Occipital lobe
Corpus callosum
Frontal lobe
Midbrain
Pons
Cerebellum
Medulla oblongata
Spinal cord

(3) THE BRAIN—from above—hemispheres

Lt. cerebral hemisphere
Rt. cerebral hemisphere
Frontal lobe
Central sulcus
Parietal lobe
Laterial sulcus
Occipital lobe
Longitudinal fissure

(4) THE BRAIN—basal view—from below

Rt. frontal lobe
Longitudinal fissure
Olfactory tract
Rt. temporal lobe
Midbrain
Pons
Diencephalon
Medulla oblongata
Cerebellum
To spinal cord

F. 21-7 GREY & WHITE SUBSTANCE (MATTER):

(1) FOREBRAIN—Frontal section

Longitudinal fissure
Corpus callosum
Grey substance forms outer part, cortex, cell bodies
Thal-amus
Third ventricle
White substance forms central-axon or dendrites.

(2) SPINAL CORD—Cross section

Grey substance (matter) forms central part of spinal cord, cell bodies
White substance (matter) outer part of cord, axon and dendrites.

The peripheral nervous system is that part outside of the skull and vertebral column, i.e. the external part. It includes the cranial nerves, spinal nerves, and the autonomic nervous system. Each of these parts has connections with the central nervous system. Detailed information concerning each is contained in Section 258, or subsequent sections.

S. 254 THE BRAIN F.21-6

(G. enkephalos = encephalon = brain) (AS. braegen)

The brain occupies the entire cerebral cranium. The human brain is well developed. It includes three parts, a large forebrain, a small midbrain, and a hindbrain. These terms may be confusing, but if the human subject is positioned like an animal on all fours, with hands and feet supporting the body, and with the face directed downwards instead of to the front these names will have a practical meaning. The forebrain or front brain will be anterior, then the mid or middle brain, and the hindbrain at the back.

1. THE FOREBRAIN F.21-6

The forebrain is the large expanded mushroom-like upper part of the brain. It occupies the entire cranium except the posterior cranial fossa. It consists of the cerebrum, and diencephalon or inter or between brain. (Fore = front; i.e. front brain).

(1) THE CEREBRUM forms the visible part of the forebrain and is by far its larger part. It obscures the diencephalon. It is composed of gray and white matter. F.21-6

Gyri or convolutions are the rounded ridges visible on the outer curved surfaces of the cerebrum. (s. gyrus)

Sulci are shallow grooves that dip down between adjacent gyri. This arrangement of alternate ridges and grooves increases the surface area of the cerebrum. (s. sulcus = a groove)

F. 21-8 NERVE FIBERS CONNECTING CONVOLUTIONS, LOBES: of cerebrum

(1) LONGITUDINAL SECTION:

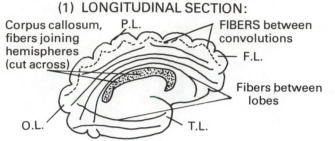

Corpus callosum, fibers joining hemispheres (cut across) — P.L. — FIBERS between convolutions — F.L. — Fibers between lobes — O.L. — T.L.

(2) FRONTAL SECTION—Fibers joining hemispheres

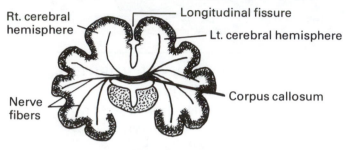

Rt. cerebral hemisphere — Longitudinal fissure — Lt. cerebral hemisphere — Nerve fibers — Corpus callosum

---

The cortex, also called the cerebral cortex, is the outer part of the cerebrum. It is composed of several layers of nerve cell bodies that form the **gray matter** of the cerebrum. Hence cell bodies surround the entire cortex. F.21-7

The white matter of the cerebrum lies inside the cortex and consists of nerve fibers (processes) arising from or ending in the cortical cell bodies. These white fibers form tracts or bundles that connect two parts of the cerebrum on the same side, or that connect the cerebrum with the other parts of the brain and spinal cord, or that connect one half of the cerebrum across the midline with the other half. F.21-7

The corpus callosum is composed of these nerve fibers that connect the two halves or hemispheres.

The cerebral hemispheres, right and left, are the two lateral halves of the cerebrum separated along the median plane by the longitudinal fissure.

FISSURES AND GROOVES   F.21-6, 7

The student may avoid confusion by realizing that the two deep grooves in the brain are named fissures while the shallow grooves that are listed here are named sulci.

The longitudinal fissure (NA) is a deep groove that extends from the front to the back of the cerebrum, along the median line, and under the saggital suture. It divides the cerebrum into right and left cerebral hemisphere. The division is incomplete below where the corpus callosum, composed of nerve fibers, connects the hemispheres.

The transverse fissure (NA) passes transversely across the upper limit of the posterior cranial fossa. It separates the inferior and posterior part of the cerebrum from the cerebellum below. F.21-6

The central sulcus (NA) is a deep groove that begins at the longitudinal fissure at about its midpoint. It extends downwards and anteriorly on the lateral surface of each

hemisphere to end in the lateral sulcus below. The motor area of the brain occupies the gyrus anterior to this central sulcus. The sensory area is in the gyrus posterior to this sulcus. F. 26-6

The lateral sulcus (NA) or lateral cerebral fissure (OT) is a definite groove that begins on the inferior surface of each cerebral hemisphere. It passes laterally, then turns upwards and posteriorly on the lateral surface of the hemisphere, towards the longitudinal fissure. It extends upwards beyond its junction with the central sulcus. F.21-6

The parietooccipital sulcus is difficult to demonstrate. It extends upwards and backwards from the inferior margin of the lateral surface of the hemisphere close to its posterior margin. It runs up and posteriorly to the longitudinal fissure where it becomes more definite.

Many other sulci are also visible on the surface of the hemispheres. These have been omitted here. The fissures and sulci divide the hemisphere into five lobes.

LOBES OF THE CEREBRAL HEMISPHERES   F.21-6

Each cerebral hemisphere has five lobes: frontal, parietal, occipital, temporal, and the insula. These lobes are in contact with the cranial bone of the same name, and the lobes are separated by the fissures and sulci as defined below. F.21-6

a. **The frontal lobe** lies adjacent to the frontal bone, and anterior to the central sulcus.

b. **The parietal lobe** lies under the parietal bone, and between the central sulcus in front, and the lateral sulcus and parietooccipital sulcus behind.

c. **The occipital lobe** lies in contact with the occipital bone posterior to the parietooccipital sulcus, and behind the parietal and temporal bones.

d. **The temporal lobe** lies adjacent to the temporal bone and below and behind the lateral sulcus. It occupies the middle cranial fossa.

e. **The insula** (central lobe) lies deep within each hemisphere, and deep to the lower part of the lateral sulcus.

It was stated in describing the white matter of the cerebrum that the many gyri and the lobes of one hemisphere are connected by a great many white nerve fibers called **association fibers.** Other white fibers pass across the midline from one hemisphere to the other below the longitudinal fissure, and are called the **corpus callosum.** A third group of fibers connects the cortex of each hemisphere with the parts of the brain below and the spinal cord — the **projection fibers.** All these may be sensory or motor. F.21-8

SPECIAL AREAS & CENTERS OF THE CORTEX

The motor area of the cortex as already noted is located in front of the central sulcus in the adjacent gyrus of the frontal lobe. Axons from this area carry nerve impulses to other parts of the brain and cord. These axons synapse with a second neuron in the brain or cord that conveys motor impulses to muscles. At the upper end of the gyrus are located centers for the muscles of the lower limb, followed by centers for the trunk, the upper limb, with centers for the head at the lower end of the gyrus. F.21-6

The sensory area which receives and interprets impulses from sensory nerves, is located posterior to the

F. 21-9 THE BRAIN — INFERIOR SURFACE:

Forebrain, midbrain, hindbrain

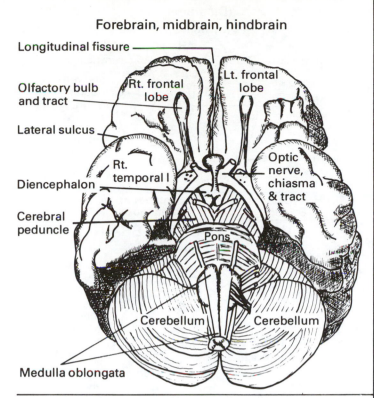

- Longitudinal fissure
- Olfactory bulb and tract
- Rt. frontal lobe
- Lt. frontal lobe
- Lateral sulcus
- Optic nerve, chiasma & tract
- Rt. temporal l
- Diencephalon
- Cerebral peduncle
- Pons
- Cerebellum
- Cerebellum
- Medulla oblongata

F. 21-10 DIENCEPHALON, MIDBRAIN, HINDBRAIN

(1) —FRONTAL VIEW—

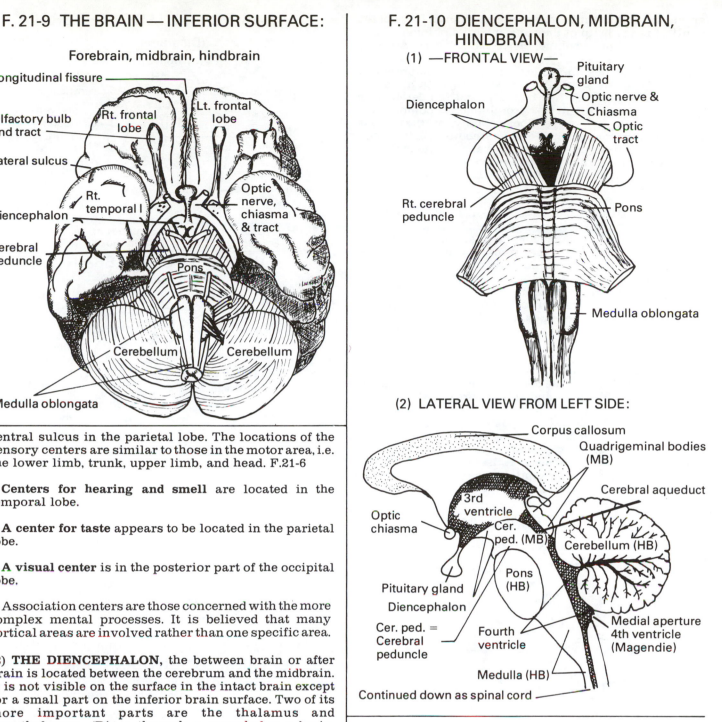

- Diencephalon
- Pituitary gland
- Optic nerve & Chiasma
- Optic tract
- Rt. cerebral peduncle
- Pons
- Medulla oblongata

(2) LATERAL VIEW FROM LEFT SIDE:

- Corpus callosum
- Quadrigeminal bodies (MB)
- Cerebral aqueduct
- 3rd ventricle
- Optic chiasma
- Cer. ped. (MB)
- Cerebellum (HB)
- Pituitary gland
- Pons (HB)
- Diencephalon
- Medial aperture 4th ventricle (Magendie)
- Cer. ped. = Cerebral peduncle
- Fourth ventricle
- Medulla (HB)
- Continued down as spinal cord

central sulcus in the parietal lobe. The locations of the sensory centers are similar to those in the motor area, i.e. the lower limb, trunk, upper limb, and head. F.21-6

Centers for hearing and smell are located in the temporal lobe.

A center for taste appears to be located in the parietal lobe.

A visual center is in the posterior part of the occipital lobe.

Association centers are those concerned with the more complex mental processes. It is believed that many cortical areas are involved rather than one specific area.

(2) THE DIENCEPHALON, the between brain or after brain is located between the cerebrum and the midbrain. It is not visible on the surface in the intact brain except for a small part on the inferior brain surface. Two of its more important parts are the thalamus and hypothalamus. (Dia = through + encephalon = brain, hence between brain). F.21-9, 10

The thalamus forms the lateral walls of the third ventricle. It has nuclei that are relay stations for sensory impulses from the body and lower brain. It appears to sort out sensations to be transmitted to the cerebrum.

The hypothalamus forms the floor of the third ventricle. It appears to have many important functions, such as regulations of secretions of the anterior and posterior lobes of the pituitary gland. It exercises some control over the autonomic nervous system. It helps to regulate body temperature, and directs the feeding habits and the quantity of food eaten. It forms the hormone oxytocin and the antidiuretic hormone. (vasopressin)

2. THE MIDBRAIN F.21-9, 10

The midbrain forms a small part of the brain and is less than one inch in length. It lies between the forebrain,

which is above, and the pons that is below it. It consists of two parts: the two cerebral peduncles, and four quadrigeminal bodies. (Mid = middle; hence middle brain).

The cerebral peduncles are visible on the ventral surface of the brain as two bulges above the pons. They consist of nerve fibers connecting the cerebrum, hindbrain, and cord.

The corpora quadrigemina (quadrogeminal bodies) cannot be seen in the intact brain as they are located in the dorsal part of the midbrain and are covered by the cerebrum. They form relay stations for auditory and visual sensations. F.21-12

3. THE HINDBRAIN F.21-9, 10

The hindbrain lies below the midbrain and extends down to the foramen magnum where it becomes the spinal cord. It occupies the posterior cranial fossa of the skull. It is made up of the pons, medulla oblongata, and cerebellum. (hind = behind; hence the behind brain.)

(1) **THE PONS** lies immediately below the midbrain and anteriorly forms a definite prominence with a median groove. The medulla lies below it and the cerebellum posterior to it. The anterior prominence is composed of white matter, i.e. nerve fibers. Many of these pass transversely from one-half of the cerebellum to the other. Other fibers pass up or down between the medulla and midbrain or forebrain. The pons also contains the nuclei of several cranial nerves. F.21-9, 10 (L. pons = a bridge, i.e. something connecting)

(2) **THE MEDULLA OBLONGATA** lies below the pons, between it and the spinal cord. It ends at the foramen magnum where the spinal cord begins. Nerve fibers pass through it connecting the spinal cord to the pons, cerebellum, midbrain and forebrain. Many of these are arranged in tracts, e.g. motor and sensory tracts which cross over within the medulla to the opposite side. This explains why paralysis or loss of sensation of a part of the body occurs on the opposite side to a brain lesion. The nuclei of several cranial nerves, including the vagus, lie in the medulla. Some vital centers such as the respiratory, cardiac, and vasomotor are located in the medulla. (L. medulla = marrow; G. myelos = marrow, i.e. something soft; hence the bone marrow, and the marrow-like medulla oblongata and spinal cord). F.21-9, 10

(3) **THE CEREBELLUM** lies posterior to the pons and upper part of the medulla oblongata. It is much larger than the other parts of the hindbrain and occupies most of the posterior cranial fossa. It is separated from the inferior surfaces of the occipital lobes of the cerebrum by the transverse fissure and the fold of dura mater, the

F. 21-11 FOREBRAIN:
—FRONTAL VIEW—

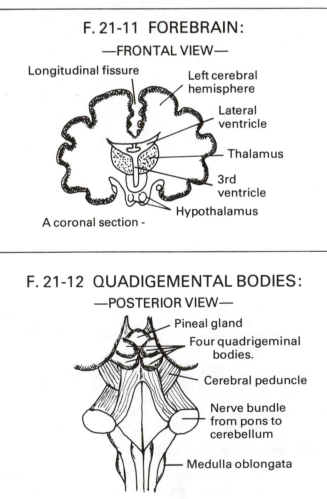

A coronal section -

Longitudinal fissure
Left cerebral hemisphere
Lateral ventricle
Thalamus
3rd ventricle
Hypothalamus

F. 21-12 QUADIGEMENTAL BODIES:
—POSTERIOR VIEW—

Pineal gland
Four quadrigeminal bodies.
Cerebral peduncle
Nerve bundle from pons to cerebellum
Medulla oblongata

F. 21-15 VENTRICLES OF THE BRAIN
(Diagrams)
(1) —Viewed from above—

Left lateral ventricle
AH
Right lateral ventricle
IH
IH
B
B
Third ventricle
Interventricular foramen (Monro)
Fourth ventricle
Cerebral aqueduct (Sylvius)
PH
PH

PH = Posterior horn   IH = Inferior horn
B = Body   AH = Anterior horn

VENTRICLE OF THE BRAIN
(2) —Right lateral view—

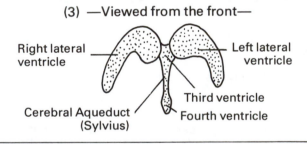

Lateral ventricle
B
AH
PH
III
Third ventricle
Interventricular foramen (Monro)
IH
Fourth ventricle
Cerebral aqueduct (Sylvius)

VENTRICLES OF THE BRAIN

(3) —Viewed from the front—

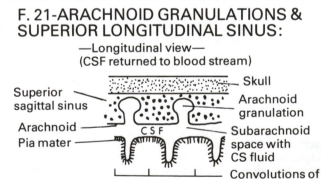

Right lateral ventricle
Left lateral ventricle
Third ventricle
Cerebral Aqueduct (Sylvius)
Fourth ventricle

F. 21-ARACHNOID GRANULATIONS & SUPERIOR LONGITUDINAL SINUS:
—Longitudinal view—
(CSF returned to blood stream)

Superior sagittal sinus
Skull
Arachnoid granulation
Arachnoid
CSF
Subarachnoid space with CS fluid
Pia mater
Convolutions of cerebrum

tentorium cerebelli within this fissure. The cerebellum consists of two lateral lobes or hemispheres joined across the midline. The cerebellum has fibers that connect it with the spinal cord, pons, midbrain, and forebrain. F.21-9.10
(Cerebellum = little brain).

The cerebellum acts as a coordinator of muscular movements, rendering them smooth and graceful rather than jerky. It is concerned with the maintenance of posture and equilibrium.

The brain stem, a term that is occasionally used, includes the midbrain, pons and medulla.

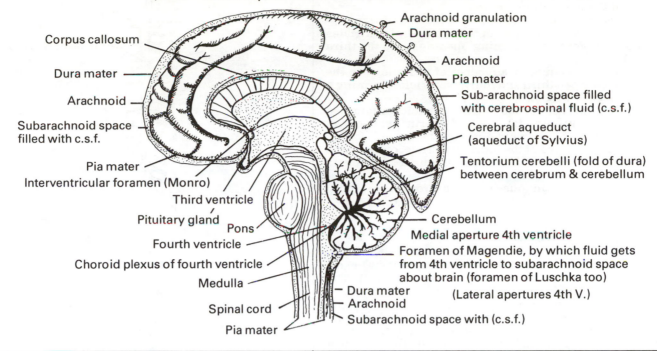

Corpus callosum
Dura mater
Arachnoid
Subarachnoid space filled with c.s.f.
Pia mater
Interventricular foramen (Monro)
Third ventricle
Pituitary gland
Pons
Fourth ventricle
Choroid plexus of fourth ventricle
Medulla
Spinal cord
Pia mater

Arachnoid granulation
Dura mater
Arachnoid
Pia mater
Sub-arachnoid space filled with cerebrospinal fluid (c.s.f.)
Cerebral aqueduct (aqueduct of Sylvius)
Tentorium cerebelli (fold of dura) between cerebrum & cerebellum
Cerebellum
Medial aperture 4th ventricle
Foramen of Magendie, by which fluid gets from 4th ventricle to subarachnoid space about brain (foramen of Luschka too)
(Lateral apertures 4th V.)
Dura mater
Arachnoid
Subarachnoid space with (c.s.f.)

S. 255 THE VENTRICLES OF THE BRAIN
F.21-13, 15

(L. ventriculus = a small cavity - a belly)

The ventricles of the brain are four cavities that lie within it: the right lateral ventricle, the left lateral ventricle, the third ventricle, and the fourth ventricle. They are filled with cerebrospinal fluid.

They are given the name ventricles of the brain rather than cerebral ventricles as they are not all located within the cerebrum.

1. **THE LATERAL VENTRICLES,** right and left, are located in the corresponding cerebral hemisphere. Each lies below the corpus collosum, and extends from front to back, corresponding somewhat in shape to that of the hemisphere within which it lies. Each has an opening into the third ventricle. Students should obtain radiographs showing the ventricles filled with air for study. The lateral view will show the lateral ventricles superimposed one upon the other. In this projection the ventricle resembles a painting of a human hand, often seen in the corridors of buildings pointing out the direction. The wrist represents the **posterior horn** in the occipital lobe. The hand is the **central part** or body lying within the parietal lobe. The index finger is the **anterior** horn, the blunt anterior end of the ventricle. The thumb is the **inferior horn** in the temporal lobe. In frontal projections the airfilled lateral ventricles resemble a butterfly with wings extended. If available a cast, or glass model filled with fluid should be utilized for this study.

The **interventricular foramen** (NA) or foramen of Monro (OT) is an opening between the medial margin of a lateral ventricle and the adjacent third ventricle. Hence there are two of them, right and left. F.21-15

2. **THE THIRD VENTRICLE,** a single cavity, lies in the midline of the diencephalon or afterbrain, between the medial margins of the thalami which form the lateral walls. The parts of the hypothalamus form its floor. In the frontal radiograph it appears as a narrow slit. In the lateral view it is four sided and appears much larger and lies within the curve formed by the parts of the lateral ventricle. It is connected by a narrow channel with the fourth ventricle. F.21-15

The cerebral aqueduct (NA), the aqueduct of Sylvius (OT) is a canal that passes between the third and fourth ventricles. It lies in the midline and extends down through the midbrain. F.21-15

3. **THE FOURTH VENTRICLE,** a single cavity lies in the hindbrain. The pons and upper medulla lie in front and the cerebellum lies posterior to it. In lateral diagrams it appears triangular in shape with the pointed end extending into the anterior surface of the cerebellum. The posterior boundary is spoken of as the roof of the fourth ventricle. There are three small openings in the roof of the fourth ventricle by which cerebrospinal fluid escapes from this cavity into the subarachnoid space surrounding the brain. See meninges below. F.21-15

The median aperture (NA) of the fourth ventricle, the foramen of Magendie (OT) is an opening between the fourth ventricle and subarachnoid space of the brain in the midline of the roof of the fourth ventricle.

The lateral apertures (NA) of the fourth ventricle, the foramina of Luschka (OT) are 2 openings between the fourth ventricle and subarachnoid space of the brain in the lateral parts of the roof of the fourth ventricle.

These median and lateral apertures allow cerebrospinal fluid to pass from the ventricular system into the subarachnoid space to circulate around the brain and subarachnoid space of the spinal cord.

The choroid plexuses are networks of veins in the wall of each ventricle. Special cells in these areas secrete the cerebrospinal fluid into the ventricles.

LOCATIONS OF VENTRICLES SUMMARIZED

The right lateral ventricle — in the right cerebral hemisphere below the corpus callosum.
The left lateral ventricle — in the left cerebral hemisphere below the corpus callosum.

The **third ventricle** — in the diencephalon of the forebrain, in the midline, between the thalami.

The **fourth ventricle** — in the hindbrain with the pons in front, and the cerebellum behind.

The **interventricular** foramina, or foramina of Monro — between the medial margins of the two lateral ventricles and the third ventricle, forming openings into the third ventricle.

The **cerebral aqueduct** or aqueduct of Sylvius — in the midbrain, between third and fourth ventricles, forming a passage between them.

The **median aperture** of the fourth ventricle, or foramen of Magendie, through the roof of fourth ventricle in cerebellum to subarachnoid space of the brain.

The **lateral apertures** of the fourth ventricle, or foramina of Luschka, two, through the roof of fourth ventricle to subarachnoid space of the brain.

They form a continuous passage from the lateral ventricles to the subarachnoid space of the brain.

S. 256   THE SPINAL CORD   F.21-14

(L. medulla spinalis = spinal marrow; the marrow or soft part within the vertebral column = spinal cord). The spinal cord, about 40 to 45 cm (16 to 18 inches) in length lies within the vertebral canal, which is formed by the vertebral foramina of the vertebrae and the ligaments between their arches. The cord is a continuation of the medulla oblongata and extends from the foramen magnum to the second lumbar vertebra. The spinal nerves below this level pass caudally as a bundle to their exits from the vertebral canal. They resemble a horse's tail, hence the name **cauda equina.**

The cord is incompletely divided into right and left lateral halves by anterior and posterior grooves that run lengthwise on the anterior and posterior surfaces. A bridge of nerve tissue in the central area connects the two halves.

The **gray substance** (matter) of the cord, composed largely of nerve cell bodies forms the inner part of the cord, an arrangement opposite to that in the cerebrum. In cross sections of the cord the gray substance forms an "H" shaped figure, one limb extending from front to back in each lateral half, and connected across the median line by the bar of nerve tissue. The anterior half of each limb is called the **anterior horn** (cornu), and the posterior half is named the **posterior horn** (cornu). Students must realize that in longitudinal sections of the cord these cornu are actually pillars of nerve cells that extend the length of the cord and are called columns. **The anterior and posterior columns** of gray substance are the pillars of nerve cells extending the length of the cord on each side, and connected across the midline. F.21-7 (2)

The **lateral horn** (lateral column) is a bulge on each limb of gray substance in the thoracic and upper lumbar segments. Cell bodies of the sympathetic nervous system are located here.

The **posterior column** (called cornu in cross section) contains sensory cell bodies, the anterior column contains motor cell bodies.

The **white substance** of the spinal cord is composed of nerve fibers (processes) and surrounds the gray substance. The "H" shaped gray substance separates this white substance into four large bundles, the anterior, posterior, and two lateral divisions. **Funiculi** are the four anterior, posterior, and two lateral divisions of white matter. (funiculus = cord). F.21-7 (2); F.21-14

Each funiculus is made up of several tracts or bundles of nerve fibers. Each tract has a specific position in the cord, a specific destination, and a specific function.

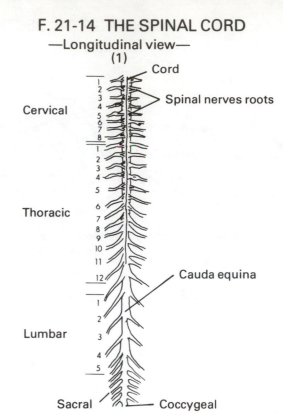

F. 21-14 THE SPINAL CORD
—Longitudinal view—
(1)

Cord

Spinal nerves roots

Cervical

Thoracic

Cauda equina

Lumbar

Sacral

Coccygeal

F. 21-14 (2) SECTION OF SPINAL CORD:

Root of spinal nerve

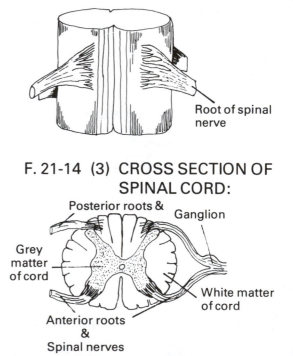

F. 21-14 (3) CROSS SECTION OF SPINAL CORD:

Posterior roots &

Ganglion

Grey matter of cord

White matter of cord

Anterior roots & Spinal nerves

The **motor tracts** in general are located in the anterior and lateral funiculi of the cord. Motor impulses to muscles, originating in cell bodies in the motor area of the cerebrum are relayed through their axons in the motor tracts through the midbrain, pons medulla and cord to various levels in the cord. These axons terminate by synapsing with motor cells in the anterior horn of gray matter. Axons of this second set of neurons pass out by spinal nerves. These motor tracts are also called **descending tracts.** Descending tracts from other parts of the brain also pass down the motor funiculi to their destinations.

The **sensory tracts** of nerve fibers, also called **ascending tracts** are located for the most part in the posterior and lateral funiculi of white matter. Sensory impulses entering the cord either end in the sensory cell bodies of the posterior horn, or are conveyed up or down the cord in the sensory tracts.

The role of the spinal cord in relation to the autonomic nervous system will be outlined in a subsequent section.

The spinal cord as suggested earlier helps to form reflex arcs at various levels of the cord.

S. 257 MENINGES OF THE BRAIN AND SPINAL CORD F. 21-13-16

(G. meninx = a membrane; pl. = meninges).

The meninges are the coverings of the brain and spinal cord. There are three of them. They enclose not only the brain but are continued down around the spinal cord. They are named the pia mater, the arachnoid, and the dura mater.

1. **THE PIA MATER** is the delicate inner membrane. It is closely applied to the outer surface of the brain and cord. It dips down into the sulci and fissures, and covers all gyri or convolutions. (L. pia = tender or clinging + mater = mother; hence a clinging membrane).

2. **THE ARACHNOID** is the middle layer and overlies the pia mater. It does not dip down into the many grooves so that a space is left between it and the pia. (G. arachne = spider + oid = like; i.e. spiderlike).

THE SUBARACHNOID SPACE is the space between the arachnoid and pia mater. The cerebrospinal fluid circulates in this space about the brain and around the cord. (Sub = under + arachnoid).

3. **THE DURA MATER** is the tough outer membrane covering the brain and cord. This membrane consists of two layers that in most areas are adherent to each other. The outer layer forms the periosteum on the inner surface of each cranial bone. The inner layer, the meningeal layer has a smooth inner surface adjacent to but not attached to the arachnoid. It becomes separated from the periosteal layer at the longitudinal and transverse fissures of the cerebrum. It dips down into these grooves to form folds that separate adjacent brain surfaces. (L. dura = hard or tough + mater = mother, i.e. tough membrane).

THE FALX CEREBRI is the fold of dura mater that dips down into the longitudinal fissure between the right and left cerebral hemisphere. 21-17

THE TENTORIUM CEREBELLI is a fold of dura mater that passes transversely across the upper end of the posterior cranial fossa in the transverse fissure. It separates the occipital lobe of the cerebrum from the upper surface of the cerebellum. F.21-17

THE EPIDURAL SPACE in the skull is the interval between the dura mater and cranial bones. In the spine it is the space between the dura mater and the vertebral canal.

THE SUBDURAL SPACE is the interval between the arachnoid and dura mater of the brain or the spinal cord.

THE ARACHNOID CISTERNS are small cavities where the pia mater and arachnoid are more widely separated from each other than usual. Cerebrospinal fluid collects in these pockets. They are found where one part of the

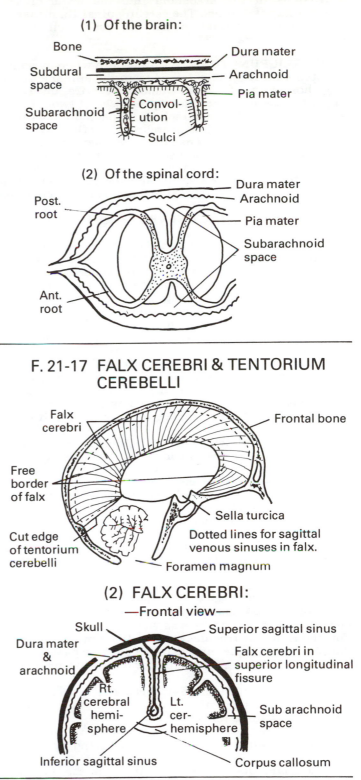

F. 21-16 THE MENINGES—Diagrammatic

(1) Of the brain:

Bone
Subdural space
Subarachnoid space
Dura mater
Arachnoid
Pia mater
Convolution
Sulci

(2) Of the spinal cord:

Post. root
Ant. root
Dura mater
Arachnoid
Pia mater
Subarachnoid space

F. 21-17 FALX CEREBRI & TENTORIUM CEREBELLI

Falx cerebri
Free border of falx
Cut edge of tentorium cerebelli
Frontal bone
Sella turcica
Dotted lines for sagittal venous sinuses in falx.
Foramen magnum

(2) FALX CEREBRI:
—Frontal view—

Skull
Dura mater & arachnoid
Rt. cerebral hemi-sphere
Lt. cer-hemisphere
Inferior sagittal sinus
Superior sagittal sinus
Falx cerebri in superior longitudinal fissure
Sub arachnoid space
Corpus callosum

brain meets another part at an angle, and the arachnoid bridges over the angle.

THE CEREBROSPINAL FLUID (C.S.F.) is a clear colorless fluid that is formed by cells in the choroid plexuses of all of the ventricles, and is secreted into their cavities. Because the ventricles open into each other the fluid passes down to the fourth ventricle and out through the median and lateral apertures of this ventricle into the subarachnoid space about the brain. It also circulates in the similar space of the spinal cord. It is reabsorbed into the blood through the superior sagittal sinus within the cranium.

ARACHNOID GRANULATIONS are small knoblike protrusions of the arachnoid into the walls of the superior sagittal sinus. The cerebrospinal fluid passes through these granulations into the venous sinus and blood. F. 21-15. Page 224.

A LUMBAR PUNCTURE is a procedure whereby some cerebrospinal fluid is obtained from the spinal cord. A long hollow needle with removable wire insert is inserted between the spinous processes of two lumbar vertebrae into the subarachnoid space to obtain a specimen of the fluid. Clear cerebrospinal fluid will drip out through the needle when the wire insert is removed. This can be examined microscopically or chemically for blood cells or microorganisms. This procedure is also utilized to administer a spinal anesthesia, or to relieve high pressure when cerebrospinal pressure is increased.

S. 258 THE CRANIAL NERVES F. 21-18

The cranial nerves form part of the peripheral nervous system. They arise directly from the base of the brain, and are not associated with the spinal cord. There are twelve pairs, each pair consisting of right and left nerves. They have their centers in the midbrain, pons, and medulla. They leave the cranium by foramina in the base of the skull. Most of them terminate in structures of the head or neck. Some are sensory, and convey sensory impulses to the brain, while others are motor and carry impulses out to muscles Some are both sensory and motor. Some of them contain neurons that form part of the autonomic nervous system. Each cranial nerve has a name, but in addition is numbered as the first cranial, second cranial, third cranial, etc., or by the Roman numerals from I to XII.

First cranial or olfactory nerve (I), the sensory nerve of smell, passes from the upper part of each nasal cavity through minute foramina in the cribriform plate of the ethmoid bone to the olfactory bulb within the cranium. From here it enters the olfactory tract to end in the temporal lobe of the brain.

Second cranial or optic nerve (II), the sensory nerve of sight, passes from the retina of the eyeball through the optic canal (foramen) to the optic chiasma within the cranium, and then by the optic tract to end in the occipital lobe of the cerebrum.

Third cranial or oculomotor nerve (III), a motor nerve to muscles of the eye, passes from the midbrain through the superior orbital fissure to the orbit. It has some sensory and automotic nerve fibers as well.

Fourth cranial or trochlear nerve (IV), a motor nerve to the superior oblique muscle of the eyeball, passes from the midbrain through the superior orbital fissure to the orbit.

Fifth cranial or trigeminal nerve (V), a motor and sensory nerve from the pons, with three divisions: (trigeminal = three roots)

(1) **ophthalmic nerve**, sensory through the superior orbital fissure to the eyeball, lacrimal gland, and upper face.

(2) **Maxillary nerve**, sensory through the foramen rotundum (round) to the face and upper teeth.

(3) **Mandibular nerve**, sensory and motor, sensory to the lower teeth, mouth, and lower part of the face, motor to the muscles of mastication (chewing), through the foramen ovale.

Sixth cranial or abducent nerve (VI), a motor nerve from the pons through the superior orbital fissure to the lateral rectus muscle of the eye.

Seventh cranial or facial nerve (VII), a motor nerve to muscles of the face, from the pons through stylomastoid foramen, and sensory nerve to taste buds on the tongue.

Eight cranial or vestibulocochlear nerve (VIII), acoustic or auditory nerve, through the internal acoustic porus

F. 21-18 TWELVE PAIRS OF CRANIAL NERVES
—Directly from brain—

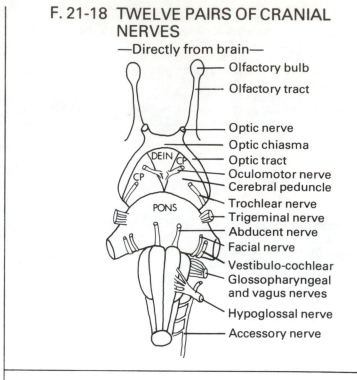

and internal acoustic meatus to the internal ear. There are two parts:

(1) **vestibular nerve**, sensory from semicircular canals of of internal ear to medulla, pons and the cerebellum, controlling equilibrium and coordination.

(2) **Cochlear nerve**, nerve of hearing, sensory, from the organ of Corti in the internal ear to the temporal lobe of the cerebrum.

Ninth cranial or glossopharyngeal nerve (IX), (glossus = tongue + pharynx = throat) sensory and motor, from medulla through jugular foramen to muscles of pharynx (swallowing) and to taste buds on posterior part of tongue — taste.

Tenth cranial or vagus nerve (X), sensory and motor from medulla through jugular foramen, and parasympathetic nerve to the thoracic and abdominal organs, visceral sensory, and motor to the larynx and pharynx.

Eleventh cranial or accessory nerve (XI), two parts: the **cranial part** from the medulla through the jugular foramen with the vagus nerve to the thoracic and abdominal organs, and the **cervical** part from the upper cervical segments of the spinal cord, motor and sensory about the neck, shoulder and upper back.

Twelfth cranial or hypoglossal nerve (XII), (hypo = below + glossus = the tongue) motor nerve to the tongue and muscles under this organ, from the medulla through the hypoglossal canal to the tongue.

Notes: the optic, oculomotor, trochlear, and abducent nerves pass to the orbit.

The oculomotor, trochlear, and abducent nerves enter the orbit through the superior orbital fissure.

The ninth, tenth, and eleventh cranial nerves leave the skull through the jugular foramen. The third, seventh, and ninth, cranial nerves contain autonomic fibers. These fibers travel with the cranial nerves to their destinations.

The student should again note that the cranial nerves form part of the peripheral nervous system. Four of them are of some concern to the radiological technician. These are: the olfactory — nerve of smell; the optic — nerve of sight; the vestibulocochlear — the nerve of hearing, and equilibrium, and the vagus — the parasympathetic nerve to the thoracic and abdominal viscera. These will be described in a subsequent section.

The spinal nerves forming part of the peripheral nervous system and consisting of 31 pairs, are connected with the spinal cord. There are 8 cervical, 12 thoracic, 5 lumbar, 5 sacral. and 1 coccygeal, right and left. The first cervical leaves the cord above the first cervical vertebrae, the others caudal to the corresponding vertebra, in the intervertebral foramen on either side. These nerves connect the spinal cord with body structures such as muscles and skin, etc. Each spinal nerve is a mixed nerve and consists of two parts, sensory and motor nerve fibers. Each spinal nerve, readily visible to the eye contains many axons and dendrites forming a cordlike structure.

The sensory fibers begin as receptors of sensory impulses in the skin, muscles, joints, etc. The dendrites do not enter the cord but end in cell bodies in the intervertebral foramina at each level. These cell bodies are called the **dorsal root ganglia.** Axons from these cell bodies enter the dorsal part of the cord on each side. These axons may end by synapsing with dendrites of cell bodies in the adjacent gray matter in the posterior column. They may also pass up or down in one of the posterior tracts of white fibers to end elsewhere.

The motor fibers or axons originate in cell bodies in the anterior column of gray substance (cornu). These axons leave the cord at each level and pass out through an intervertebral foramen to join the sensory fibers and form a spinal nerve. The motor fibers end as endorgans on muscle fibers, or about secreting glands.

Each spinal nerve is connected to the spinal cord by two roots. These unite to form the spinal nerve.

The ventral root of a spinal nerve consists of the bundle of axons that leaves the cord at its anterior surface — motor fibers F.21-14 (3)

The dorsal root consists of a bundle of dendrites that enters each intervertebral foramen and contains sensory nerve fibers.

S. 260 DISTRIBUTION OF SPINAL NERVES

After emerging from the spinal cord, each spinal nerve, right or left, divides into dorsal and ventral divisions or branches. The dorsal division supplies the skin and muscles of the posterior part of the trunk in the area where it emerges. The ventral division of each spinal nerve follows a different course to be outlined below.

In the introduction to this chapter a plexus of nerves was defined as a network of branching nerve fibers, formed by several nerves. The ventral divisions of spinal nerves (except the thoracic) form right and left cervical, lumbar, and sacral plexuses. The ventral divisions of

F. 21-19 SPINAL NERVE—THORACIC
—Motor & Sensory parts—

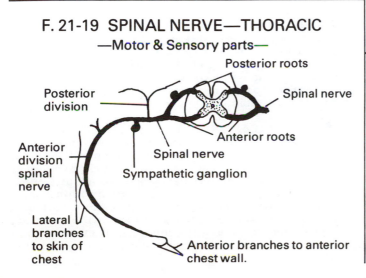

F. 21- PLEXUSES OF SPINAL NERVES

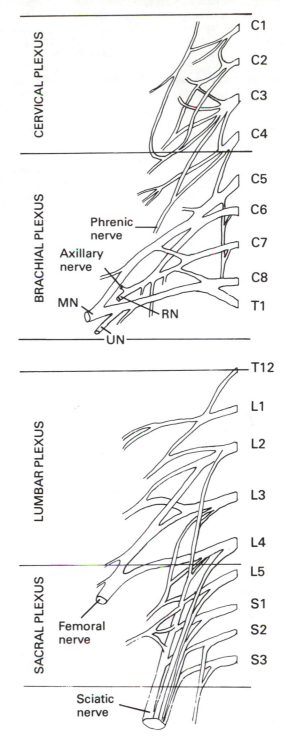

several adjacent spinal nerves unite to form trunks, which redivide into branches containing a different arrangement of nerve fibers, and finally form several nerves with components of several original trunks. A nerve plexus could be compared to an aerial view of a railway center with its complicated system of tracks.

The cervical plexus on each side of the body is made up of the ventral divisions of the first, second, third and fourth cervical nerves. This plexus is located high up in the muscles on each side of the neck. Nerves emerging from it supply the skin and muscles of the scalp, neck, and part of the face. **The phrenic nerve** to the diaphragm comes from this plexus and descends in the neck and thorax. F.21-19a

The brachial plexus is composed of ventral divisions of the lower four cervical and first thoracic nerves. It lies posterior to the clavicle. Nerves emerging from it supply the shoulder and upper limb. The median, radial, ulnar, and musculocutaneous nerves are important divisions. (L. Brachium = arm) — brachial. F.21-19a.

The thoracic nerves do NOT form a plexus, but the ventral divisions pass around the thoracic wall as intercostal nerves between adjacent ribs. The lower six, after leaving the costal cartilages of the lower chest wall, emerge in the anterior abdominal wall. Each successive nerve is directed more obliquely down and forward so that the anterior end of the twelfth terminates close to the symphysis pubis. The anterior abdominal wall is therefore enervated by the thoracic spinal nerves. F.21-19

The lumbar plexus is composed of the ventral divisions of the first, second, third, and part of the fourth lumbar spinal nerves. This plexus lies in the posterior abdomen in the area of the psoas major muscle. Its largest branch is the **femoral nerve** that passes into the thigh with the femoral artery. F.21-19b.

The sacral plexus is formed from the ventral divisions of the fourth and fifth lumbar, and the first second and third sacral spinal nerves. It is located within the pelvis on the posterolateral wall. The largest branch, **the sciatic nerve** leaves the pelvis posteriorly through the greater sciatic notch to reach the posterior thigh. F.21-19b.

The lumbar and sacral plexuses, by their many nerves including the femoral and sciatic enervate the buttocks, external genitals, and lower limb, including the muscles and skin.
Each spinal nerve supplies a different area of skin of the body or limbs, so that localized pain or numbness of an area indicates a lesion of a certain nerve. The medial surface of the thigh for example is supplied by the ventral branches of the second and third lumbar nerves.

S. 261   THE AUTONOMIC NERVOUS SYSTEM
F.21-20

(G. autos = self + nomos = a law; i.e. a law unto itself - self regulating).
The autonomic nervous system is classified as a part of the peripheral nervous system. Its controlling centers however lie within the central nervous system, in the midbrain, pons, medulla and spinal cord. These centers are not regulated by the cerebrum, the centers of consciousness, but are independent or involuntary. The subject is usually unaware of what is taking place in the organs controlled by this system, and cannot by conscious effort influence its activities. For instance, it is not possible to consciously cause contraction of the wall of the stomach, nor to influence the secretion of a salivary gland. The name autonomic restricts it to motor or efferent nerves. There are also sensory or afferent neurons from the viscera included in the cranial and spinal nerves as well as visceral sensory fibers that accompany motor visceral nerves. The motor neurons, to be studied here control:
1. The contraction and dilation of all organs with visceral (involuntary) muscles in their walls.
2. The rate of the heart beat.
3. The activity of many secreting exocrine glands.
The structures supplied include the pupillary muscles of the eyes, salivary and lacrimal glands, sweat glands, capillary blood vessels of the skin, other blood vessels, the heart, bronchi, digestive organs and accessory digestive organs, (peristalsis and secretions), urinary system, and reproductive organs.

There are two divisions of this autonomic nervous system. Each one is often referred to as "a system" or in NA as "a part". The two parts frequently supply nerves to the same organ, and these nerves often run together. The two parts are antagonostic and cause opposite responses, e.g. where one will cause contraction of a visceral muscle the other will cause dilatation and if one stimulates glands to secrete the other inhibits their secretion. In both parts two neurons are required to convey impulses from the center in the lower brain or spinal cord to the viscus, gland or skin. The two divisions of the autonomic system include:
1. The sympathetic nervous system or part
2. The parasympathetic nervous system or part

Before studying the two parts of this system the autonomic nerve plexuses and ganglia must be defined. A plexus of nerve fibers is a network of them similar to the complicated network of railway tracks in a railway freight yard. Bilateral plexuses involving some groups of spinal nerves have already been defined as the cervical brachial, lumbar, and sacral plexuses in S. 260. The **autonomic nerve plexuses** on the other hand are networks of nerve fibers formed by axons from both parts of this division. These are located at strategic points in the thorax, abdomen and pelvis, with subplexuses originating from the major networks. The cardiac and pulmonary plexuses lie in the mediastinum, the celiac (solar) plexus upon the abdominal aorta at the origin of the celiac trunk. Its offshoots are the renal, abdominal aortic, mesenteric, and hypogastric plexuses. The pelvic plexus lies in the pelvis and sends nerve fibers to the pelvic organs. The nerve fibers that eventually leave these plexuses frequently travel on branches of the aorta to the organs that are supplied by these vessels.
Autonomic ganglia are groups of nerve cell bodies located in the main plexuses defined above, or upon or close to the organs supplied, or the trunk chain sympathetic ganglia to be described below in S. 262. They are the cell bodies of the second neurons required to carry impulses to organs.

S. 262   THE SYMPATHETIC NERVOUS SYSTEM OR PART, OR THE THORACOLUMBAR SYSTEM   F.21-20(1)

This part of the autonomic nervous system has:
— nerve centers in the spinal cord
— communicating rami, i.e. branches — axons
— sympathetic trunks and trunk ganglia
— autonomic plexuses and ganglia
— nerve fibers (axons) to the organs supplied.
These structures make up two neurons from the cord to the organ. They are paired structures. To render the study intelligible a brief discription of each component is included here.

THE NERVE CENTERS of this sympathetic part are the groups of cell bodies that lie in the lateral horns of gray matter of the spinal cord in the 12 thoracic and upper lumbar segments of the cord on each side. **Thoracolumbar**, an alternate name, is appropriate as it defines the location. These paired centers control the activities of the sympathetic division.

THE WHITE COMMUNICATING RAMI are axons from the cell bodies of the sympathetic centers of the cord defined above. They join the motor roots of the spinal nerves at the same level, i.e. 12 thoracic and upper lumbar. The other spinal nerves do not acquire these rami. These axons leave the spinal nerves within a short distance to enter the trunk ganglion beside a vertebral body and at the same level. They may end in this ganglion by synapsing with a ganglion cell. They may pass through the ganglion, or may pass up or down the sympathetic trunk.

F. 21-20 THE AUTONOMIC NERVOUS SYSTEM:

(1) Centers for sympathetic part:

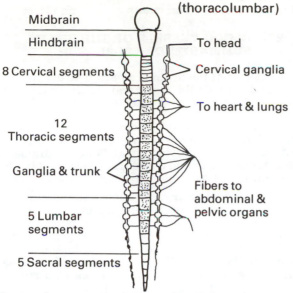

(thoracolumbar)

- Midbrain
- Hindbrain
- 8 Cervical segments
- To head
- Cervical ganglia
- To heart & lungs
- 12 Thoracic segments
- Ganglia & trunk
- Fibers to abdominal & pelvic organs
- 5 Lumbar segments
- 5 Sacral segments

Centers for sympathetic located in 12 thoracic and upper lumbar segments of spinal cord; form ganglia and trunks.

(2) Centers for parasympathetic part:

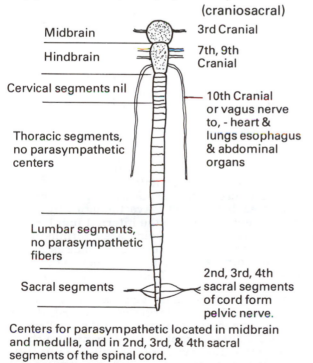

(craniosacral)

- Midbrain — 3rd Cranial
- Hindbrain — 7th, 9th Cranial
- Cervical segments nil
- 10th Cranial or vagus nerve to, - heart & lungs esophagus & abdominal organs
- Thoracic segments, no parasympathetic centers
- Lumbar segments, no parasympathetic fibers
- Sacral segments — 2nd, 3rd, 4th sacral segments of cord form pelvic nerve.

Centers for parasympathetic located in midbrain and medulla, and in 2nd, 3rd, & 4th sacral segments of the spinal cord.

THE SYMPATHETIC TRUNK GANGLIA consist of about 25 small groups of nerve cell bodies lying on either side of the vertebrae. They extend from the base of the skull to the sacrum on either side. The number varies possibly with 3 cervical, 12 thoracic, 5 lumbar, and 5 sacral ganglia bilaterally. They are connected to form a chain by the sympathetic trunks. White rami reach them through the 12 thoracic and upper two lumbar spinal nerves only. The cell bodies synapse with these white rami, and have axons that are discussed below. These cells form the second neuron.

THE SYMPATHETIC TRUNKS are made up of the trunk ganglia and nerve fibers (axons) that connect them, or run through them to higher or lower levels. The white rami from the upper thoracic centers, for instance pass up these trunks to reach the cervical trunk ganglia, as these have no white rami of their own. Similarly some white rami pass down by the sympathetic trunk to reach the lower lumbar and sacral trunk ganglia. These also have no white rami of their own.

PLEXUSES & NERVE FIBERS. Axons from the trunk ganglia plus some axons that simply pass through the trunk ganglia help to form the autonomic plexuses or nerve networks on the abdominal aorta — the celiac, mesenteris, etc., the cardiac and pulmonary plexuses in the thorax, and other plexuses in the head, etc.

GANGLIA OTHER THAN TRUNK GANGLIA are associated with the celiac and other plexuses. These consist of cell bodies that synapse with axons of those cell bodies in the centers in the cord that did not end in trunk ganglia but passed through them, e.g. the splanchnic nerves. They form the cell bodies of the second neurons. Their axons end in the organs supplied by them.

Frequently the nerve fibers or axons from trunk ganglia, the plexuses, etc., accompany the branch of an artery to the organ supplied. From the superior cervical trunk ganglion, for instance, the nerve fibers join the internal carotid and vertebral arteries and their branches to reach the brain, or the external carotid to reach the scalp, face and neck. Fibers from the plexuses such as the celiac follow along with the branches of the aorta to reach their termination in organs supplied by these arteries. Similarly nerve fibers accompany the arteries to the upper and lower limbs.

From the trunk ganglia some axons rejoin the spinal nerves to be distributed to sweat glands, skin capillaries, etc., along with the spinal nerves. These are the gray communicating rami.

To summarize: two neurons are required — a cell body within the spinal cord with its axon leaving the cord; a second cell body in a trunk ganglion or other ganglion, the axon of which reaches the organ supplied.

S. 263 THE PARASYMPATHETIC NERVOUS SYSTEM OR PART OR THE CRANIO-SACRAL NERVOUS SYSTEM OR PART F.21-20(2)

The parasympathetic part of the autonomic nervous system provides nerve fibers to many but not all of the same organs supplied by the sympathetic part. Stimulation of these nerves produces reactions opposite to those induced by stimulation of the sympathetic nerves. Most of the structures are paired, right and left. Two neurons are required to convey impulses to the organs. There are:

Nerve centers within the central nervous system.
Axons or nerve fibers from these centers
Ganglia in the walls of the organs supplied
Axons from these ganglia in the supplied organs

THE NERVE CENTERS are groups of cell bodies in the cranial or sacral parts of the central nervous system. **The cranial centers** are located in the midbrain, pons, and medulla oblongata on each side.

The sacral centers are located in the 2nd, 3rd, and 4th sacral segments of the spinal cord.

AXONS OR NERVE FIBERS FROM THE CRANIAL CENTERS in the brain join the 3rd, 7th, and 9th cranial nerves in order to reach the organs supplied by them, i.e., the muscles of the pupils of the eyes, the lacrimal, parotid, submandibular, and sublingual glands. They end in ganglia close to or within the organs supplied.

The vagi nerves (s. vagus) or 10th cranial nerves originate in centers in the medulla on both sides. Their motor fibers or axons pass out of the base of the skull through the jugular foramina. They pass down through the neck to reach the thorax and abdomen. They form the esophageal plexus along the esophagus. They help to form the pulmonary, cardiac, celiac, and other plexuses of nerve fibers. In the abdomen they follow branches of the abdominal aorta along with sympathetic fibers to reach the organs supplied. In additition to the accessory digestive organs, the esophagus, stomach, and small intestine they supply the colon to the splenic flexure. Students should note particularly that the vagus nerve supplies the thoracic and abdominal organs with parasympathetic nerve fibers.

AXONS OF NERVE CENTERS IN THE SACRAL SEGMENTS of the spinal cord join the pelvic plexus and end in ganglia close to pelvic organs and descending colon and rectum.

THE GANGLIA containing the cell bodies of the second neurons are located in the walls of the organs supplied. They synapse with the axons from the first neurons defined above. They have axons that are distributed to organs.

THE AXONS FROM THE GANGLIA are distributed to the organs supplied.

To summarize: two neurons are required — a cell body within the base of the brain or sacral segment of the spinal cord — with its axon passing to a ganglion close to or upon the organ supplied — from this ganglion (cell body) a second axon passes to the organ supplied.

It should be noted that although the cerebrum has no direct control of the autonomic system, subconsciously the hypothalamus, a part of the diencephalon, exerts definite control upon this system.

S. 264 FUNCTIONS OF THE AUTONOMIC NERVOUS SYSTEM

The simplest way to explain the actions of the two parts of the autonomic nervous system is to assume that the sympathetic part is the system that is active in emergencies. Fright, flight, and fight are involved. Those body functions that are essential to survival are activated. The heart rate is increased, the blood pressure becomes elevated, the bronchi dilated, and the digestive processes become arrested, etc. The action is similar to the responses to stimulation of the medulla of the suprarenal glands.

The parasympathetic part on the other hand, is concerned with normal body functions, usually with reactions opposite to those of the sympathetic part. The heart is slowed, blood vessels dilate, blood pressure falls, and the digestive processes and glandular secretions are stimulated. The following list demonstrates some of the responses:

Organ	Sympathetic	Parasympathetic
Pupil	dilates	constricts
Bronchi	dilates	constricts
Heart	speeds up	slows down
Digestive —		
stomach	decrease	increases
intestine	peristalsis &	
GB & Bile ducts	contractions	
Glands —		
sweat	sweating	nil
salivary	decreases secretions	increases secretions
gastric	decreases secretions	increases secretions
intestinal		
Arteries of skin & viscera	constricts	no effect

S. 265 BLOOD VESSELS OF THE BRAIN SUMMARIZED F.21-21, 22, 23

The arteries of the brain are paired structures, right and left, and include the right and left internal carotid and right and left vertebral arteries.

1. **The internal carotid artery**, a branch of the common carotid, enters the brain through the carotid canal in the petrous part of the temporal bone. It divides into an anterior cerebral, and a middle cerebral artery, to supply the brain. F.21-21

2. **The vertebral artery**, a branch of the subclavian artery, reaches the skull by ascending through the foramina transversaria of the cervical vertebrae and by the foramen magnum. The two vertebral arteries, left and right, join to form the basilar artery on the upper surface of the basilar part of the occipital bone. This vessel gives off cerebellar branches and divides into right and left posterior cerebral arteries. F.21-22

3. **The middle meningeal artery**, a branch of the maxillary part of the external carotid artery, enters the skull through the foramen spinosum. Its branches do not supply the brain but pass upward in grooves on the inner surface of the skull to supply the adjacent bones and dura mater. These grooves are visible in radiographs of the skull in the lateral projection.

The veins of the brain empty into large trunk vessels called the venous sinuses of the dura mater. F.21-23

1. **The superior sagittal sinus, and the inferior sagittal sinus** run from front to back in the falx cerebri that separates the cerebral hemispheres. The superior drains into a transverse sinus, and the inferior becomes the straight sinus which drains into the other transverse sinus.
2. **The transverse sinuses**, right and left, pass transversely across the inner surface of the occipital bone in grooves that are visible in radiographs of the occipital region. They end posterior to the petrous parts of the temporal bone in the sigmoid sinuses.
3. **The sigmoid sinuses** pass down in the posterior cranial fossa on each side and leave the skull by the jugular foramina, emptying into the internal jugular veins.
4. **The internal jugular veins** are continuations of the sigmoid sinuses of the dura. Each passes down in the neck to join a subclavian vein and form the brachiocephalic vein (OT, innominate vein).

There are other small venous sinuses within the cranium as well as the larger ones described above.

It is of some interest to know that if a needle is inserted through the anterior fontanelle of an infant in the midline it will enter the superior saggital sinus. Fluid or medication may be introduced by this route when necessary, and blood samples may be obtained.

S. 266 PROTECTION OF THE BRAIN AND SPINAL CORD

The skull protects the brain from injury. Blows to the head do not usually affect the brain.

The vertebrae and ligaments between them afford considerable protection to the spinal cord. In most injuries vertebrae must be fractured or dislocated to produce cord damage.

The meninges and cerebrospinal fluid also give some protection to both the brain and spinal cord.

F. 21-21 LEFT ANTERIOR & MIDDLE CEREBRAL ARTERIES:

(1) Frontal view—as in carotid arteriogram-

(2) Left lateral view

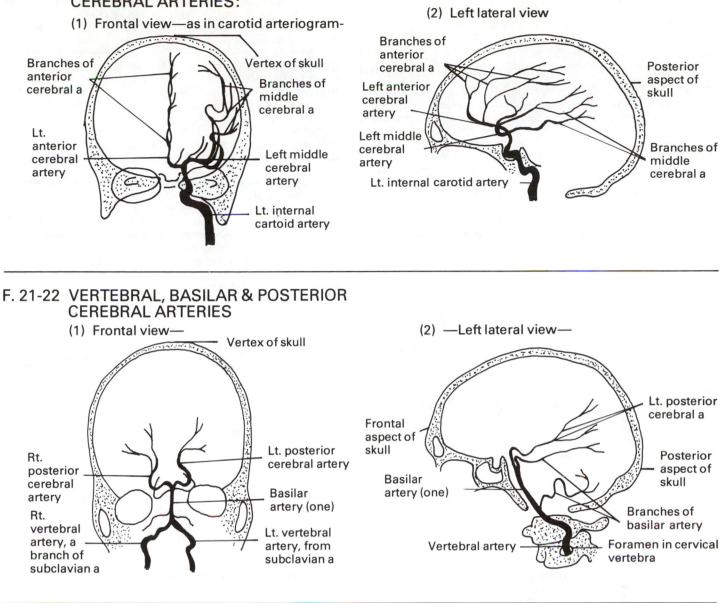

Frontal view labels:
- Branches of anterior cerebral a
- Vertex of skull
- Branches of middle cerebral a
- Lt. anterior cerebral artery
- Left middle cerebral artery
- Lt. internal cartoid artery

Left lateral view labels:
- Branches of anterior cerebral a
- Posterior aspect of skull
- Left anterior cerebral artery
- Left middle cerebral artery
- Branches of middle cerebral a
- Lt. internal carotid artery

F. 21-22 VERTEBRAL, BASILAR & POSTERIOR CEREBRAL ARTERIES

(1) Frontal view—

(2) —Left lateral view—

Frontal view labels:
- Vertex of skull
- Rt. posterior cerebral artery
- Lt. posterior cerebral artery
- Basilar artery (one)
- Rt. vertebral artery, a branch of subclavian a
- Lt. vertebral artery, from subclavian a

Left lateral view labels:
- Frontal aspect of skull
- Lt. posterior cerebral a
- Posterior aspect of skull
- Basilar artery (one)
- Branches of basilar artery
- Vertebral artery
- Foramen in cervical vertebra

F. 21-23 INTRACRANIAL VENOUS SINUSES: (of DURA MATER)

(1) Left lateral view

(2) Viewed from interior of base of skull.

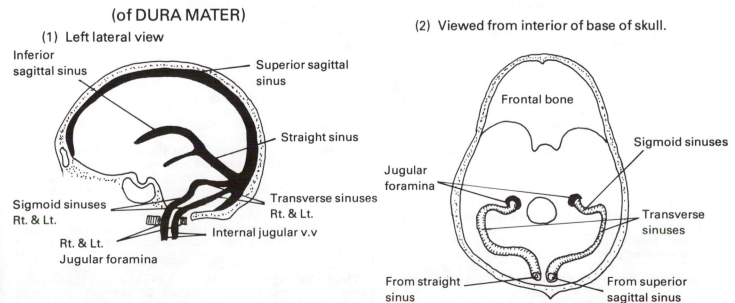

Left lateral view labels:
- Inferior sagittal sinus
- Superior sagittal sinus
- Straight sinus
- Sigmoid sinuses Rt. & Lt.
- Transverse sinuses Rt. & Lt.
- Internal jugular v.v
- Rt. & Lt. Jugular foramina

Base of skull view labels:
- Frontal bone
- Sigmoid sinuses
- Jugular foramina
- Transverse sinuses
- From straight sinus
- From superior sagittal sinus

General — system of communication and control

Sensory neurons conduct nerve impulses from the skin, muscles, joints, etc. into the spinal cord or up to the brain. These include sensations of touch, pressure, pain, cold, heat, and position.

Motor neurons conduct nerve impulses from the brain to other parts of the brain or spinal cord or from the spinal cord to muscles or secreting glands, causing movement or secretion.

Dendrites convey nerve impulses towards the cell body.

Axons conduct nerve impulses away from the cell body.

Receptors are sensory end organs, structures at the distal ends of dendrites that pick up sensations in the skin, muscles, joints, etc. and transmit these nerve impulses to sensory neurons.

Effectors are structures at the distal ends of the axons that transmit nerve impulses from the motor neuron to muscle fibers or **secreting glands.**

A synapse is a very minute gap between the end of one axon and the receptor at the distal end of a dendrite of an adjacent neuron. The nerve impulse is carried across the gap by some chemical means.

A reflex consists of a stimulus causing a nerve impulse to pass from the skin, etc. through a sensory neuron to the spinal cord, then through a connecting neuron (internuncial neuron) to a motor neuron, the axon of which conducts a nerve impulse to a group of muscles, causing them to contract and cause movement. For example, when a finger is pricked by a pin the nerve impulse causes the finger to be pulled away from the pin.

Gray substance consists of the cell bodies of neurons and direct the activity of the nerve cell.

White substance includes dendrites and axons and conducts nerve impulses. They are also named nerve fibers.

Neuroglia are connective tissue cells that lie among the neurons and support the adjacent neurons.

The cerebrum

The motor area of the cerebrum is located in the frontal lobe of each hemisphere anterior to the central sulcus. This area controls all movements of the skeletal muscles of the body.

The sensory area is located in the parietal lobe of each hemisphere posterior to the central sulcus. This area receives and interprets sensory nerve impulses from all parts of the body.

Auditory centers are located in the temporal lobes. These are concerned with hearing.

Visual centers lie in the occipital lobes of the cerebrum, and are concerned with sight.

Olfactory centers are in the temporal lobes, and are concerned with smell.

Gustatory centers are probably located in the parietal lobes. They are concerned with taste. Higher mental functions such as memory, speech, etc. are said to be scattered through the cerebrum rather than localized in any one area.

The corpus callosum, composed of a broad band of white nerve fibers conveys nerve impulses from one cerebral hemisphere across the midline to the other cerebral hemisphere.

Association fibers conduct nerve impulses from one gyrus to another, or from one lobe to another in the same hemisphere.

Projection fibers convey impulses from one part of the brain to another part, or to the spinal cord.

The diencephalon — the thalamus and hypothalamus.

The thalamus, part of the diencephalon, lies below the cerebrum, and in the midline adjacent to the third ventricle. It is a relay station for sensory nerve impulses and sorts them out before passing them on to the cerebrum. It is also a relay station for motor impulses on their way to other parts of the brain or cord.

The hypothalamus manufactures oxytocin and antidiuretic hormones that are then stored in the posterior lobe of the pituitary gland. These are released as required. It also controls the release of the trophic hormones secreted by the anterior lobe of the pituitary gland.

It exerts some control over the autonomic nervous system and emotional states such as rage. It controls thirst, hunger, body temperature and the waking state.

The midbrain — cerebral peduncles and quadrigeminal bodies (colliculi).

The midbrain contains some nerve centers. It has ascending and descending tracts that convey nerve impulses up or down to other parts of the brain or cord. The cerebral peduncles convey motor impulses from the forebrain to the hindbrain or spinal cord. The quadrigeminal bodies are way stations concerned with sight and hearing.

The hindbrain — pons, medulla, cerebellum.

The pons, a bridge, has nerve fibers, passing from one part of the cerebellum across the midline to the other side of it.

It has centers for several cranial nerves. It has ascending and descending tracts between the midbrain above and medulla below. These conduct nerve impulses up and down to adjacent parts.

The cerebellum is concerned with coordination of muscular movements causing them to be smooth and graceful rather than jerky.

The cerebellum also controls equilibrium. It receives sensory impulses from the semicircular canals in the ear, and from other parts of the body. It directs contraction of some muscles, and partial relaxation of others to maintain the upright position.

The medulla oblongata contains centers that control some of the cranial nerves.

Other centers, that are necessary to maintain life, control cardiac, respiratory, and vasomotor activity. The vasomotor are concerned with the contraction of blood vessels, and hence blood pressure. The medulla oblongata has also centers concerned with coughing, sneezing, vomiting and swallowing.

The medulla has ascending (sensory) and descending (motor) tracts by which nerve impulses are conducted up or down to other parts of the brain and spinal cord. As most of these tracts cross over to the opposite side in the medulla, one side of the brain communicates with the opposite side of the body.

The spinal cord

Ascending (sensory) tracts in the spinal cord transmit sensory impulses received from posterior roots of spinal nerves to various parts of the brain. Descending (motor) tracts conduct motor impulses from various parts of the brain to synapse with anterior roots of spinal nerves.

The spinal cord also has reflex arcs that are responsible for many reflex acts such as the knee jerks. The thoracic and upper lumbar segments have control centers for the sympathetic nervous system.

The sacral segments of the cord have centers that control the parasympathetic functions of the pelvic organs through the pelvic nerves.

The spinal nerves

The sensory part of each spinal nerve conducts sensory impulses from the skin, muscles, joints, etc. to synapse with neurons in the posterior horns of gray substance of the spinal cord, for transmission by ascending tracts to various parts of the brain. The motor part of each spinal nerve conducts impulses from descending tracts of the cord out to skeletal muscles or secreting glands.

The cranial nerves summarized

Sensory;
- 1st. olfactory — smell — nasal cavities
- 2nd. optic — vision — eyeball
- 8th. vestibulocochlear: two parts —
  cochlear — hearing — internal ear
  vestibular — equilibrium — ear.

Motor;
- 3rd. oculomotor — muscle of eye, muscles of iris and ciliary body
- 4th. trochlear — muscle of eye
- 6th. abducent — muscle of eye
- 12th. hypoglossal — muscles of tongue and under tongue

Sensory and Motor
- 5th. trigeminal: three parts — opthalmic, maxillary, mandibular. motor — muscles of mastication, sensory — eyeball, face, part of scalp, nasal cavities, mouth, teeth, anterior tongue
- 7th. facial — motor to facial muscles, sensory — taste — anterior tongue, secretion of saliva — submandibular and sublingual glands and secretory to lacriman gland (parasympathetic)
- 9th. glossopharyngeal — motor for swallowing, sensory — posterior tongue, pharynx, secretion of saliva by parotid gland, (parasympathetic) taste — posterior tongue,
- 10th. vagus — motor to larynx, pharynx, thoracic and abdominal organs (parasympathetic, peristalsis, secretory glands) sensory to thoracic and abdominal organs
- 11th. accessory — two parts — (1) with vagus, (2) skeletal muscles of shoulder and neck — motor.

The ventricles of the brain secrete cerebrospinal fluid into their cavities. This then circulates in the subarachnoid space about the brain and spinal cord.

Cerebrospinal fluid forms a fluid cushion that surrounds the brain and cord, and provides some protection for these organs from physical trauma.

The meninges form three coverings that enclose the brain and spinal cord affordings some protection. They also provide a space for the circulation of cerebrospinal fluid about the brain and spinal cord.

The autonomic nervous system

Centers in the midbrain, hindbrain, and spinal cord control the activity of many organs with visceral muscle in their walls, as well as of many secreting glands. Activities of these organs may therefore be carried on independent of any control by the cerebrum. Thus we breathe while asleep, and the heart pumps. Also included are contraction and dilatation of the iris (pupil), changes in convexity of the lens, rate of heart beat, constriction of bronchi and blood vessels, peristalsis of digestive organs, ureters, etc. and secreting activity of salivary glands, lacrimal glands, secretion of digestive enzymes, of mucus, etc.

The parasympathetic part is concerned with maintaining the normal activity while the sympathetic part slows or stops normal activity, and is active in emergent situations — fright — flight — fight. The student should review the table included in S. 264.

S. 268 SOME ANOMALIES OF THE NERVOUS SYSTEM

Anencephalus — absence of a brain: adj. anencephalic.

Microcephalus — a small head: adj. microcephalic.

Hydrocephalus — a large head due to a block in the apertures in the roof of the fourth ventricle resulting in an accumulation of cerebrospinal fluid in the ventricles. These enlarge and compress the brain against the skull. (Hydro — water — cephalus, head; i.e. water in the head).

Meningocele — a swelling in the midline of the subject's back due to a hernia of the spinal cord and meninges through a defect in a vertebral arch. (cele = a swelling, i.e. a swelling of the meninges).

S. 269 SOME PATHOLOGICAL CONDITIONS

Encephalitis — an inflammation of the brain.

Meningitis — an inflammation of the meninges, the membranes covering the brain and spinal cord.

Poliomyelitis — infantile paralysis — an inflammation of the motor cell bodies in some part of the spinal cord, caused by a virus. Some cell bodies are destroyed along with their axons to skeletal muscles, resulting in paralysis of the involved muscles.

S. 270 APPLICATION TO RADIOGRAPHY F.21-24

Radiographs demonstrating normal cerebral ventricles and blood vessels, as well as the subarachnoid space of the brain and cord are utilized in teaching the normal anatomy. For this reason a very brief introduction to the names of examinations employed and a definition of each is included in the paragraphs to follow.

The various parts of the brain, the fluid filled ventricles, the blood vessels and the subarachnoid space and the fluid within it, are not visible as separate images in radiographs of the skull or vertebrae. They must be filled with air or an opaque medium in order to render them visible.

In the examination of patients with suspected brain abscess, hemorrhage, tumors, or diseases of the blood vessels of the brain, these examinations are used. In routine skull films these conditions are not outlined. They may displace a calcified pineal gland from its normal position in the midline. They may be calcified and so become visible. A tumor of the pituitary gland may enlarge or erode the sella turcica or clinoid process.

Since the ventricles of the brain and blood vessels may be deformed or displaced from their normal position by these conditions, the examination of these structures may help in obtaining a diagnosis.

In **VENTRICULOGRAPHY** a small hole is made in the cranium, and a small needle is introduced through this into a ventricle. Cerebrospinal fluid is withdrawn and air injected into the ventricle as a contrast medium to outline

F. 21-24 VENTRICLES OF BRAIN:

(1) —Sagittal section

Lateral ventricle shown with black dots

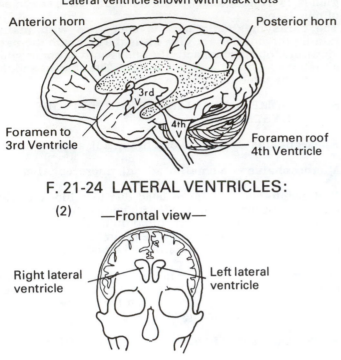

Anterior horn
Posterior horn
Foramen to 3rd Ventricle
Foramen roof 4th Ventricle
3rd V
4th V

F. 21-24 LATERAL VENTRICLES:

(2) —Frontal view—

Right lateral ventricle
Left lateral ventricle

F. 21-25 MYELOGRAM:

- Dark shadow represents opaque medium injected.-

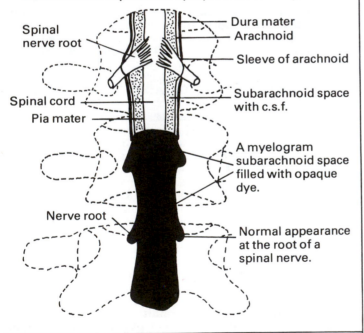

Spinal nerve root
Dura mater
Arachnoid
Sleeve of arachnoid
Spinal cord
Subarachnoid space with c.s.f.
Pia mater
A myelogram subarachnoid space filled with opaque dye.
Nerve root
Normal appearance at the root of a spinal nerve.

the ventricles for radiography. The film is a ventriculogram.

In **ENCEPHALOGRAPHY** a lumbar puncture is done with the patient sitting upright and some cerebrospinal fluid is withdrawn. It is replaced by air which will ascend in the subarachnoid space of the cord to surround the brain. Some will pass through the apertures in the roof of the 4th ventricle and fill the cerebral ventricles. Radiographs will show the parts filled with air. The film is an encephalogram. F.21-24

In **CEREBRAL ARTERIOGRAPHY** an opaque medium is injected into one of the internal carotid arteries in the neck. Films exposed in rapid succession will demonstrate arteries within the cranium. The vertebral artery may be similarly injected to outline the remaining intracranial arteries. The film is a cerebral or vertebral arteriogram. F.21-21

In **MYELOGRAPHY** a small quantity of an opaque medium is injected through a lumbar puncture needle into the subarachnoid space of the spinal cord. This is done under fluoroscopic vision, with the x-ray table horizontal. By tilting the table the fluid may be made to move up or down in the subarachnoid space. Tumors and displaced intervertebral discs that press upon the subarachnoid space will show deformities in the column of medium. The film is a myelogram. F.21-25

S. 271   ANATOMICAL TERMS — NERVOUS SYSTEM

BASIC TERMS

Neuron;
  dendron, dendrites
  axon
sensory neuron or
afferent neuron
motor neuron or
efferent neuron
connecting neuron
synapse
reflex arc
nerve fiber
nerve
endorgans;
  receptor
  effector
ganglion, ganglia
center (nerve)
nucleus (nerve)
plexus
gray substance (matter)
white substance (matter)
neuroglia
funiculus, funiculi
tract
irritability (nerves)
conductivity (nerves)

Central Nervous System; C.N.S.
brain
spinal cord
Peripheral Nervous System; P.N.S.
cranial nerves
spinal nerves
autonomic nervous
  system

BRAIN: Encephalos

FOREBRAIN:
Cerebrum:
gyrus, gyri
convolutions
sulcus, sulci
fissures
longitudinal fissure
transverse fissure
central sulcus
lateral sulcus
parietooccipital
  sulcus

cortex
  white substance
  gray substance
  cerebral hemispheres

Lobes;
frontal lobe
parietal lobe
occipital lobe
temporal lobe
insula
motor area
sensory area

Diencephalon; inter or
after brain
thalamus
hypothalamus

MIDBRAIN
corpora quadrigemina
cerebral peduncles

HINDBRAIN:
Pons
Medulla oblongata
cerebellum
cerebellar, hemisphere

VENTRICLES OF THE BRAIN
Rt. lateral ventricle
Lt. lateral ventricle
Third ventricle
Fourth ventricle
Interventricular foramen
  (2) right & left
  or Foramina of Monro
Cerebral aqueduct or
  aqueduct of Sylvius
Median aperture of 4th
  ventricle, or the
  foramen of Magendie
Lateral apertures of 4th
  ventricle, or the
  foramina of Luschka
Choroid plexuses
Cerebrospinal fluid, CSF

SPINAL CORD OR MEDULLA SPINALIS
Cauda equina
Gray substance (matter);

* cornua or horns; anterior, posterior, lateral
* columns; anterior, posterior & lateral
White substance (matter);
* funiculi — anterior, posterior lateral
* tracts; ascending or sensory, descending or motor

MENINGES OF BRAIN & CORD
Meninx — membrane, meninges — pl.
meningeal, adj.
Pia mater
Arachnoid
Dura mater
subarachnoid space
falx cerebri
tentorium cerebelli
subdural space

PERIPHERAL NERVOUS SYSTEM

CRANIAL NERVES
12 pairs
originate from brain
Note:
   olfactory — smell
   optic — sight
   vestibulocochlear or accoustic — hearing
vagus

SPINAL NERVES
31 pairs
ventral root or anterior or motor root
dorsal root or posterior or sensory root

cervical plexus
   phrenic nerve
brachial plexus;
   median, ulnar, radial, and musculocutaneous
lumbar plexus
   femoral nerve
sacral plexus
   sciatic nerve
thoracic nerves;
   intercostals

AUTONOMIC NERVOUS SYSTEM
Sympathetic part or
Sympathetic nervous s. or thoracolumbar part
   sympathetic centers
Sympathetic ganglia

sympathetic trunks
preganglionic fibers
postganglionic fibers
splanchnic nerves
great plexuses
Parasympathetic part or parasympathetic nervous s. or craniosacral part

BLOOD VESSELS OF BRAIN
internal carotid aa (2)
vertebral aa (2)
venous sinuses of the dura mater;
   superior sagittal sinus
   inferior sagittal sinus
   straight sinus
   transverse sinuses (2)
   sigmoid sinuses (2)
   internal jugular veins (2)
      right and left
middle meningeal aa (2)

NOTES

238

22. THE SENSE ORGANS

Sensory neurons of the cranial and spinal nerves conduct impulses to the central nervous system. These sensations may be divided into two groups. The first, a general group, includes sensations of heat, cold, touch, pain, pressure, and position. A second group is made up of the special senses: sight, hearing, smell and taste. In all of these, stimuli affect receptors at the outer ends of dendrites of sensory nerves.

The special senses have sense organs that contain the receptors for stimuli:
1.   The eye — organ of sight and vision
2.   The ear — organ of hearing
3.   The nose — olfactory — organ of smell
4.   The taste buds — gustatory — organ of taste.

S. 272   THE PARTS OF THE EYE   F. 22-1

The visual apparatus includes the eyes, i.e. the organs of sight, the right and left optic nerves, the optic chiasma, the optic tracts, and centers within the brain.

The parts of the eyeball include:

1. **Coverings of eyeball** (Tunics)
   (1) Fibrous or outer layer
      a.   sclera
      b.   cornea with conjunctiva
   (2) Vascular or middle layer
      a.   choroid
      b.   ciliary body
      c.   iris and pupil
   (3) Nervous or internal layer
      a.   retina
2. **Refracting media** of the eyeball
   (1) Aqueous humor
   (2) Lens, capsule & suspensory ligament
   (3) Vitreous humor (body)
   (4) Cornea
3. **Muscles within the eyeball**
   (1) Of iris = constrictor of pupil
               dilator of pupil
   (2) Of lens = ciliary circular
                radiating

S. 273   DETAILED STUDY OF THE EYE   F.22-1

(L. oculus; A.S. eage; eyeball, bulb of eye.) The eyeball occupies the anterior part of the orbit. It is spherical in shape except anteriorly where it bulges forward. It measures about one inch in diameter. It has three coverings, coats, or tunics, a fibrous or outer layer, a vascular or middle layer, and the nervous layer or retina, which is the inner layer.

1. The coats, coverings, or tunics of the eye:
(1) The fibrous covering is composed of fibrous tissue:
(a) **The sclera** covers the posterior five-sixths of the eyeball, and is white. The student should examine his/her eye in a mirror, and verify the facts listed here.

(b) The **cornea** covers the ramaining one-sixth of the eyeball. This layer is transparent so that light may pass through it. The cornea is covered by the transparent conjunctiva. This is a thin membrane that not only covers the cornea but lines the eyelids.
(2) **The vascular coat** or middle layer is rich in blood vessels and is composed of:
(a) **The choroid covering** that includes the posterior two-thirds of this middle layer, and lies inside of the fibrous layer.
(b) **The ciliary body,** a thickened ring of tissue that encircles the eyeball in front of the choroid proper. Anteriorly it is attached to the margin of the lens by the suspensory ligament. The ciliary body contains the ciliary muscle. This muscle regulates the convexity of the lens by pulling on the suspensory ligament attached to the lens.

(c) **The iris** is a colored membrane that surrounds the pupil. It is attached along its outer margin to the ciliary body. Pigment in the iris determines the color of the eyes. See F.22-1 (3)

The pupil is the circular opening in the center of the colored iris, and appears black.

The iris contains a circular layer of muscle fibers, the sphincter of the pupil, and radiating fibers, the dilator of the pupil. When the circular fibers contract they cause the iris to cover more of the pupil which seems to become smaller. The radiating fibers when they pull on the free inner margin of the iris render the opening larger. The pupil therefore appears to contract and dilate.

(3) **The retina** is the inner coat. It is made up of several layers of cells, but only the visual layer will be considered here. F.22-2

This visual layer is made up of rods and cones, named from their resemblance to these two objects. These are the receptors for light and color stimuli. The cones are adapted to bright light and color. The rods are sensitive to dim light. The cones are most numerous in that part of the retina directly opposite to and behind the pupil. The rods are more numerous than the cones and lie in the peripheral part of the retina. These visual cells contain a substance, visual purple, that fades with exposure to light and regenerates with rest. Waves of light or color cause some chemical changes in the visual cells. Resulting impulses transmitted along the optic nerves to the visual center in the occipital lobe of the brain are interpreted by this center.

Several other layers of cells are present in the retina, these are not included in this study.

The macula lutea is a **yellow spot** on the posterior and inner surface of the retina opposite the pupil. It is composed of cones with no rods and is the most sensitive area of the retina to light. (Macula = a spot + lutea = yellow).

The fovea contralis (central pit) is a small depression at the center of the macula, opposite the pupil and very sensitive to light.

The blind spot or **optic disc** is a small circular area on the inner surface of the retina where the optic nerve enters the retina. It is located slightly medial to the fovea centralis, towards the nasal cavity. It has neither rods nor cones so is insensitive to light — blind spot.

2. THE REFRACTING MEDIA   F.22-1, 2

The refracting media fill the cavity of the eyeball. They are transparent so that light may pass through them to reach the retina. Their function is to change the direction of light waves so as to focus the image on the inner surface of the retina. This image is actually upside down, but this is disregarded. There are four refracting media, if the cornea is included: the aqueous humor, the lens, the vitreous humor or body, and the cornea.
(1) **The aqueous humor** lies posterior to the cornea in the anterior part of the eyeball. It is a thin clear fluid.
(2) **The lens** and its capsule lie vertically in the anterior part of the eyeball, behind the iris and the aqueous humor. It is attached along its margin to the suspensory ligament, and by this to the ciliary body and ciliary muscle. The lens is biconvex with anterior and posterior convex surfaces. It is transparent. The ciliary muscles in the ciliary body vary the convexity of the lens by pulling upon or relaxing the suspensory ligament. The lens provides accommodation to focus light upon the retina.

F. 22-1 THE RIGHT EYEBALL
(1) VIEWED FROM ABOVE

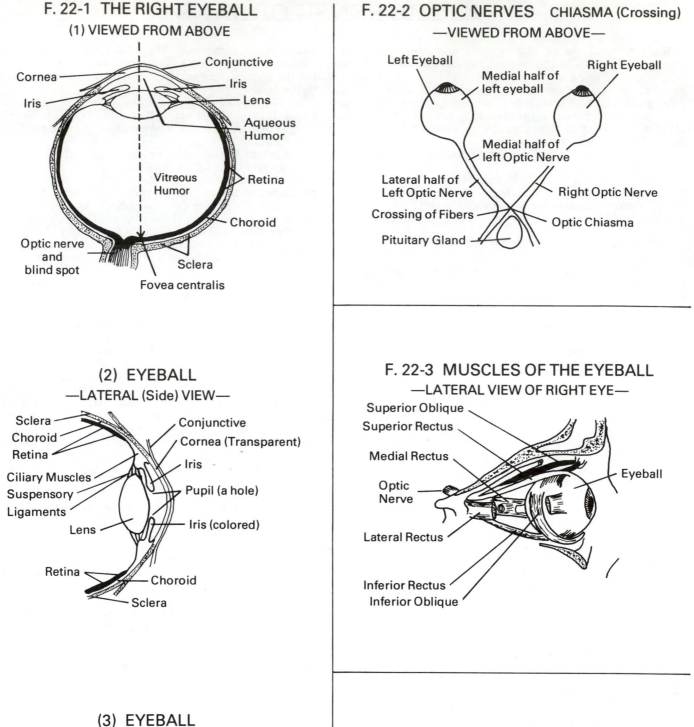

Cornea
Iris
Conjunctive
Iris
Lens
Aqueous Humor
Vitreous Humor
Retina
Choroid
Optic nerve and blind spot
Sclera
Fovea centralis

(2) EYEBALL
—LATERAL (Side) VIEW—

Sclera
Choroid
Retina
Ciliary Muscles
Suspensory
Ligaments
Lens
Retina
Conjunctive
Cornea (Transparent)
Iris
Pupil (a hole)
Iris (colored)
Choroid
Sclera

(3) EYEBALL
—FRONTAL VIEW—

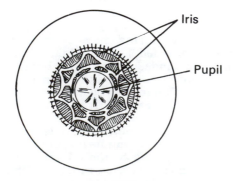

Iris
Pupil

F. 22-2 OPTIC NERVES CHIASMA (Crossing)
—VIEWED FROM ABOVE—

Left Eyeball
Medial half of left eyeball
Right Eyeball
Medial half of left Optic Nerve
Lateral half of Left Optic Nerve
Right Optic Nerve
Crossing of Fibers
Optic Chiasma
Pituitary Gland

F. 22-3 MUSCLES OF THE EYEBALL
—LATERAL VIEW OF RIGHT EYE—

Superior Oblique
Superior Rectus
Medial Rectus
Optic Nerve
Lateral Rectus
Eyeball
Inferior Rectus
Inferior Oblique

F. 22-4 LACRIMAL OR TEAR GLAND & DUCT
—FRONTAL VIEW OF RIGHT EYE—

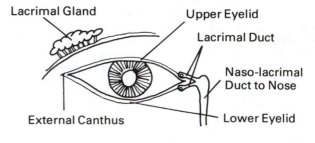

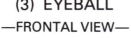

Lacrimal Gland
Upper Eyelid
Lacrimal Duct
Naso-lacrimal Duct to Nose
External Canthus
Lower Eyelid

(3) **The vitreous humor** or body is a transparent jellylike material that fills the cavity behind the lens.

3. THE MUSCLES WITHIN THE EYEBALL
(1) **The muscles of the iris,** the constrictor and dilator have been mentioned.
(2) **The ciliary muscles** of the lens were also defined in the preceding paragraph.

THE OPTIC NERVES, OR SECOND CRANIAL NERVES are the nerves of sight. There are two, a right and a left. Nerve fibers from the rods and cones i.e. the receptors in the eyeball join to form these nerves. From each eyeball an optic nerve passes posteriorly through the optic canal to enter the cranial cavity below the brain. Behind the pituitary gland the nerve fibers from the medial half of each eyeball cross the midline to join the fibers from the lateral half of the other eyeball. Hence the nerve fibers from the right half of each eyeball join together, and likewise those from the left half of each eyeball join together.

The optic chiasma is the point of crossing of the medial nerve fibers. It resembles the letter "X", a letter in the Greek alphabet called chi, so — chiasma. F.22-2

The optic tracts are the two bundles of nerve fibers that emerge from the optic chiasma and pass obliquely across the cerebral peduncles to enter the brain. They have connections in the lower brain centers and end in the occipital lobes of the cerebral hemispheres.

Diagrams of the base of the brain show the optic nerves, chiasma, and optic tracts posterior to the olfactory tracts. F.22-13

S. 274 EXTRINSIC MUSCLES OF THE EYE F.22-3

Movements of the eyeball are controlled by six extrinsic muscles of the eyeball. Each of these is attached by one end to the apex of the orbit close to the optic canal. The other end inserts into the sclera of the eyeball. There are three opposing pairs of these muscles in each orbit. There is a superior rectus attached to the upper surface of the eyeball. Contraction of this muscle turns the eye upwards. The opposing inferior rectus is attached to the under surface and turns the eye downwards. The medial rectus muscle on the medial side turns the eye in, while the lateral rectus on the lateral side turns the eyeball outwards. The superior oblique muscle, attached to the top of the eyeball rolls the eye inwards. The inferior oblique below the eyeball rolls the eye outwards. A further muscle, the elevator of the upper eyelid is attached to the upper lid to open the eyelid. (L. rectus = straight)

These muscles receive motor nerve fibers from the oculomotor (3rd), or trochlear (4th), or abducent (6th) cranial nerves.

This nerve control is arranged in such a manner that both eyes move in the same direction at the same time, up or down, or to the right and left.

This does not apply to convergence when both eyes turn medially to view some close object.

Strabismus — squint or crossed eyes, is a condition in which both eyes do not move in the same direction at the same time.

S. 275 THE EYELIDS

The eyelids, upper and lower protect the eyes and cover them during sleep.

The **medial canthus** or internal canthus is the point at the medial border of the eyeball where the upper and lower eyelids meet.

The **lateral canthus** or external canthus is the point where the upper and lower eyelids meet at the lateral margin of the eye.

The **conjunctiva** is a thin transparent membrane that covers the inner surface of each eyelid, and is reflected over the anterior surface of each eyeball. It is therefore in front of the cornea.

S. 276 THE LACRIMAL APPARATUS F.22-4

(L. lacrima = a tear; G. dakryon = a tear). The lacrimal apparatus is responsible for the formation, circulation, and disposal of the tears. Each eye has a lacrimal gland with its excretory ducts, two lacrimal ducts, a lacrimal sac, and a nasolacrimal duct.
1. **The lacrimal gland** lies behind the outer part of the supraorbital margin of the orbit. It secretes tears through several small ducts that open upon the conjunctiva.
2. **The lacrimal ducts** one for each eyelid begin as small openings on the free margins of the eyelids close to their medial ends, the punctae. Each passes medially within the eyelid and joins the lacrimal sac close to the nose.
3. **The lacrimal sac** lies at the medial end of each eye in a lacrimal groove formed in the lacrimal bone and frontal process of the maxilla. The sac has openings from the lacrimal ducts.
4. **The nasolacrimal duct** passes down in the nasolacrimal groove and opens into the nasal cavity. The lacrimal sac opens into it above.

Tears secreted by the lacrimal gland flow across the surface of the cornea to reach the lacrimal. They enter the lacrimal ducts and pass into the lacrimal sac and nasolacrimal duct and then into a nasal cavity. The tears lubricate the surface of the eyeball.

S. 277 APPLICATION TO RADIOGRAPHY

Deformities of the optic canals may be accompanied by diplopia (double vision) or by blindness. The state of these canals may be demonstrated radiographically. The student should locate the optic canals through the opening of the orbit and determine what position of the skull shows this canal to best advantage.

Enlargement of the pituitary gland may cause pressure upon the sella turcica and the clinoid processes and may cause some destruction of these structures. This erosion may be demonstrated in radiographs. Also with enlargement the pituitary may press upon the optic nerves or chiasma and may cause blindness.

The lacrimal ducts, sac or nasolacrimal ducts **may become obstructed** so that tears cannot pass into the nasal cavity but must run down the cheek instead.

Dacryocystography is a procedure whereby an opaque medium is injected through a fine needle into one of the lacrimal ducts to outline it, the lacrimal sac, and the nasolacrimal duct. (Dacryo = tear + cyst = a small cyst or bladder).

S. 278 APPLICATION TO FLUOROSCOPY
or DARK ADAPTATION

Image intensification in fluoroscopy has largely replaced conventional fluoroscopy in a darkened room. By image intensification the illumination (lighting up) of the fluoroscopic screen has been increased many times.

Images may now be seen on a screen in a room with ordinary lighting. However as conventional fluoroscopy is still utilized in some departments a brief study of what it entails is still pertinent.

A fluoroscopic screen contains calcium tungstate crystals. These crystals light up or illuminate when the undertable x-ray tube is energized. The degree of illumination is very faint if compared to that of an ordinary desk lamp. This may be demonstrated by having someone step on the fluoroscopic foot switch, while the overhead light is left on. It will be impossible to detect any lighting up of the screen. When a patient's body is interposed between the tube and screen the illumination is still further reduced. It is also further reduced by a thick body or a dense part.

A radiologist or assistant going from a well lighted room into a darkened fluoroscopic room will see images on the screen only very faintly if at all. The image is there but the eyes are not capable of seeing the light. The eyes must be accommodated by dark adaptation. In faint light the rods take over sight. Cones are insensitive to this degree of light.

The preparation at one time consisted of a period of resting the rods, the longer the resting period the more acute the vision. This was accomplished by sitting in a room in complete darkness, along with the wearing of dark goggles. Later when it was decided that the rods of the retina were insensitive to colored light red goggles were worn for a period of time. The radiologist could sit in a room with ordinary lighting. The red goggles filtered out all but the red rays to which the rods were not sensitive. Since it is possible that the rods are not entirely insensitive to red light a short period in a darkened room was utilized as well.

If the red goggles are removed between examinations in a lighted room, and particularly if films are studied on viewing boxes the dark accommodation is lost. The adaptation process must be repeated. It is therefore the duty of assistants to protect the radiologist from any interruptions wherein a request is made to have films examined between fluoroscopic examinations.

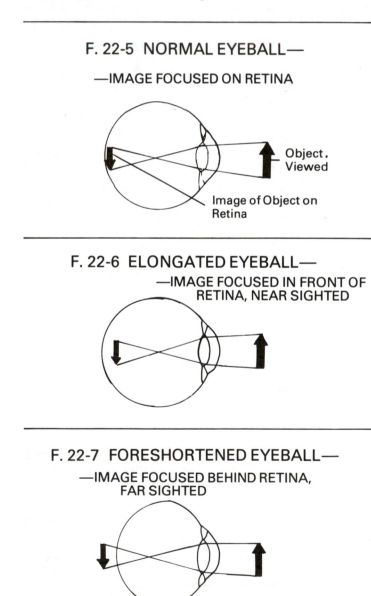

F. 22-5 NORMAL EYEBALL—

—IMAGE FOCUSED ON RETINA

Object. Viewed

Image of Object on Retina

F. 22-6 ELONGATED EYEBALL—

—IMAGE FOCUSED IN FRONT OF RETINA, NEAR SIGHTED

F. 22-7 FORESHORTENED EYEBALL—

—IMAGE FOCUSED BEHIND RETINA, FAR SIGHTED

S. 279 THE EAR — A LIST OF ITS PARTS

The ear has three parts: the external, the middle, and the internal ear.

1. THE EXTERNAL EAR
   (1) auricle or pinna = cartilages
   (2) opening (porus) of external acoustic meatus
   (3) external acoustic (auditory) meatus

2. THE MIDDLE EAR
   (1) tympanic membrane or ear drum
   (2) auditory ossicles — 3 pairs
     a. malleus or hammer
     b. incus or anvil
     c. stapes of stirrup
   (3) openings from middle ear
     a. to external acoustic meatus
     b. tympanic opening of auditory tube or Eustachian tube
     c. to mastoid antrum (tympanic antrum) and mastoid cells
     d. to cochlea = round opening
     e. to vestibule = oval opening

3. INTERNAL EAR OR LABYRINTH
   (1) Osseous labyrinth — 3 parts:
     a. Vestibule with round & oval openings
     b. Semicircular canals, 3 ampullae
     c. Cochlea
   (2) Membranous labyrinth — 3 parts
     a. utricle & saccule in vestibule
     b. semicircular ducts in semicircular canals
     c. cochlear duct in cochlea
       2 partitions:
         vestibulae membrane
         basilar membrane — Organ of Corti
       3 canals:
         vestibular canal (scala)
         tympanic canal (scala)
         cochlear duct (scala media)
           Organ of Corti on basilar membrane, has:
           hair cells
           tectorial membrane (roof)
           dendrites of sensory neurons
           Spiral ganglia — to cochlear nerve

S. 280 A DETAILED STUDY OF THE EAR F.22-8

(A.S. eare, L. auris, aural; organ of hearing). The ear is the organ of hearing and balance or equilibrium. It is contained within the petrous part of the temporal bone. In it sound waves are transformed into nerve impulses. These are conveyed by the cochlear part of the vestibulocochlear (acoustic) nerve to the auditory centers in the occipital lobes of the cerebrum. This nerve also transmits sensations of the position of the head in space and of movements of the head. These are conveyed to the cerebellum and cerebrum by the vestibular part of the vestibulocochlear nerve, or the eighth cranial nerve. The body equilibrium is adjusted. The ear consists of three parts: the external, middle and internal ear.

1. THE EXTERNAL EAR OR AURIS EXTERNA F.22-8

The external ear includes the auricle, the external acoustic meatus, and its opening, i.e. the external acoustic opening or porus.

F. 22-8 THE EAR EXTERNAL, MIDDLE, AND INTERNAL PARTS

—FRONTAL VIEW—

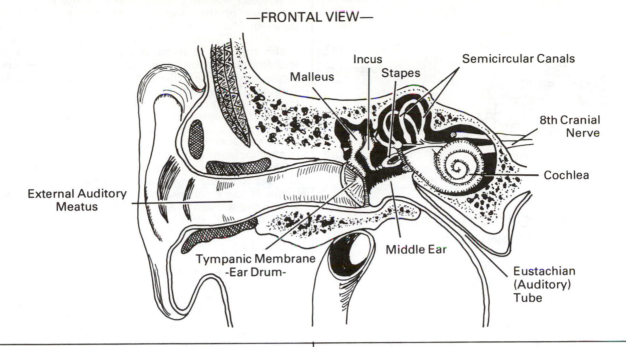

2. **THE MIDDLE EAR OR TYMPANIC CAVITY**
   F.22-8

The middle ear, auris media, also named the tympanic cavity is a small hollow space between the external and internal ear. It lies in the petrous part of the temporal bone. It is separated from the external acoustic meatus by the tympanic membrane or ear drum. It is filled with air and contains three minute bones, the auditory ossicles. It has five openings.

The **tympanic membrane** or ear drum is a thin membrane stretched across the opening between the inner end of the external acoustic meatus and the middle ear. Sound waves carried through the meatus cause it to vibrate.

The **auditory ossicles** are three very minute bones in the middle ear. They are named from objects they resemble: the malleus or hammer, the incus or anvil, and the stapes or stirrup. F.22-9

The **auricle** or pinna is the trumpetlike cartilaginous part that protrudes from the side of the head. It helps to direct sound waves into the ear canal.

The **external acoustic meatus** (OT. auditory) is the canal extending from the opening in the auricle to the ear drum. It is about 2.5 cm (1 inch) in length. It is lined with skin that contains modified glands — ceruminous glands — that secrete wax into the canal. This passage conducts sound waves to the ear drum.

The **malleus** or hammer is attached by its handle to the inner surface of the tympanic membrane i.e. the ear drum.

The **incus** or anvil is attached at one end to the malleus, and by its other end to the stapes.

The **stapes** or stirrup extends from the incus to an opening into the inner ear, the vestibular opening or the fenestra ovalis. This opening is closed by the footpiece of the stapes.

These bones form a system of levers that convey vibrations from the ear drum to the internal ear, magnifying them in transit.

Openings from the middle ear:

The opening between the external acoustic meatus, and the middle ear, covered by the eardrum has been noted.

The **tympanic opening** of the auditory tube, is its opening into the middle ear. The auditory tube is a passage extending from the nasal pharynx into the middle ear. Air pressure on the outer and inner surfaces of the ear drum is kept equal by this pathway.

The **aditus** (opening) of the mastoid or tympanic antrum is an opening from the middle ear into this cavity or antrum. This cavity opens into the mastoid cells within the mastoid process. The opening is sometimes termed "the aditus ad antrum". (opening into antrum.)

The **fenestra rotunda** or round opening or the cochlear opening is located between the middle ear and cochlea of the internal ear. It is covered by a memberane. (L. fenestra = a window)

The **fenestra ovalis** or oval opening or vestibular opening is located between the middle ear and vestibule of the internal ear. The footpiece of the stapes fits into this opening.

3. **THE INTERNAL EAR OR LABYRINTH** — auris interna F.22-11, 12

The internal ear also called the labyrinth lies medial to the middle ear. It is located in the petrous part of the temporal bone. There is a bony labyrinth hollowed out from bone, and within this a membranous labyrinth. A labyrinth is a series of intercommunicating cavities. The bony labyrinth has three parts: a **vestibule**, three **semicircular canals**, and the **cochlea**.
(G. labyrinthos = a maze or complicated series of communicating cavities).

F. 22-9 THE AUDITORY OSSICLES
—GROSSLY ENLARGED—

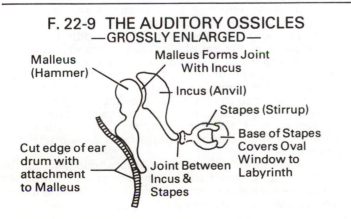

F. 22-10 CROSS SECTION OF COCHLEAR DUCT

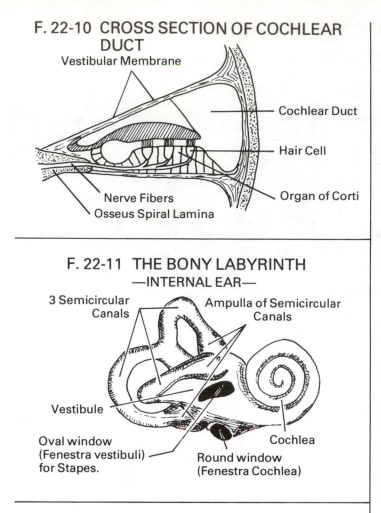

Vestibular Membrane

Cochlear Duct

Hair Cell

Organ of Corti

Nerve Fibers
Osseus Spiral Lamina

F. 22-11 THE BONY LABYRINTH
—INTERNAL EAR—

3 Semicircular Canals

Ampulla of Semicircular Canals

Vestibule

Oval window
(Fenestra vestibuli)
for Stapes.

Cochlea

Round window
(Fenestra Cochlea)

F. 22-12 THE MEMBRANOUS LABYRINTH

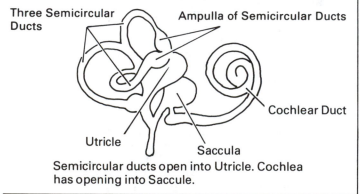

Three Semicircular Ducts

Ampulla of Semicircular Ducts

Cochlear Duct

Utricle

Saccula

Semicircular ducts open into Utricle. Cochlea has opening into Saccule.

Before proceeding with the anatomy of the bony labyrinth the student should review the anatomy of the petrous part of the temporal bone. Posterior to the ridge on the petrous part there is the internal acoustic opening and the internal acoustic meatus. These carry the vestibulocochlear or acoustic or eighth cranial nerve into the internal ear. This nerve has two parts, a cochlear part concerned with carrying nerve impulses caused by sound waves to the brain. The vestibular part carries impulses from the semicircular canals to the cerebellum and cerebrum, and is concerned with the position of the head in space and movements of the head.

THE BONY LABYRINTH F.22-11

The vestibule (F.22-11) is a small cavity with walls of bone lying between the semicircular canals and the cochlea. The cochlea is in front and the semicircular canals behind it. An opening the fenestra ovalis is located between the vestibule and middle ear, hence vestibular opening. This oval window is covered by the footpiece of the stapes. There are also five openings between the ends of the semicircular canals and the vestibule. (Vestibulum = ante-chamber.) See below.

The three semicircular canals are curved passages within the petrous bone, each forming more than a half circle, and shaped like the letter "C". They open at each end into the vestibule. Since two of them join together at one end there are five openings into the vestibule, not six. One end of each canal is expanded to form an ampulla. The three semicircular canals lie in different planes, one in the vertical, one in the vertical posterior, and the other in the horizontal plane.

The cochlea consists of a spiral passage or canal similar to a spiral staircase, with almost three complete turns. It has a central core of bone, the modiolus, around which the windings are built. The cochlea is surrounded by a bony wall. It has an opening into the anterior part of the vestibule. (F.22-11). (L. cochlea = a snail shell; i.e. snail-like.)

THE MEMBRANOUS LABYRINTH (F.22-12) is contained within the bony labyrinth and has a membranous covering that separates it from the osseous labyrinth. It includes three **semicircular ducts**, a **cochlear duct**, (not canals) and two sacs, the **utricle**, and the **saccule**. The semicircular ducts lie within the semicircular canals, and the cochlear duct within the cochlea. These ducts conform in shape to the bony canals but are much smaller. The sacs, utricle and saccule lie inside the vestibule but differ in shape from it. The space surrounding each duct and sac, i.e. between it and the bony wall is filled with a fluid — perilymph. The ducts and sacs are also filled with a fluid — endolymph. The membranous labyrinth is thereby surrounded by **perilymph**, and is filled with **endolymph**.

The three semicircular ducts, one inside of each semicircular canal, have five openings into the utricle within the vestibule. Each has an expanded end that lies in an ampulla of a bony semicircular canal. (F.22-12)

The cochlear duct lies within the spiral bony cochlea and follows its course from base to apex. Two membranes are stretched across the spiral canal of the bony cochlea along the length of the canal. These divide the cochlea into three passages, the cochlear duct is in the middle, the scala vestibuli above, and the scala tympani below. (F.22-12)

The lower or **basilar membrane** forms the floor of the cochlear duct. The organ of hearing, i.e. **the organ of Corti** lies along the length of this membrane. Specialized cells extend up from this basilar membrane. These end in hairs, and are named hair cells. A gelatinous membrane, the tectorial membrane, covers the hairs. Dendrites of the cochlear nerve are in contact with the hair cells. Sound vibrations transmitted to this membrane through the surrounding fluid are said to pull the hairs, creating nerve impulses that are carried by the cochlear nerve to the temporal lobe of the brain. F.22-12

Sound waves entering the external acoustic meatus cause vibrations of the ear drum. These are transmitted by the auditory ossicles through the middle ear to the oval window of the vestibule. They cause vibrations in the perilymph and endolymph and are carried to the cochlear duct and organ of Corti. The hair cells here produce nerve impulses that are conveyed by the cochlear nerve to the temporal lobe for interpretation.

The utricle is a small membranous sac that lies inside the vestibule of the bony labyrinth. It is filled with endolymph and has five openings from the semicircular ducts.

The saccule is a smaller membranous sac in the vestibule. It is also filled with endolymph. It and the utricle are surrounded by perilymph. The saccule has an opening into the cochlear duct.

Small structures named **maculas** are located within the utricle, and saccule, and similar organs also lie in the ampullae of the semicircular ducts. These have hair cells and are covered by gelatinous membranes. The hair cells have dendrites of the vestibular part of the vestibulocochlear nerve about their bases. These dendrites receive nerve impulses from changes in pressure within the sacs and ducts, that affect the hair cells. Movements of the head, and the position of the head in space are transmitted to the vestibular nerve, the cerebellum and cerebrum. As a result of these impulses equilibrium is maintained by the body.

S. 281 ANATOMICAL PECULIARITIES & RADIOGRAPHY

The petrous part of the temporal bone contains the middle and internal ear. This lies obliquely in the base of the skull. When viewed from the front it appears foreshortened similar to the neck of the femur that also is obliquely placed. In order to place it parallel to the table top the head must be turned somewhat obliquely.

The mastoid cells, mastoid antrum, and middle ear contain air. The internal acoustic meatus contains the vestibulocochlear nerve. All of these structures are less dense than the surrounding bone, and have separate images in radiographs more translucent than the dense petrous pyramid itself.

The auditory ossicles are composed of bone, and lie in the air-filled middle ear, and may be radiographed particularly when body section methods are employed.

A continuous passage extends from the nasal pharynx through the auditory tube to the middle ear, then from the middle ear through the mastoid antrum to the mastoid cells. This arrangement is important medically, as an infection of the nasal pharynx (sore throat) may travel into the middle ear, and also out into the mastoid cells.

Otitis media = an infection of the middle ear. Mastoiditis = an infection of the mastoid cells.

A bulge on the upper surface of the petrous pyramid denotes the position of a semicircular canal.

S. 282 ORGANS OF SMELL: OLFACTORY ORGANS F.22-13

Olfactus = smell. The olfactory apparatus is concerned with the sense of smell. It has right and left olfactory organs with receptors or endograns within the nasal cavities. These continue as olfactory nerves, olfactory bulbs, olfactory tracts, and olfactory nerve centers in the brain.

The receptors or endorgans are tufts of hairs in the lining membrane of the upper part of each nasal cavity. They are the olfactory organs. About 20 olfactory nerves pass from these hairs up through the cribriform plate of the ethmoid bone and end in the olfactory bulbs.

The olfactory bulbs are the expanded anterior ends of the two olfactory tracts, two club-shaped structures that lie on the inferior surfaces of the frontal lobes of the cerebrum. They are visible in diagrams of the base of the brain. These bulbs, and the cordlike tracts posterior to the bulbs help to form the neurons that carry impulses from the receptors in the nasal cavities to the centers in the temporal lobes of the brain.

The hairlike cells in the upper part of each nasal cavity are stimulated by gas particles (in solution) coming into contact with them during respiration, particularly in sniffing. Nerve impulses thus generated are conveyed by the olfactory nerves, bulbs, and tracts to the temporal lobes for interpretation. The receptors are quickly fatigued which may explain why an obnoxious odor, though intollerable at first, does not continue to upset one after a period of time.

F. 22-13 THE BRAIN — BASAL VIEW
— Olfactory and optic nerves —

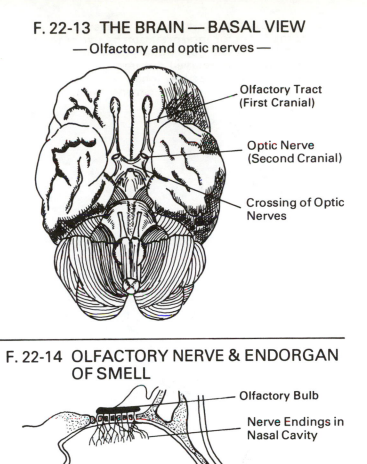

Olfactory Tract (First Cranial)

Optic Nerve (Second Cranial)

Crossing of Optic Nerves

F. 22-14 OLFACTORY NERVE & ENDORGAN OF SMELL

Olfactory Bulb

Nerve Endings in Nasal Cavity

F. 22-15 TASTE BUDS OF TONGUE
—ENDORGANS OF TASTE—

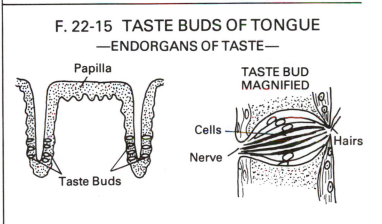

Papilla

TASTE BUD MAGNIFIED

Cells

Nerve

Hairs

Taste Buds

S. 283 ORGANS OF TASTE: GUSTATORY ORGANS F.22-15

(L. Gustatio = to taste, hence gustation and gustatory.)

The end organs of taste are the taste buds. These are located on papillae on the upper surface and sides of the tongue, as well as on other parts of the mouth.

The papillae are minute elevations on the tongue that are the cause of the roughness of this organ. (L. papilla = nipple or nipplelike)

The taste buds (F.22-15) are oval shaped bodies located on the sides of the papillae. They have elongated central cells with hairs at their surfaces. Stimulation of these hairs results in impulses that are carried to the parietal lobes of the brain for interpretation by two of the cranial nerves, the facial and glossopharyngeal nerves. Actually the impulses are relayed through the medulla, pons, and thalamus to the parietal lobe.

There are only four taste sensations: salty, bitter, sweet, and acid. Taste differentiation is limited, and is usually a combination of taste and smell. The sensation of taste often experienced may be of smell rather than taste. Patients with head colds frequently complain that they cannot taste their food. In fact it is their sense of smell that is likely to be affected by the head cold.

23. THE CONTENTS OF BODY CAVITIES

S. 284 **A LIST OF STRUCTURES IN BODY CAVITIES**

See body cavities S. 15 of Handbook

I. **THE HEAD:** CEREBRAL AND VISCERAL CRANIUM
1. The brain
2. Pituitary gland
3. Pineal gland
4. The organs of hearing
5. Orbits and eyeballs
6. Nasal cavities
7. Paranasal sinuses
8. The mouth
9. The nasal pharynx

II. **THE NECK**
1. Pharynx or throat
2. Larynx or voice box
3. Upper trachea
4. Thyroid cartilage (Adam's apple)
5. Thyroid gland
6. Parathyroid glands
7. Blood vessels, lymphatics
8. Nerves — cervical, brachial plexuses

III. **CONTENTS OF THE CHEST OR THORAX**
1. Right lung
2. Left lung
3. Mediastinum:
   1) Circulatory system;
     (1) heart
     (2) aorta, ascending, arch, descending thoracic
     (3) superior vena cava + tributaries
     (4) Inferior vena cava, (part)
     (5) pulmonary trunk
     (6) pulmonary arteries, r & l
     (7) pulmonary veins, r & l
     (8) thoracic duct
     (9) bronchial arteries + veins
   2) Other structures;
     (1) trachea, lower part
     (2) main bronchi, r & l
     (3) esophagus
     (4) thymus gland
     (5) lymph nodes (glands)
     (6) nerves — vagus, r & l
             phrenic, r & l

IV. **CONTENTS OF THE ABDOMEN**
1. Digestive tract;
   1) stomach
   2) small intestine
   3) large intestine, except rectum
   4) appendix, usually
2. Accessory digestive organs;
   1) pancreas
   2) liver
   3) gall bladder
   4) bile ducts
3. Urinary system;
   1) kidneys, r & l
   2) renal pelves, r & l
   3) upper ureters, r & l
4. Glands;
   1) suprarenal or adrenal glands
5. Vessels;
   1) abdominal aorta and branches
   2) inferior vena cava + tributaries
   3) portal vein + tributaries

   4) cisterna chyli, thoracic duct, lymph nodes
6. Nerves;
   1) sympathetic
   2) parasympathetic

V. **CONTENTS OF THE PELVIS**
1. Digestive system;
   1) sigmoid (pelvic) colon
   2) rectum
   3) ileum, part of
   4) cecum and appendix, sometimes
2. Urinary system;
   1) lower ureters, r & l
   2) urinary bladder
   3) prostatic urethra
3. Female genitals;
   1) broad ligament
   2) uterus
   3) uterine tubes, r & l
   4) ovaries, r & l
   5) vagina
4. Male genitals
   1) part of deferent ducts, r & l
   2) seminal vesicles + ducts, r & l
   3) ejaculatory ducts, r & l
   4) prostate gland

VI. **LOCATIONS OF ABDOMINAL ORGANS**

See S. 187, Factors determining position of the abdominal organs;
See S. 188, Divisions of the abdomen

1. **THE RIGHT UPPER QUADRANT**
R U Q or U R Q
   1) pyloric part of stomach
   2) duodenum, superior, descending + horizontal parts
   3) parts of small intestine
   4) upper ascending colon, cecum rarely
   5) right half of transverse colon
   6) liver, greater part
   7) gall bladder, usually
   8) bile ducts
   9) head of pancreas
   10) end of pancreatic duct
   11) right suprarenal gland
   12) right kidney
   13) renal pelvis + upper ureter, right,
   14) blood and lymph nodes

2. **THE RIGHT LOWER QUADRANT**
R L Q or L R Q
   1) lower ascending colon
   2) cecum + appendix, usually
   3) right ureter midpart
   4) terminal ileum, usually
   5) blood and lymph vessels

3. **THE LEFT UPPER QUADRANT**
L U Q or U L Q
   1) stomach, fundus and body
   2) duodenum, ascending part
   3) parts of small intestine
   4) left half of transverse colon
   5) upper part of descending colon
   6) body and tail of pancreas
   7) spleen
   8) left suprarenal gland
   9) liver, a small part
   10) left kidney

11) renal pelvis + upper ureter, left
12) blood vessels, lymph vessels
13) cisterna chyli + lower thoracic duct

4. **THE LEFT LOWER QUADRANT**
   L L Q
   1) lower part of descending colon
   2) part of sigmoid colon, sometimes
   3) small intestine, part of ileum
   4) left ureter, midpart
   5) blood and lymph vessels

Students should compare the contents of:
the right and left upper quadrants
the right and left lower quadrants

ORGANS VISIBLE IN THORAX & ABDOMEN
FOLLOWING REMOVAL OF ANTERIOR WALL

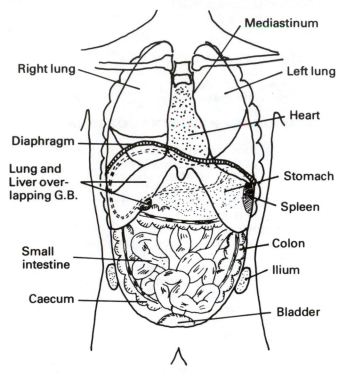

Right lung — Mediastinum — Left lung — Heart — Diaphragm — Lung and Liver overlapping G.B. — Stomach — Spleen — Small intestine — Colon — Ilium — Caecum — Bladder

BLANK OUTLINES —fill in—

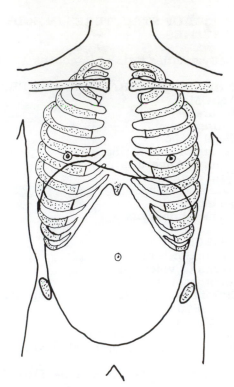

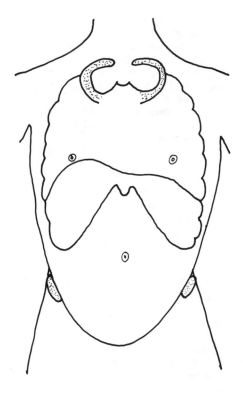

24. APPENDICES

APPENDIX I

UNUSUAL ANATOMICAL TERMS AND DERIVATIVES

Cytos	= kytos = a cell, leukocyte
Os, ossa	= bone & bones, ossicle = little bone osseous = adj. refers to bone ossify = to form bone osteitis = inflammation of bone osteoma = tumor of bone
Chondros	= cartilage chondral, adj. refers to cartilage chondritis = inflammation of
Costa	= a rib costal = adj. refers to a rib costochondral = a rib cartilage
Arthron	= a joint arthritis = inflammation of a joint arthritic = inflamed joint
Arthrosis	= a joint arthrography = radiography of joint
Articulation	= a joint articular = adj. refers to a joint pertiarticular = surrounding a joint
Phren	= the diaphragm phrenic = adj. refers to diaphragm subphrenic = under the diaphragm
Angeion	= angion = a vessel or a duct angeitis or angitis = inflammation of angiography = radiography of a blood vessel or bile duct, etc.
Cor	= heart, eg. cor bovinum = bull's heart very large; cor pulmonale — type of heart
Cardia	= kardia = heart cardiac = adj. refers to heart, endocardium, myocardium, pericardium
Phleb	= a vein, phlebitis = inflammation of phlebography = radiography of
Vena	= a vein; adj. venous, refers to a vein vena cava
Pulmo & pulmones	=lung & lungs pulmonary = adj. refers to lung pulmonic = adj. refers to lung
Pneumon	= lung; pneumonia = specific inflammation of; pneumonic = adj. refers to specific inflammation, or to the lung
Pneumia	= air; pneumatic = refers to air containing object; pneumatic tire, pneumatic cyst = contains air.
Sialon	= saliva; sialography = radiography of salivary glands
Gaster	= stomach; adj. gastric, refers to stomach gastritis = inflammation of stomach epigastric = upon the stomach
Enteron	= intestine, any part of; enteric = adj. refers to bowel, enteritis = inflammation of; enterolith = a stone in
Hepar	= the liver; hepatic = adj. refers to the liver; hepatitis = inflammation of liver
Chole	= bile; choledochus = bile duct cholelithiasis = stone in bile tract
Cyst	= kystis = bladder, any bladder, cholecyst = gall bladder;

	cholecystitis = inflammation of gall bladder cholecystography = radiography of gall bladder; cystitis = inflammation of urinary bladder
Ren	= the kidney; renal — adj. refers to kidney; perirenal = around the
Nephros	= a kidney; nephritic = adj. refers to kidney; nephritis = inflammation of a kidney nephroma
Nephron	= a renal tubule + glomerulus
Ure, uri, uro	= urine = (ouron) ureter = a passage for urine urinary adj. refers to urine
Pyelos	= pelvis = a trough; pyelitis = inflammation of renal pelvis; pyelonephritis, of kidney & pelvis pyelography = radiography of
Vesica	= bladder, any bladder or blister vesical = adj. refers to bladder ureterovesical = ureter + bladder
Hystera	= uterus, hysterography = radiography, hysterectomy = removal of
Metra	= uterus; endometrium = lining of uterus; myometrium = muscle of
Salpinx	= uterine tube; salpingeal = adj. salpingography = radiography of salpingitis = inflammation of salpingectomy = removal of
Oophoron	= ovary (egg bearer) oophorectomy = removal of ovary
Mastos	= a breast; mastitis = inflammation mastectomy = removal of breast
Mamma	= a breast; mammary = adj. mammalia = animals with breasts mammography = radiography of
Myelos	= marrow or middle; something soft, so bone marrow or spinal cord myelitis = inflammation of cord myelography = radiography of cord osteomyelitis = inflammation of a bone & its bone marrow
Medulla	= marrow or middle part; similar to myelos. 1. refers to spinal cord — myelos 2. refers to bone marrow — myelos 3. central part of any structure, medulla of kidney, intramedullary pin; medulla oblongata.
Cephale	= kephale = head, cephalic = adj. refers to head cephalad = towards the head microcephalic = small head macrocephalic = large head
Encephalos	= enkephalos = brain anencephaly = no brain encephalography = radiography of
Meninx	= a membrane; meninges = plural form, = coverings of brain & spinal cord. meningeal = adj. refers to these meningitis = inflammation of covers meningioma = a tumor of

APPENDIX II

ANATOMICAL TERMS APPLIED TO METHODS OF EXAMINATION

Intravenous = into a vein; material injected by syringe & needle.

Subcutaneous = under the skin; injection by syringe & needle through skin.

Percutaneous = through the skin; an injection through skin into artery, muscle, vein, organ.

Intramuscular = injection into a muscle using a syringe and needle.

Lumbar = the lumbar region of the back, injection through the lumbar muscles; eg. lumbar aortography

Presacral = in front of the sacrum, air injected into tissues in front of sacrum, between coccyx & anus, air passes up to surround the kidneys, etc.

Excretory = to be excreted by an organ, e.g. excretory urography, medium injected into a vein, to be excreted by and to outline urinary tract.

Operative = performed during an operation, e.g. operative cholangiography bile ducts injected at operation.

T-tube = injection through a T-tube that was inserted during an operation for gall bladder; i.e. T-tube cholangography.

Pneumo = air; air injected, pneumothorax = air injected into pleural cavity; pneumoperitoneum = into abdomen, pneumoarthrogram = into a joint.

Retrograde = backwards, against the normal flow; retrograde pyelography, retrograde percutaneous femoral arteriography

Oral = by mouth, administer something through the mouth.

Parenteral = administration by any means other than by mouth; e.g. intramuscular intravenous, intraspinal, etc. Literally — para = beside, i.e. not into the gut.

APPENDIX III

ANATOMICAL TERMS APPLIED TO RADIOGRAPHIC PROCEDURES

Arthrography = radiography of joints (arthron).

Pneumoarthrography = radiography of joints by the injection of air or gas.

Discography = outlining nucleus pulposus of intervertebral joint with opaque medium injection + radiography.

Cardiography = outlining chambers of heart by opaque medium + radiography.

Arteriography = outlining an artery & its branches with opaque medium + radiography.

Venography = outlining a vein + tributaries by injection + radiography.

Phlebography = as above for venography (phleb).

Lymphangiography = outlining lymphatic vessels & nodes with opaque medium + radiography.

Angiography = outlining any vessel, blood vessel, or duct with medium + radiography.

Angiocardiography = outlining chambers of heart + connected vessels + radiography.

Aortography = outlining aorta with opaque medium + radiography.

Splenography = outlining internal structure of spleen + radiography.

Hepatography = outlining structure of liver by injection of opaque medium + radiography (percutaneous).

Splenoportography = outlining spleen, splenic vein, & portal vein with opaque medium + radiography.

Bronchography = outlining trachea, main bronchi, lobar, segmental & smaller bronchi with medium + radiography.

Sinography = (1) outlining paranasal sinus with opaque medium + radiography.
(2) outlining a sinus tract, such as from bone to skin with opaque medium + radiography.

Sialography = outlining a salivary duct and its branches with an opaque medium + radiography.

Cholecystography = outlining the gall bladder with an opaque medium + radiography, may be by mouth, intravenous injection; ducts often are outlined as well.

Cholangiography = outlining hepatic & common and cystic ducts & gall bladder if not removed + radiography, by the oral, intravenous, T-tube, or operative methods.

Choledochography = as cholangiography above.

Esophagraphy = outlining esophagus by opaque medium + radiography.

Pyelography = (intravenous) = outlining kidneys, renal pelves, ureters & bladder by intravenous injection of opaque medium + radiography.

Pyelography = (excretory) as above for intravenous or by intramuscular injection of opaque medium + radiography.

Pyelography = (retrograde) — inserting catheters by urethra, & bladder & through ureters, and injection of an opaque medium into them, outlining the calyces, pelves and sometimes the ureters + radiography.

Urography = same as pyelography above, and may be intravenous (excretory) or retrograde.

Nephrogram = presence of opaque medium in renal blood vessels causes an increased opacity of kidneys, seen in intravenous examinations.

Cystography = outlining urinary bladder with an opaque medium + radiography, or may be retrograde injection of medium into the urinary bladder.

Urethrography = outlining urethra with opaque medium, as part of intravenous or retrograde injection + films.

Voiding cystography = radiography during act of voiding of opaque medium from the bladder outlining bladder & the urethra.

Pyelography = (infusion type) = a slow intravenous injection of diluted opaque medium to outline pelves, calyces, and ureters + radiography.

Urea clearance = injection of urea with a series of films taken to determine delay in excretion by kidneys.

Dacryocystography = outlining lacrimal ducts, lacrimal sac, + nasolacrimal duct by injection of opaque medium into a lacrimal duct through small needle.

Hysterosalpingography = injection of opaque medium into uterus through vagina to outline cavity of uterus, & uterine tubes + radiography.

Uterosalpingography = as in hysterosalpingography.

Hysterography = similar procedure for uterus.
Uterography = as above for hysterography.
Salpingography = outlining uterine tubes + films.
Placentography = outlining placenta by injection of opaque medium in pregnant female + radiography.
Mammography = outlining structure of female breast + radiography by a soft tissue technic.
Metrography = same as uterography above.
Myelography = outlining subarachnoid space of spinal cord with opaque medium injected through lumbar puncture needle into it + radiography.
Encephalography = outlining subarachnoid space of brain and ventricles of brain by injection of air or gas through a lumbar puncture needle in the subarachnoid space of spinal cord.
Ventriculography = outlining the ventricles by injection of air through burr holes in the skull into lateral ventricles + radiography.
Presacral air insufflation = injecting air through perineum between coccyx and anus, in front of sacrum. Air passing up behind peritoneum to surround kidneys, etc. + radiography.
Perirenal air insufflation = injection of air through lumbar muscles into area behind peritoneum to outline the kidneys, etc. + radiography.
Pneumothorax = injection of air into the pleural cavity as a contrast medium to outline a mass, etc. + films.
Pneumoperitoneum = injection of air into abdominal cavity as contrast medium to outline a mass, enlarged organ, etc. + radiography.

APPENDIX IV

ANATOMICAL TERMS APPLIED TO SOME SPECIAL NONRADIOGRAPHIC EXAMINATIONS

G. skepeo = scope; to view, to look at.

Endoscopic = viewing the interior of a hollow organ, through an endoscope inserted into that organ.
Bronchoscopic = viewing the trachea and larger bronchi by inserting bronchoscope into the trachea and peering in.
Gastroscopic = viewing the inner surface of the stomach by inserting a gastroscope through the mouth and esophagus into it.
Esophagoscopic = peering into the esophagus through an esophagoscope inserted into it.
Proctoscopic = examining the inside of the rectum and sigmoid colon through a proctoscope inserted through the anus into it.
Sigmoidoscopic = as for proctoscopic examination.
Peritoneoscopic = viewing the structures in the abdomen through a peritoneoscope inserted through a small opening made in the abdominal wall.
Cystoscopic = viewing the urinary bladder through a cystoscope inserted through the urethra into the bladder.
Urethroscopic = using a urethroscope inserted into the urethra to view this passage.
Otoscopic = examination of the external acoustic meatus or canal by an otoscope inserted into this passage. (Ous or ot = ear).
Ophthalmoscopic = looking through the pupil into the eyeball using an ophthalmoscope. (ophthalmos = eye).

squamous part of:
  frontal bone, 93
  occipital bone, 94
  temporal bones, 95
stapes, 101, 243
starches, 167
stenosis — definition, 169
Stensen's duct, 171
sternomastoid muscle, 117
sternal angle, 82
sternum, 82
steroids, 214
stomach, 172
stratified squamous epithelium, 27
stratum corneum, 27
striated muscle, 17, 111
structure of blood vessel, 130
styloid process — definition, 33
styloid process of:
  radius, 47
  temporal bone, 94
  ulna, 47
subarachnoid space, 227
subclavian artery, 137
subclavian vein, 138
subcostal plane, 6
subcutaneous tissue, 29
subdural space, 227
sublingual ducts, 171
sublingual glands, 171
submandibular ducts, 171
submandibular glands, 171
sucrase, 179
sucrose, 167
sudoriferous glands, 29
suffixes, 9
sugars, 167
sulci of brain, 222
sulci, interventricular, 134
sulcus — definition, 4
sulcus, central of brain, 222
sulcus, coronary of heart, 132
sulcus, lateral of brain, 222
superficial — definition, 5
superficial veins, 142
superior — definition, 5
superior orbital fissure, 103
superior vena cava, 138
supinate — definition, 37
supine — definition, 5
supraorbital margin, 93
suprarenal arteries, 137
suprarenal glands, 213
suprasternal notch, 82
sutures of skull, 103
swallowing, 167, 178
sweat glands, 29
sympathetic ganglia, 231
sympathetic nervous system, 230
sympathetic trunks, 231
symphysis pubis, 67
synapse, 220
synovial joints, 37
synovial joints, types, 38
synovial membrane, 16
synovitis, 112
systemic circulation, 136
systems of body, 14, 18
systole, 136

talus, 63
tarsal bones, 63
taste buds of tongue, 245
taste, organ of, 245

tear ducts, 241
tear glands, 241
teeth, 106
teeth, structure, 106
telophase, 23
temporal bones, 94
temporal lobes of brain, 222
temporomandibular joints, 103
tendon, 112
tendon sheath, 112
tenia coli, 175
tentorium cerebelli, 227
terminal — definition, 5
terminal ileum, 174
testicular artery, 137
testicle, 203
testis, 203
testosterone, 203
thalamus, 223, 234
thigh bone, 59
third ventricle of brain, 225
thoracic cavity, 13
thoracic duct, 145
thoracic nerves, 230
thoracolumbar system (part), 230
thorax, 81
throat (pharynx), 157, 171
thrombocytes of blood, 125
thrombin, 125
thrombus, 150
thumb, bones of, 50
thymus gland, 147
thyroid cartilage, 159
thyroid gland, 212
thyrotrophic hormone, 212
thyroxine, 212
tibia, 61
tibial tuberosity, 61
tidal air, 162
tissues — defined, 14
tissues, types of, 15
toes, bones of, 65
tongue, 171
tonsils, 147
trachea, 159
translucent layer of skin, 27
transpyloric plane, 6
transverse — definition, 5
transverse colon, 174
transverse fissure of brain, 222
transverse plane, 6
trapezoid bone, 49
trapezium, 49
trapezius muscle, 119
triangular bone, 49
triceps brachii muscle, 119
tricuspid valve of heart, 134
trigeminal nerve, 228
trigone, 188
triquetral bone, 49
trochanter — definition, 33
trochanter, greater, 61
trochanter, lesser, 61
trochlea of humerus, 47
trochlea of talus, 63
trochlear nerve, 228
trochlear notch, 47
trophic hormones, 212
true pelvis, 13
true ribs, 83
trypsin, 179
trypsinogen, 179
tubercle — definition, 33
tuberosity — definition, 33

tuberosity of:
  ischium, 58
  radius, 47
  tibia, 61
turbinate bones, (conchæ), 97
twins, 208
tympanic cavity, 243
tympanic membrane, 243
tympanic part temporal bone, 95
tympanum, 243

ulna, 47
ulnar artery, 142
ulnar nerve, 230
ulnar notch, 47
ulnar styloid process, 47
umbilical arteries, 148
umbilical cord, 148
umbilical hernia, 120
umbilical vein, 148
umbilicus, 148
upper limb, bones of, 43
upper limb, joints of, 43
urea, 179
ureters, 187
urethra, 188
urethral openings, 188
urethral sphincter, 188
urinary bladder, 188
urinary system, 187
urinary system functions, 191
urine, 192
urogenital system, 187
urogram (pyelogram), 193
urography, 193
uterine arteries, 200
uterine opening, 198
uterine tubes, 198
uterogram, 202
uterography, 202
uterosalpingogram, 202
uterosalpingography, 202
uterus, 198
utricle, 244

vagina, 198
vagina, functions of, 200
vagus nerves, 228, 232
valves of heart, 134
valves of veins, 134
vas deferens, 207
vascular system, 129
vasopressin, 212
Vater, ampulla of, 177
veins — definition, 130
veins of:
  brain, 140
  heart, 134
  lower limb, 143
  upper limb, 142

vena cava, superior, 138
vena cava, inferior, 139
venography, 150
ventral — definition, 4
ventricles of brain, 225
ventricles of heart, 132
ventriculogram, 235
ventriculography, 235
venules, 130
vermiform appendix, 174
vertebrae, 71, 72
vertebrae, cervical, 73
vertebrae, lumbar, 74
vertebrae, thoracic, 74
vertebral arteries, 140
vertebral canal (spinal), 13
vertebral column, 71
vertebral column, divisions, 71
vertebral curves, 71, 72, 78
vertical — definition, 5
vesicles, seminal, 204
vesicular ovarian follicles, 197
vestibule of ear, 244
vestibule of mouth, 178
vestibular nerve, 288, 242, 244
vestibulocochlear nerve, 228, 242
villi of small intestine, 174
visceral — definition, 5
visceral cranium, 90
visceral muscle, 17
visceral pericardium, 132
visceral peritoneum, 168
visceral pleura, 161
visual purple, 239
vital capacity, 162
vitamins, 167
vitamin B 12, 124
vitreous body (humor), 241
vocal cords, 159
vocal folds, 159
volar — definition, 5
voluntary muscles, 111
vomer, 99

weak abdominal areas, 120
white blood cells, 124
white substance (matter), 220, 222
Willis circle of, 140
wings of:
  sacrum, 75
  sphenoid bone, 96
wrist, bones of, 49
wrist, joints of, 52

xiphoid process of sternum, 82
xiphisternal joint, 85

zygomatic bone, 98
zygomatic process, 98
zygote, 197, 207